QUANTITATIVE CHARACTERIZATION AND PERFORMANCE OF POROUS IMPLANTS FOR HARD TISSUE APPLICATIONS

A symposium
sponsored by ASTM
Committee F-4 on Medical and
Surgical Materials and Devices
Nashville, TN, 18–19 Nov. 1985

ASTM SPECIAL TECHNICAL PUBLICATION 953
Jack E. Lemons, University of Alabama
at Birmingham, editor

ASTM Publication Code Number (PCN)
04-953000-54

 1916 Race Street, Philadelphia, PA 19103

Library of Congress Cataloging-in-Publication Data

Quantitative characterization and performance of porous implants for
hard tissue applications: a symposium/sponsored by ASTM Committee
F-4 on Medical and Surgical Materials and Devices, Nashville, TN,
18–19 Nov. 1985; Jack E. Lemons, editor.
  (ASTM special technical publication; 953)
  "ASTM publication code number (PCN) 04-953000-54."
  Includes bibliographies and indexes.
  ISBN 0-8031-0965-2
  1. Orthopedic implants—Testing—Congresses. 2. Porous materials—
Testing—Congresses. 3. Biocompatibility—Congresses.
4. Protective coatings—Mechanical properties—Congresses.
I. Lemons, Jack E. II. ASTM Committee F-4 on Medical and Surgical
Materials and Devices. III. Series.
  [DNLM: 1. Alloys—congresses. 2. Arthroplasty—congresses.
3. Biocompatible materials—congresses. 4. Implants, artificial—
congresses. 5. Materials testing—congresses. QT 34 Q15 1985]
RD755.5.Q36   1987
617'.4710592'028—dc19
DNLM/DLC
for Library of Congress   87-33430
                                                                    CIP

NOTE

The Society is not responsible, as a body,
for the statements and opinions
advanced in this publication.

Printed in West Hanover, MA
December 1987

DEDICATION

This volume is dedicated to the memory of Emmett M. Lunceford, Jr., M.D.

Dr. Lunceford, as a professor of orthopaedic surgery at the University of South Carolina School of Medicine and senior member of the Moore Clinic, fulfilled all aspects of education, research, and service associated with his chosen discipline. His activities in basic, applied, and clinical research earned him the respect of all colleagues. Within ASTM Committee F-4 on Medical and Surgical Materials and Devices, his activities were many, and his last active responsibility was the chairmanship of the Section on Arthroplasty. Most important to all who met Dr. Lunceford, was his nature as a gentleman who always had time to discuss his programs, provide guidance to colleagues on many topics, and provide credit where due and encouragement where indicated. Many of the results presented during the symposium and published in this volume are an outgrowth of his contributions. We will all miss Dr. Lunceford and, with deep appreciation, we dedicate this book to him.

FOREWORD

The symposium on Quantitative Characterization and Performance of Porous Implants for Hard Tissue Applications was held 18–19 Nov. 1985, in Nashville, TN. The event was sponsored by ASTM Committee F-4 on Medical and Surgical Materials and Devices, in cooperation with the American Academy of Orthopaedic Surgeons. Jack E. Lemons, of the University of Alabama at Birmingham, presided as chairman of the symposium and also served as editor of this publication.

Related
ASTM Publications

Vascular Graft Update: Safety and Performance, STP 898 (1986), 04-898000-54

Corrosion and Degradation of Implant Materials: Second Symposium, STP 859 (1985), 04-859000-27

Cell Culture Test Methods, STP 810 (1983), 04-810000-54

Medical Devices: Measurements, Quality Assurance, and Standards, STP 800 (1983), 04-800000-54

Titanium Alloys in Surgical Implants, STP 796 (1983), 04-796000-54

Corrosion and Degradation of Implant Materials, STP 684 (1979), 04-684000-27

A Note of Appreciation
to Reviewers

The quality of the papers that appear in this publication reflects not only the obvious efforts of the authors but also the unheralded, though essential, work of the reviewers. On behalf of ASTM we acknowledge with appreciation their dedication to high professional standards and their sacrifice of time and effort.

ASTM Committee on Publications

ASTM Editorial Staff

Helen Mahy
Janet R. Schroeder
Kathleen A. Greene
William T. Benzing

Contents

MODELING AND IMPLANT FIXATION

Overview

This volume provides papers presented in November 1985 at a symposium on the characterization and performance of porous biomaterials used for surgical reconstructions of musculoskeletal systems.

The symposium was organized to bring together contributors from private practice, government and industrial research laboratories, universities, and other fields with related interests, in an attempt to describe what has been done in this field, what we are doing now, and what our plans are for the future. This book reflects the state-of-the-art-and-science information on porous biomaterials publicly available during late 1985. Clearly, all basic and applied research and development activities are not included; however, the omissions were not intentional. The efforts of the symposium session chairmen, S. Brown, E. Frisch, and J. Parr, contributed significantly to the initiation, organization, and finalization of the program activities. These efforts are applauded.

Applications of porous biomaterials for the fixation of surgical implant prostheses through tissue ingrowth have expanded significantly during the 1980s. A great need for expanded information exchange and standardization of evaluation methods continues to exist, in part because of the number of materials, manufacturing techniques, designs, surface modifications, variations in porosity dimensions, and device applications.

This volume, like the symposium, is separated into topic areas to combine interrelated investigations. These topics include the mechanical properties of porous coatings, mechanical properties of substrate and coating, characterization of pore dimensions, biodegradation and biological analyses, performance in humans and laboratory animals, modeling and implant fixation, and systems for future applications.

The various standards and recommended practices of ASTM Committee F-4 on Medical and Surgical Materials and Devices can be separated into those on materials, those on test methods, and those on performance. Within the materials area, most ASTM standards provide details on the chemical analyses, basic properties, and surface finishes. A significant need exists for similar standards for porous biomaterials and devices. One result of the symposium was the development of several task force activities to address key needs related to the applications of devices utilizing porous biomaterials. A number of recommendations and task force activities evolved during the symposium discussions.

An overview of the findings in these papers, summarized by topic area, follows:

Mechanical Properties of Coatings and Substrates

The papers in the first two sections, which are on basic material characteristics, coatings, and coating-substrate properties, provide detailed reviews on mechanical performance measurements as a function of manufacturing, specimen and test design, and materials. Interestingly enough, and critically, the relative strengths of the porous-surfaced alloys were between 0.25 and 0.8 of the alloys' ultimate tensile strengths. Tension, shear, and fatigue test data showed a very strong dependence on the test and specimen designs. This phenomenon, combined with the authors' strong recommendations on the need for standardized quality analysis and assurance test methods for porous coatings, emphasizes the need for considerable additional research and development in this area. The necessity of careful

1

design and application to minimize fatigue fracture is stated by many of the authors. This, certainly, is a critical concern of all involved. An ASTM task force continues work in this area.

One aspect of the materials characterization focuses on alloy chemical analysis for porous and nonporous conditions, gradients, and localized regions of impurity concentrations and on how these might influence mechanical and biological performance. Although limited data exist, strong recommendations for additional studies are made with regard to both cobalt-based and titanium-based alloy systems.

Characterization of Pore Dimensions

Most analyses of properties depend on structural characterizations to establish structure versus property relationships as a function of the manufacturing process. The papers in this section provide basic guidelines on methods to characterize the pore size, distribution, and connectiveness. Systems and techniques for hand and automated quantitative microscopy provide various measurements of the pore mean intercept, volume fraction, surface area in the volume, areal fraction, genus, and other parameters. The importance of these measurements as a function of the material and device application is emphasized. Clearly, true three-dimensional characterizations will be required in order to interpret all types of *in vitro* and *in vivo* information. An ASTM task force continues work in this area.

Biodegradation and Biological Analyses/Performance in Humans and Laboratory Animals

Laboratory, laboratory animal, and human investigation results are presented in an attempt to correlate corrosion, biodegradation, measurements of elemental distributions within laboratory animals, and excretions from humans currently implanted with porous biomaterial prostheses. The biological significance of metallic ions is also reviewed from laboratory animal results.

During the symposium, a number of concerns were expressed after the laboratory and laboratory-animal-related presentations with regard to local and systemic tissue responses. However, the early human data have produced mixed opinions and a position that no clear-cut correlations exist between the human and laboratory results. There continues to be a very significant need for additional clinical and laboratory data in this area. At this time, it is not possible to define dose-response-time relationships clearly for elements released from existing porous implant systems. Since this is a multifactorial area of research, additional studies are strongly recommended.

Modeling and Implant Fixation

One objective of the research community has been better definition of the implant characteristics associated with force transfer through mechanical and biomechanical properties, while independently considering the transfer of elemental species through chemical and biochemical property evaluations. The laboratory animal and human studies completed within the past three to five years have shown important correlations between ingrowth characteristics. However, the need for standardized evaluation criteria is supported for both porous and nonporous devices. Laboratory and laboratory animal modeling studies provide key insights into device limitations; however, considerable discussion was associated with the extrapolation of existing data. Although no ASTM activities were suggested at the symposium, additional basic research is strongly recommended.

Systems for Future Applications

A number of metallic, ceramic, and polymeric materials in the form of porous implant devices are reported on in the last section of this volume. Lengthy discussions at the symposium presentation followed several of the papers with regard to the interpretation of the information provided. These reactions showed clearly that the research and development activities related to porous biomaterials and their applications remain a dynamic field of worldwide interest.

Future research, development, and applications of prostheses that are fixed through tissue growth into porosities certainly provide an area for continued activity. All of the contributing organizations and authors are to be complimented for bringing this information to presentation and publication. The volume should provide worthwhile references. The editor strongly recommends the continued and expanded involvement of the industrial sector in this type of public disclosure and information transfer.

Jack E. Lemons

University of Alabama at Birmingham, Birmingham, AL 35294; symposium chairman and editor.

Mechanical Properties of Porous Coatings

Phillip Andersen[1] *and Danny L. Levine*[1]

Adhesion of Fiber Metal Coatings

REFERENCE: Andersen, P. and Levine D. L., **"Adhesion of Fiber Metal Coatings,"** *Quantitative Characterization and Performance of Porous Implants for Hard Tissue Applications, ASTM STP 953,* J. E. Lemons, Ed., American Society for Testing and Materials, Philadelphia, 1987, pp. 7–15.

ABSTRACT: The strength of attachment of fibrous titanium pads to solid Ti-6Al-4V substrates was measured in tension and shear using both static and fatigue strength experiments. The static tensile strength ranged from 29.3 to 65.6 MPa, and the median 2-million-cycle tensile fatigue strength was 18.3 MPa. The static shear strength ranged from 51.7 to 83.2 MPa, and the median 2-million-cycle shear fatigue strength was 13.8 MPa. The tension specimens failed within the bulk of the pad, while the shear test specimens failed at the pad/substrate interface. Increases in either the processing temperature or the pressure applied to the specimen during diffusion bonding were found to increase the strength of the assembly.

KEY WORDS: porous implants, implant materials, porous metal coatings, interface strength, Ti-6Al-4V alloy, commercially pure titanium

Whether they are used with or without bone cement, porous-surfaced hard tissue implants offer a tremendous improvement in the strength of the metal-to-bone attachment. Since the early days of porous implant development for hard tissue, many investigators [1–13] have studied such devices, using histology to demonstrate ingrowth and show the pore morphology and using mechanical tests to show the interfacial strength. While these research reports have answered many questions, a large number of new questions has arisen, particularly with regard to implant strength and long-term reliability. With the currently available profusion of material-morphology combinations for porous surfaces, many of the questions and discussions relate to the relative merits of these systems [14].

The majority of mechanical studies on interface strength [1–3,5,9,12,13] have focused on the attachment of the porous-surfaced implant to bone or bone cement. While the implant/bone interface strength is of obvious clinical importance, the porous-surface/substrate attachment is also significant in terms of the reliability of the implant, yet there has been relatively little published on this subject. Ducheyne et al. [6] have presented results for titanium-fiber-coated specimens and obtained fiber-to-substrate attachment shear strengths up to 14.12 MPa. Pilliar [10] has presented results for beaded cobalt-chromium alloy specimens and reported static shear strength in excess of 21 MPa and fatigue shear strengths in excess of 7 MPa for the bead/substrate interface.

In practical terms, we expect the coating/substrate interface to be at least as strong as the coating/bone interface. Test results for the latter, available in the literature, have come from experiments with dogs and are limited to interfacial shear strength, since this seems to be most relevant clinically. Static shear strength values for the bone/porous-metal interface ranged from 3.6 MPa (519 psi) [3] to 22.6 MPa (3273 psi) [13]. Pilliar et al. [7] have given

[1] Manager, Metallurgical Research, and group manager, Analytical Services, respectively, Zimmer, Inc., Warsaw IN 46580.

an estimated fatigue strength for the bone/coating interface of 5 MPa (725 psi).

The purpose of this study was the measurement of the static and dynamic (fatigue) strengths of the interface between commercially pure titanium fibers and a Ti-6Al-4V substrate. Both tension and shear experiments were performed.

Experimental Procedure

These experiments on fiber-metal/substrate interface strength utilized all-metal specimens, consisting only of a porous coating and a solid substrate. Some experimenters, including Pilliar [10] have applied load to the porous coating by means of some hard polymer, such as epoxy or polymethyl methacrylate, which has been formed in place and which has infiltrated the porosity of the coating. The chief drawback of such an approach is that specimen failure most often occurs at the polymer/porous-coating interface and therefore does not give a valid strength for the porous-coating-to-substrate bond. The experimenter can only report that the strength of the latter must be greater than the measured value.

Tension Testing

The tension test specimens consisted of two "mushroom-shaped" substrate pieces of implant-grade Ti-6Al-4V with a flat disk of commercially pure porous fiber metal sandwiched between. The components of the assembly were joined by diffusion bonding, a process that combines elevated temperature exposure in a vacuum furnace with the application of pressure, thus creating a metallurgical bond. The temperature cycle employed was the same as that used for the processing of implants.[2] Because of the application of pressure it is not necessary to exceed the beta transus temperature of the Ti-6Al-4V substrate and therefore no radical microstructural changes are produced.

Two thermal processing conditions, hereinafter referred to as A and B were used in the bonding of porous fiber coatings on all the specimens described in this report. Condition B is the higher temperature condition of the two and thus is closer to the beta transus temperature of the Ti-6Al-4V alloy.

The completed specimen (Fig. 1) was loaded by steel dowel pins placed through the holes at each end. The quasi-static tension tests were run on an Instron Universal Test Machine. The crosshead speed used for all tests was 2.54 mm/min (0.1 in./min). The cross-sectional area of the substrate surface to which the fiber pad was bonded was 645.16 mm^2 (1 in.^{2}). This area was used for all stress calculations.

Fatigue tests with the "double mushroom" specimen were run on an MTS servo-hydraulic fatigue machine at a rate of 1 Hz with an R value (ratio of minimum stress to maximum stress) of 0.1.

A means was also devised to measure the bond strength of fiber pads on implants. A smaller mushroom-shaped piece was diffusion bonded to the exterior surface of the pad to be tested (as shown in Fig. 2). The small button was threaded to permit attachment of a load application coupling. The surface area of the button face in contact with the fiber metal was 71.256 mm^2 (0.11 in.^{2}). Static tension tests were run on the Instron machine at a crosshead speed of 2.54 mm/min (0.1 in./min).

---

[2] Note that the time, temperature, and pressure used for implant production are considered to be proprietary, and therefore the authors are not at liberty to present exact details.

Shear Testing

The lap shear specimens, like the tension test pieces, had a sandwich structure consisting of a fiber metal pad between two Ti-6Al-4V substrate pieces. As was the case for the tension specimens, the diffusion-bonding furnace cycle was one used for implant production. The lap shear specimen (shown in Fig. 3) was designed to fit into a pair of loading fixtures (shown in Fig. 4).

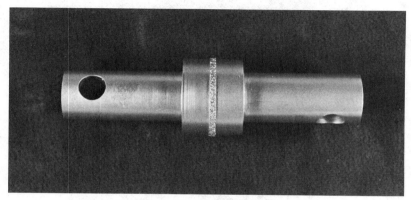

FIG. 1—*Tension test specimen.*

FIG. 2—*Small button tension specimen.*

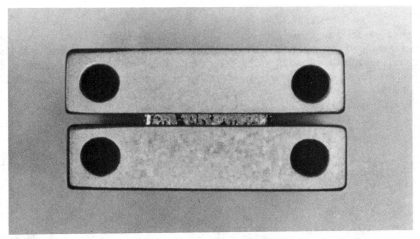

FIG. 3—*Lap shear specimen.*

FIG. 4—*Loading arrangement for the lap shear tests.*

Quasi-static lap shear tests were run on the Instron machine at a crosshead speed of 2.54 mm/min (0.1 in./min). The setup for such tests is shown in Fig. 4.

Lap shear fatigue tests were run on an MTS servo-hydraulic fatigue machine at the accelerated rate of 30 Hz with a load ratio R equal to 0.1.

Results and Discussion

The characterization of the mechanical performance of fiber-metal/substrate interfaces has focused on two aspects. In addition to the magnitude of the breaking stress, the mode of fracturing was also examined.

Tension Testing

In the course of running static tests, specimen preparation was done using a range of values for the processing parameters, which included the furnace temperature and the pressure applied to the substrate/fiber interface.

The tests showed that changes in the thermal processing conditions can substantially alter the tensile strength of the assembly. Tension tests on the large double-mushroom specimens prepared using Processing Condition A ($n = 4$) gave a static tensile strength of 29.3 ± 8.2 MPa (4230 ± 1188 psi), while those prepared using Processing Condition B ($n = 14$) gave a static tensile strength of 47.8 ± 13.4 MPa (6926 ± 1948 psi).

The mode of fracture failure of the tension specimens was partially adhesive (at the fiber/substrate interface) and partially cohesive (within the fiber pad) for those specimens processed with lower temperature and pressure. As pressure and temperature were increased, the mode of fracture became entirely cohesive, indicating that the fiber-to-substrate bonds were stronger than the wire that made up the pads.

The terms adhesion and cohesion are being used here in the sense that they are used to describe the bonding of two substrate layers by an interposed "glue" layer. If we think of the fiber metal as being the glue, the term adhesive defines the circumstance in which the interposed layer is more adherent to the substrate than to itself. Conversely, cohesion describes the situation wherein the glue layer is more adherent to itself than to the substrate. Furthermore, adhesive failure is a failure of adhesion, and cohesive failure is a failure of cohesion. Clearly, this terminology does not address the microscopic effects of wire-to-wire bonding in the fibrous layer, but does describe the character of the failures on a macroscopic scale.

Tests with the small test button bonded to an implant were done to show that results achieved in controlled laboratory experiments on specimens of simple geometry could be reproduced in products intended for clinical use. Tests on a series of porous-surfaced hip stem prostheses (Processing Condition A) gave an average strength ($n = 23$) of 57.6 ± 5.1 MPa (8349 ± 739 psi), and similar tests on a group of knee tibial implants (Processing Condition B) gave an average strength ($n = 4$) of 65.6 ± 14.2 MPa (9511 ± 2064 psi). In all cases, the tensile button pulled away from the top surface of the pad leaving the fiber metal still bonded to the substrate, so we may assume that these numbers are the lower bound strength values for fiber-to-implant adhesion.

Fatigue tests on the large double-mushroom-shaped tension specimens (Processing Condition A) produced a median 2-million-cycle fatigue strength of 18.3 MPa (2650 psi). The 2-million-cycle fatigue life was used primarily because of the lack of available testing time on the company's fatigue machines, which are primarily devoted to tests of implants. All implant tests are run using the accepted 10-million-cycle lifetime criterion. Fatigue experiments were run on six specimens. The method of assessment of the median fatigue strength was a fairly subjective one, involving extrapolation of the plotted *S-N* curve. The fatigue fractures were entirely cohesive, which suggests that the interface fatigue strength was stronger than that of the wire. For one experiment, a transparent plastic bag was mounted around the specimen to catch fiber fragments (if any) released at fracture. None were found.

Shear Testing

Static shear tests were run on several different geometrical configurations. A design feature often used on fiber metal implants is the presence of one or more surface "pockets" into which the fiber metal pad is set and bonded. With such a design, the top surface of each pad is thus nearly flush with the rest of the implant surface. The lap shear experiments used the presence or absence of a pocket as a parameter to be investigated. One other consideration was whether the pad fit snugly into the pocket or if there was a gap between the pad and the pocket. A gap of 0.3175 mm (0.0125 in.) was used in testing.

In the making of the specimens used for static shear tests, only one temperature cycle was used, corresponding to Processing Condition A.

The static shear experiments did not show a decrease in shear strength for the specimens without pockets. The specimens with pockets ($n = 11$) had a shear strength of 51.1 ± 18.7 MPa (7412 ± 2712 psi), while those without a pocket ($n = 2$) had a shear strength of 83.2 ± 33.1 MPa ($12,067 \pm 4808$ psi). Of the specimens with pockets, those having a gap

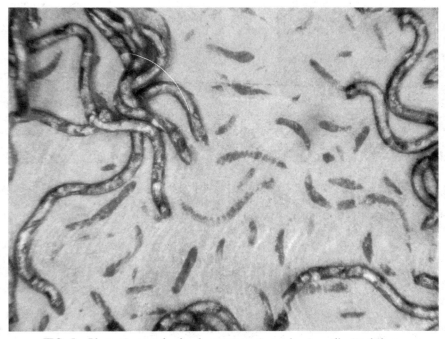

FIG. 5—*Photomicrograph of a shear test specimen showing adhesive failure.*

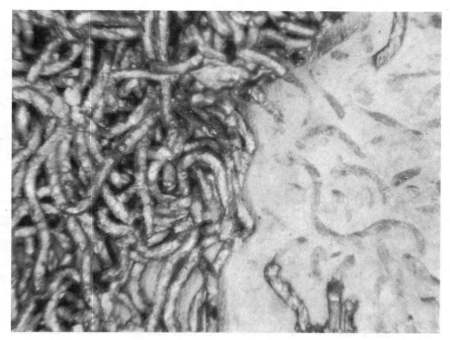

FIG. 6—*Photomicrograph of a shear test specimen showing mixed adhesive and cohesive failure.*

between the pad and pocket (n = 4) gave a static shear strength of 51.7 ± 6.9 MPa (7497 ± 996 psi), while those with no gap (n = 7) gave a shear strength of 58.7 ± 9.9 MPa (8507 ± 1442 psi). The overall average shear strength (n = 13) was 60.3 ± 16.3 MPa (8744 ± 2358 psi). The mode of failure was adhesive for 11 of the 13 specimens. The pad-to-substrate bonds were broken on one side, while the pad remained securely bonded to the second substrate piece. Two of the specimens showed mixed adhesive and cohesive failure so that some wire-to-substrate bonds were broken and some wire-to-wire bonds were broken within the bulk of the pad. Figures 5 and 6 show examples of adhesive and mixed-mode fractures.

The lap shear fatigue tests all employed specimens of the same geometrical configuration. The specimens used had substrate pieces with milled pockets and no gap between pad and pocket. Only one thermal cycle was used for diffusion bonding, and it was the same cycle used for the static lap shear tests.

The fatigue experiments gave a median 2-million-cycle fatigue strength of 13.8 MPa (2000 psi). Shear fatigue experiments were run on 20 specimens and showed a high degree of repeatability. In eight experiments, the test ran out (that is, the specimen did not fail at 2 million cycles) at the 13.8-MPa (2000-psi) stress level. The assessment of the median fatigue strength was subjective and was based on extrapolation of the S-N curve. The mode of failure was adhesive in every case. The wire-to-substrate bonds were broken on one side of the specimen, while the bonds remained secure on the other side. Fatigue striations were easily observed on all the wire-to-substrate bond sites. An example of the striations is shown in Fig. 7.

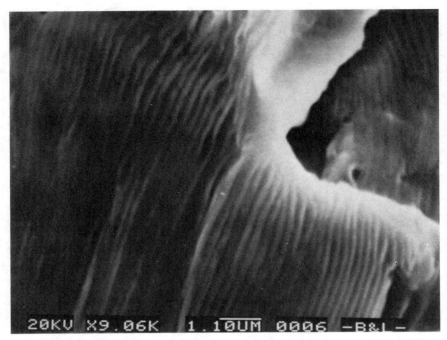

FIG. 7—*Scanning electron photomicrograph of fatigue striations on a wire-to-substrate bond.*

Conclusions

The strength of the attachment between fibrous titanium and solid Ti-6Al-4V was examined in tension and shear using both static and fatigue strength experiments. The average tensile strength was found to range from 29.3 MPa (4250 psi) to 65.6 MPa (9514 psi), and the shear strength ranged from 51.7 MPa (7498 psi) to 83.2 MPa (12 067 psi). The 2-million-cycle fatigue tests gave strengths of 18.3 MPa (2654 psi) in tension and 13.8 MPa (2002 psi) in shear. During tension testing, failure occurred through the fiber metal pad. In shear tests, failure occurred at the pad/substrate interface.

While variations in specimen processing parameters (the thermal cycle used, pressure) did affect the assembly strength, the use of a pocket on the substrate did not appreciably affect the static shear strength.

The mode of failure in tension is cohesive (within the pad), while the mode in shear is adhesive (at the pad/substrate interface). Fiber pad specimens do not release small fragments when they fail by fatigue.

Acknowledgments

This research was supported by Zimmer, Inc., Warsaw, Indiana. The authors would like to acknowledge the contributions and helpful comments offered by R. Crowninshield, G. Delli Santi, and C. Jacobs. The static tests were run by D. Waites, and the fatigue tests were run by D. Slone and M. Sparks.

References

[1] Hahn, H. and Palich, W., *Journal of Biomedical Materials Research,* Vol. 4, 1970, pp. 571–577.
[2] Galante, J., Rostoker, W., Lueck, R., and Ray, R. D., *Journal of Bone and Joint Surgery,* Vol. 53-A, No. 1, 1971, pp. 101–114.
[3] Galante, J. and Rostoker, W., *Journal of Biomedical Materials Research Symposium,* No. 4, 1973, pp. 43–61.
[4] Homsy, C. A., Cain, T. E., Kessler, F. B., Anderson, M. S., and King, J. W., *Clinical Orthopaedics and Related Research,* No. 89, 1972, pp. 220–235.
[5] Nilles, J. L. and Lapitsky, M., *Journal of Biomedical Materials Research Symposium,* No. 4, 1973, pp. 63–84.
[6] Ducheyne, P., Martens, M., Aernoudt, E., Mulier, J., and DeMeester, P., *Acta Orthopaedica Belgica,* Vol. 4, 1974, pp. 799–805.
[7] Pilliar, R. M., Cameron, H. V., and Macnab, I., *Biomedical Engineering,* April 1975, pp. 126–131.
[8] Murray, G. A. W. and Campbell Semple, J., *Journal of Bone and Joint Surgery,* Vol. 63-B, No. 1, 1981, pp. 138–141.
[9] Clemow, A. J. T., Weinstein, A. M., Klawitter, J. J., Koeneman, J., and Anderson, J., *Journal of Biomedical Materials Research,* Vol. 15, 1981, pp. 73–82.
[10] Pilliar, R. M., *Clinical Orthopaedics and Related Research,* No. 176, 1983, pp. 42–51.
[11] Cook, S. D., Georgette, F. S., Skinner, H. B., and Haddad, R. J., *Journal of Biomedical Materials Research,* Vol. 18, 1984, pp. 497–512.
[12] Manley, M. T., Stern, L. S., and Gurtowski, J., *Journal of Biomedical Materials Research,* Vol. 19, 1985, pp. 563–575.
[13] Cook, S. D., Walsh, K. A., and Haddad, R. J., *Clinical Orthopaedics and Related Research,* No. 193, 1985, pp. 271–280.
[14] Lunceford, E. M., *Contemporary Orthopaedics,* Vol. 9, No. 6, 1984, pp. 53–89.

James A. Davidson,[1] *Hugh U. Cameron,*[2] *and Michael Bushelow*[3]

Determining the Cantilever-Bend Fatigue Integrity of a Porous-Coated Tibial Component

REFERENCE: Davidson, J. A., Cameron, H. U., and Bushelow, M., **"Determining the Cantilever-Bend Fatigue Integrity of a Porous-Coated Tibial Component,"** *Quantitative Characterization and Performance of Porous Implants for Hard Tissue Applications, ASTM STP 953,* J. E. Lemons, Ed., American Society for Testing and Materials, Philadelphia, 1987, pp. 16–27.

ABSTRACT: Two currently popular porous-coated tibial plates, made of cobalt-chromium-molybdenum alloy, were tested in cantilever-bend fatigue loading to determine the fatigue integrity of these plates under partially unsupported *in vivo* conditions. The test method was carefully established based on an *in vitro* strain gage analysis using an embalmed cadaver tibia. Fatigue test loads were selected, based on results of the cadaver study, to represent knee forces in excess of six times the body weight of a 91-kg (200-lb) person. Results showed both types of tibial plates to be capable of withstanding more than 10 million cycles under these extreme cantilever-bend fatigue load conditions.

KEY WORDS: porous implants, porous coatings, tibial plate, cobalt-chromium-molybdenum alloy, fatigue design, *in vitro* testing

With the increasing popularity of porous-metal-coated implant devices, associated test methods are needed to assess the integrity of these devices under anticipated worse-case *in vivo* conditions. Such test procedures must also address concerns of the surgeon. One concern of surgeons is the effect of cantilever loading on the integrity of the tibial plate component of the porous-coated total knee system. The literature has recently reported [1] a case in which a tibial plate cracked following cantilever loading *in vivo*.

In addition to demonstrating the integrity of porous-coated tibial components subjected to cantilever loading, this study recognizes the important fact that, unlike the porous coatings on tension-loaded surfaces of a femoral component, the porous coating on a tibial plate is located on a compression-loaded surface. Thus, notch effects associated with the presence of a sintered metallic porous coating are not operative. Nonetheless, it is still important to verify that the presence of a porous metal coating and the thermal treatment or treatments to the substrate associated with it do not compromise the intended functional integrity of the device under anticipated service (*in vivo*) loading.

In this study, two popular porous-metal-coated tibial plates were tested under severe cantilever-bend fatigue loading. Such loading may occur from loss of bone support under

[1] Director, Materials Research, Richards Medical Co., Memphis, TN 38116.
[2] Orthopedic surgeon, Orthopaedic and Arthritic Hospital, Toronto, Ontario, Canada M4Y 1H1.
[3] Research engineer, Howmedica Corp., Rutherford, NJ 07070.

the tibial plate. Both the beads used to form the porous coating and the substrate material were made of a cobalt-chromium-molybdenum alloy meeting the requirements of the ASTM Specification for Cast Cobalt-Chromium-Molybdenum Alloy for Surgical Implant Applications (F75-82). Results of the fatigue tests were compared with assumed *in vivo* fatigue loading characteristics for six times body weight. Because of the high cost of these devices, only a limited number of tests could be performed.

Procedure

Two different types of porous-coated tibial plates of similar size (Fig. 1) were fatigue loaded in cantilever bending to evaluate the fatigue integrity in comparison with the anticipated *in vivo* loading, based on a strain gage study. Plate Type T was 45 by 76 mm with a thickness of 3.35 mm. The polyethylene articulating surface was 6 mm thick. The other type of tibial plate, Type P, was 45.6 by 72.5 mm with a thickness of 3.2 mm. The polyethylene surface of Plate Type P was 10 mm in thickness. Neither plate incorporated a central post for fixation into the tibia.

Stress Calibration

Axial strain gages were attached to the tension surface of Plate Type T, as shown in Fig. 2. The gages were oriented to measure the tensile bending strains due to the cantilever loading. Because the gages were used for calibration purposes, it was not necessary to determine strains in other directions during loading. The plate was then properly placed in an embalmed cadaver tibia with bone removed under half of the plate, as illustrated in Fig. 3. Such a test method is appropriate for tibial plates without a central post. Removal of the bone enabled cantilever bending stresses to develop, which are characteristic of a potentially extreme *in vivo* condition, when the tibia was loaded by the matching femoral component. A universal joint was used to ensure that the vertical load was evenly divided between the two femoral condyles. A calibration curve was then developed relating the applied load to the resultant bending strain. To provide the best representation of the anatomical condition [2,3], both the calibration and the subsequent fatigue tests were performed with the tibial plate supported at a 7° anterior-posterior angle.

Because the cadaver bone used to develop the calibration curve could not support loads characteristic of several times the body weight (BW) of an average adult, the identical calibration test was repeated using a steel block, as illustrated in Fig. 4.

Fatigue Testing

Once the calibration curve was established, cantilever-bend fatigue tests were performed on both types of tibial plates. The plates were supported as shown in Fig. 5. The strain-gaged plate used in the cadaver bone calibration study was used to calibrate the fatigue loading setup, and fatigue load ranges were selected to produce strain ranges in excess of that characteristic of six times BW, based on the cadaver calibration.

Both the calibration and fatigue tests were performed on a 9072-kg (20 000-lb)-capacity MTS closed-loop servo-hydraulic test machine. For the fatigue tests, the ratio of the minimum load divided by the maximum load (R ratio) was equal to 0.1. The cyclic frequency was 10 cycles per second (10 Hz).

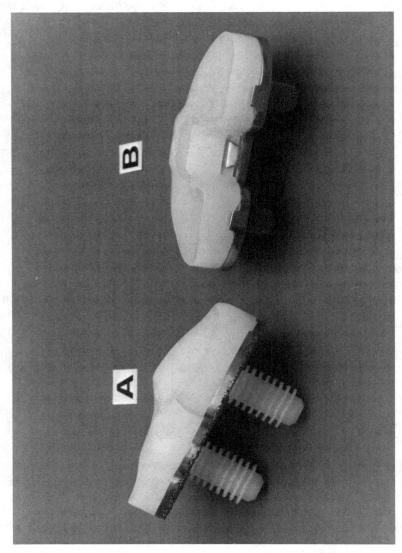

FIG. 1—*Photograph of (a) tibial Plate Type T and (b) tibial Plate Type P evaluated in this study.*

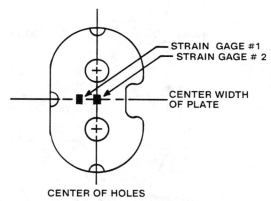

FIG. 2—*Location of the two strain gages placed on tibial Plate Type T (under the polyethylene surface) to record the cantilever bending strains.*

Results

The calibration curves, based on both the cadaver tibia and the steel block, are shown in Fig. 6. Both strain gages showed similar results in the tests on the cadaver bone and on the steel block. The results were roughly equal at an applied load of 454 kg (1000 lb). Results from the cadaver tibia were only obtainable to a load just over 178 kg (400 lb). At a load of 200 kg (440 lb), the tibia crushed under the applied force. However, the results of the calibration using the steel block are in very good agreement with those obtained from the cadaver tibia.

Calibration results of the cantilever fatigue loading setup are also shown in Fig. 6. This calibration curve was used to establish loads for the fatigue tests so that the load range

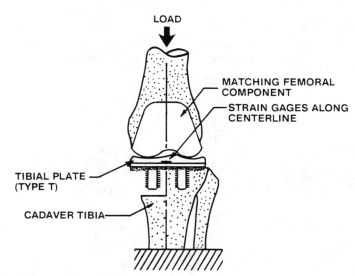

FIG. 3—*Schematic illustration showing the cadaver tibia test setup used to obtain the load-strain calibration curve.*

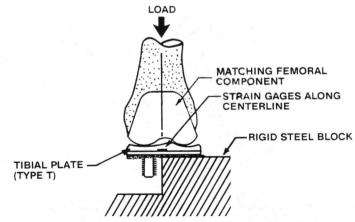

FIG. 4—*Schematic showing the rigid steel block and load configuration for determining the cantilever bending strains at high loads.*

would be representative of several times body weight, based on the cadaver calibration. For instance, a maximum fatigue load of 58 kg (128 lb) on the fatigue specimen would be characteristic of a 545-kg (1200-lb) *in vivo* load, based on the cadaver calibration. Because the fatigue tests were performed at an R ratio of 0.1, the minimum fatigue load would be equal to 0.1 of 545 kg, or 54.5 kg. Thus the load range, defined as the difference between the maximum and minimum fatigue loads, would be equal to 545 kg minus 54.5 kg, or 491.5 kg. The equivalent load range for the fatigue test specimen condition would be equal to 58 kg minus 5.8 kg, or 52.2 kg (115 lb).

Results of the cantilever fatigue tests are summarized in Table 1. Both the number of cycles to fatigue initiation, and the crack initiation site are included in the table. The locations of the fatigue crack origins summarized in Table 1 are shown in Fig. 7. Cracks initiated at both the anterior and posterior sides of Plate Type T and at the posterior side adjacent to a raised metal support in Plate Type P.

Figure 8 is a plot of the applied cantilever-bend fatigue load range versus cycles to the first indication of a fatigue crack for both plate types. Also shown in Fig. 8 is a dashed line

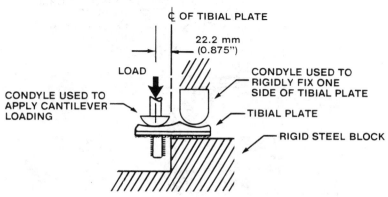

FIG. 5—*Schematic showing the test setup used for the strain-load calibration of tibial Plate Type T and f̶ subsequent cantilever-bend fatigue testing.*

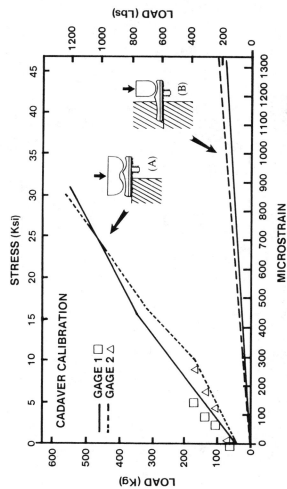

FIG. 6—*Results of the cantilever strain calibrations for* (a) *the unsupported tibial plate under anatomical loading (upper curves shown by Test Setup A) and* (b) *the unsupported tibial plate fatigue test condition (lower curves shown by Test Setup B). The open squares and triangles indicate the load-strain relationship observed during the cadaver tibia test prior to fracture of the tibia.*

TABLE 1—*Results of the cantilever-bend fatigue tests of two types of porous-coated tibial plates.*

Plate Type	Load Range, kg (lb)	No. of Cycles Before Crack Was First Detected	Crack Initiation Site	No. of Cycles Before Crack Was Observed on Both Sides	Notes
T	109 (240)	5×10^6	posterior	8.2×10^6	polyethylene able to maintain load after cracking
T	100 (220)	5×10^6	anterior	8.9×10^6	polyethylene able to maintain load after cracking
T	86 (190)	1.1×10^7	posterior	test terminated at 1.1×10^7 (very small crack)	...
T	73 (160)	no crack	...	no crack	test terminated at 1.0×10^7
P	86 (190)	7.9×10^5	posterior	1.9×10^6	polyethylene able to support load after cracking
P	86 (190)	5.1×10^6	posterior	9.1×10^6	polyethylene able to maintain load after cracking

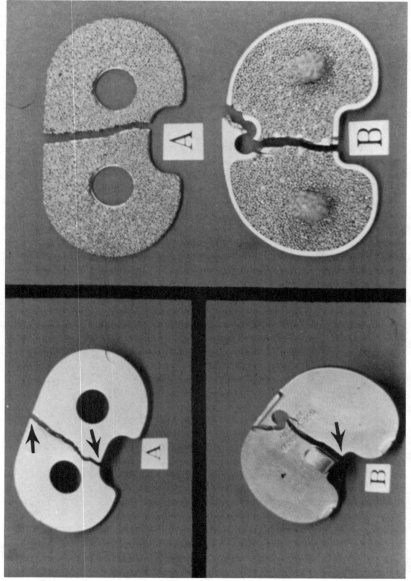

FIG. 7—Photographs of the fatigue cracks that developed during the cantilever-bend fatigue tests on (a) tibial Plate Type T and (b) tibial Plate Type P. The arrows indicate the location of crack initiation.

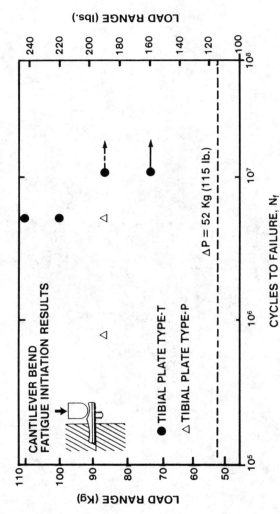

FIG. 8—*Results of the cantilever-bend fatigue tests for both types of tibial plates. The dashed line at a load range of 52 kg (115 lb) represents expected in vivo conditions at six times BW.*

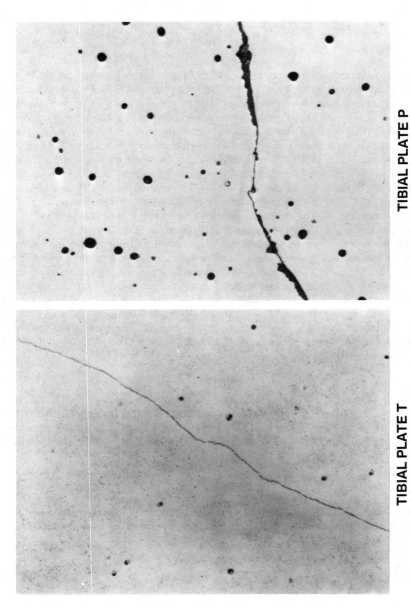

TIBIAL PLATE T   **TIBIAL PLATE P**

FIG. 9—*Photomicrographs showing the microstructural comparisons between tibial Plate Type T (left) and tibial Plate Type P (right) cast cobalt alloy substrate material: ×320 magnification, 2% chromium trioxide and water electrolytic etch for 10 s.*

representing the load range of 52 kg (115 lb). As mentioned earlier, this load range is characteristic of an *in vivo* peak load of 545 kg (1200 lb) at an R ratio of 0.1, based on the cadaver calibration results given in Fig. 6. A peak load of 545 kg is equivalent to six times the body weight of a 91-kg (200-lb) person.

Discussion

Looking at Fig. 8, one can see that, although the fatigue life of porous-coated tibial Plate Type P is less than that of Plate Type T, both types of plate can be expected to perform *in vivo* at loads characteristic of over six times BW of a 91-kg (200-lb) person. The lower fatigue life of Plate Type P is partially attributed to the slightly smaller thickness of the porous-coated substrate, and is thus a reflection primarily of a particular choice of design.

Another factor that may have contributed to the difference in the fatigue performance is the microstructure. Plate Type T had received a hot-isostatic-pressing (HIP) treatment, which has the effect of closing much of the microporosity resulting from the casting process and reducing grain-boundary carbides. Studies [4–7] have shown that HIP can improve the fatigue strength of ASTM Specification F75 cast cobalt alloy. The microstructure of Plate Type T is shown in Fig. 9. The microstructure of Plate Type P is also shown in Fig. 9 for comparison. Plate Type P did not receive a HIP cycle, as is indicated by the increased porosity and grain boundary carbides. Thus, this microstructural aspect may have also contributed to the relatively lower cantilever-bend fatigue performance of Plate Type P. Regardless of the fatigue mechanism, the loading condition used in this study represents an extreme case, in which a dramatic loss in bone support is proposed to have occurred.

The fatigue endurance limit of cobalt alloy is defined at 10 million cycles. If, when subjected to repetitive cyclic loads, a material or device can withstand 10 million cycles, it is assumed that it will perform at that load level indefinitely. In Fig. 8 it is apparent that both plates will be able to perform for more than 10 million cycles at cyclic loads above the dashed line representing six times BW.

Conclusion

By using a severe *in vivo* loading model, a test method for determining the fatigue strength of tibial plates has been developed which indicates that two currently popular porous-coated ASTM Specification F75-82 tibial implant devices can withstand 10 million load cycles under cantilever bending.

Acknowledgment

The authors wish to thank Gary Lynch, supervisor of the Mechanical Test Laboratory at Richards Medical Co., for his assistance with the equipment scheduling and the test fixtures for this study.

References

[1] Mendes, D. G., Brandon, D., Galor, L., and Roffman, M., "Breakage of the Metal Tray in Total Knee Replacement," *Orthopaedics*, Vol. 7, No. 5, May 1984.
[2] Laskin, R. S., Denham, R. A., and Apley, A. G., *Replacement of the Knee,* Springer-Verlag, New York, 1984.
[3] Townley, C. O., "The Anatomic Total Knee Resurfacing Arthroplasty," *Clinical Orthopaedics and Related Research,* No. 192, January/February 1985, pp. 82–96.

[4] Hodge, F. G. and Lee T. S. III, "Effects of Processing on Performance of Cast Prosthesis Alloys," *Corrosion,* Vol. 31, No. 3, March 1975.

[5] Georgette, F. S. and Davidson, J. A., "The Effect of HIP'ing on the Fatigue and Tensile Strength of a Cast, Porous-Coated Co-Cr-Mo Alloy," *Journal of Biomedical Materials Research,* Vol. 20, No. 8, October 1986.

[6] Galante, J. O., Rostoker, W., and Doyle, J. M., "Failed Femoral Stems in Total Hip Prostheses," *Journal of Bone and Joint Surgery,* Vol. 57A, 1978, pp. 230–651.

[7] Dobbs, H. S. and Robertson, J. L. M., "Heat Treatment of Cast Co-Cr-Mo for Orthopaedic Implant Use," *Journal of Materials Science,* Vol. 18, 1983, pp. 391–404.

DISCUSSION

L. C. Jones[1] (written discussion)—I question the use of such a severe test for cantilever bending of the tibial tray. The magnitude of the load applied, determined by an extrapolation from cadaver specimen tests, appears to be slightly above that experienced normally by patients. Furthermore, this test allows unsupported bending, which may not be relevant to the clinical situation. It is possible that a tibial tray that fails this test may still be adequate clinically.

J.A. Davidson, H.U. Cameron, M. Bushelow (authors' closure)—The authors wish to thank Lynne Jones for reemphasizing what has been pointed out in the text. That is, this test method does indeed reflect a severe, worse-case cantilever-bend fatigue loading condition for a tibial plate without a central tibial post. On the other hand, the occurrence of this type of unsupported bending *in vivo* cannot be ruled out for this type of tibial plate. If off-axis loading by the femoral condyles occurs, the design of the plate is such that one side will rise while the opposite side is loaded, creating a bending condition at the edge of the unsupported location. However, as Dr. Jones has pointed out, such a loading condition would not be expected to be maintained for 10 million load cycles. Thus, the authors agree that a tibial tray that fails this test may still be adequate clinically.

[1] Johns Hopkins University School of Medicine, Department of Orthopaedic Surgery, Baltimore, MD 21239.

Mechanical Properties of Substrate and Coating

Frederick S. Georgette[1]

Effect of Hot Isostatic Pressing on the Mechanical and Corrosion Properties of a Cast, Porous-Coated Co-Cr-Mo Alloy

REFERENCE: Georgette, F. S., **"Effect of Hot Isostatic Pressing on the Mechanical and Corrosion Properties of a Cast, Porous-Coated Co-Cr-Mo Alloy,"** *Quantitative Characterization and Performance of Porous Implants for Hard Tissue Applications, ASTM STP 953,* J. E. Lemons, Ed., American Society for Testing and Materials, Philadelphia, 1987, pp. 31–46.

ABSTRACT: The corrosion properties of a cast Co-Cr-Mo alloy were evaluated using anodic and cathodic polarization techniques. These data were compared with those for a Co-Cr-Mo alloy which had been exposed to a porous coating thermal cycle, both with and without a subsequent hot isostatic pressing (HIP) procedure. The thermal exposure during sintering was found to result in the formation of grain boundary precipitates and porosity, accompanied by a reduction in elemental segregation. This resulted in a reduction in mechanical properties, as well as an increase in corrosion resistance. The use of a HIP cycle, subsequent to sintering, provided an increase in both mechanical properties and corrosion resistance. The improvement in corrosion resistance was attributed to a further reduction in elemental segregation, in comparison with the as-sintered condition.

KEY WORDS: sintering, corrosion, hot isostatic pressing, porosity, Co-Cr-Mo alloy, implant materials, porous implants

Microstructural inhomogeneities observed in the as-cast condition of the Co-Cr-Mo alloy include casting porosity, coring, oxides, sigma phase formation, large grains, and grain boundary precipitates [1–4]. The thermal treatment commonly employed to apply a porous coating to this alloy results in a redistribution of these microstructural features. It is therefore important to understand the nature of these transformations, and how they relate to the resulting properties of the alloy.

Research concerning the effect a porous coating has on the fatigue properties of implant materials has concentrated on two areas: the effect that the porous coating has on the substrate, and the effect that the thermal treatment has on the substrate properties. For the Co-Cr-Mo alloy, it has been shown that the thermal process required to sinter a porous coating degrades the fatigue strength of the alloy, with or without a porous coating [5] (see Table 1). By applying a hot isostatic pressing (HIP) treatment to sintered materials, an improvement in their fatigue properties can be achieved. The benefits of a HIP treatment are further shown by an increase in tensile properties (see Table 1). These mechanical property improvements were attributed to the microstructural variations induced by thermal cycling.

[1] Manager, Advanced Technologies, Richards Medical Co., Memphis, TN 38116.

TABLE 1—*Mechanical properties of a Co-Cr-Mo alloy in the as-cast and sintered conditions.*

Condition Treatment	Ultimate Tensile Strength, MPa	Yield Strength, MPa	Elongation, %	Reduction of Area, %	Fatigue Strength, MPa		
					Uncoated	Three-Layer Porous Coating	One-Layer Porous Coating
A sintering plus hot isostatic pressing plus heat treatment	731	495	13.0	16.5	255	234	179
B sintering plus heat treatment	659	474	9.3	12.8	177	193	...
C as cast	711	471	11.9	10.8	267	...	...

TABLE 2—*Chemical composition of the base material.*

Element	Weight %
Cr	27.85
Mo	5.98
Ni	0.71
Fe	0.28
C	0.22
Si	0.80
Mn	0.72
Co	balance

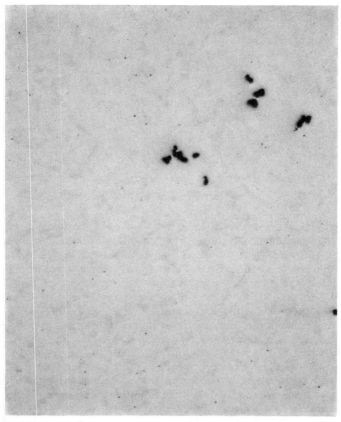

FIG. 1a—*Microstructure of the as-cast condition, in the as-polished state (×100).*

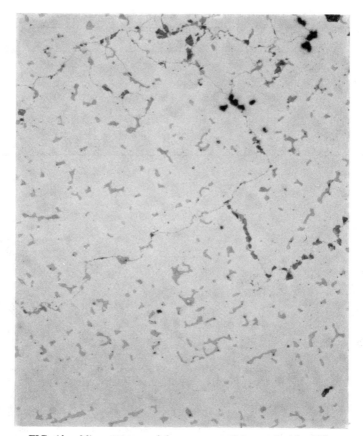

FIG. 1b—*Microstructure of the as-cast condition, etched (×100).*

Previous investigators have shown that as-cast Co-Cr-Mo alloy exhibits inferior corrosion resistance when compared with the heat-treated condition. Specifically, Hollander and Wulff [6] and Hodge and Lee [1] have observed an improvement in crevice corrosion resistance as a result of the use of a HIP process. Further, Acharya et al. [7] have theorized that a microcouple effect might be present in cast Co-Cr alloys, occurring between precipitates and the matrix. In general, the improved corrosion resistance has been attributed to a reduction in the segregation of alloying elements, inherent in the as-cast material condition. Because the high-temperature application of a porous coating can result in precipitate dissolution, as well as a reduction in segregation, it is likely that there may be a change in corrosion resistance.

This study evaluates the corrosion behavior of the Co-Cr-Mo alloy in three conditions: as cast; sintered without a porous coating and heat-treated; and sintered without a coating, HIP-treated, and heat-treated. The resistance to pitting corrosion was evaluated using potentiodynamic anodic polarization techniques. A relative corrosion rate for each condition was calculated from the intersection of the anodic and cathodic polarization curves. The resulting corrosion behavior was correlated with corresponding microstructural features. Chemical analyses were performed on the surface of the material to give an indication of the distribution of alloying elements.

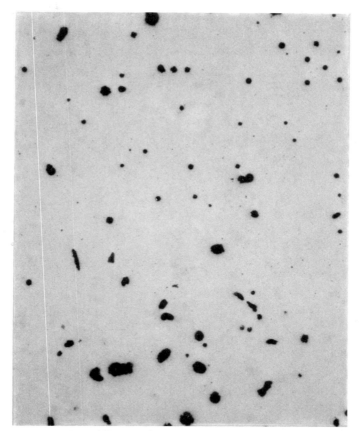

FIG. 2a—*Microstructure of the sintered and heat-treated condition, in the as-polished state* (×*100*).

Experimental Procedure

Materials

All materials utilized in this investigation were found to conform to the chemical requirements outlined in the ASTM Specification for Cast Cobalt-Chromium-Molybdenum Alloy
for Surgical Implant Applications (F 75-82) (see Table 2). The corrosion specimens were
machined from 15.9-mm-diameter cast Co-Cr-Mo bars and tested in the following conditions:

(a) exposed to a sintering cycle, HIP-treated, and subsequently heat-treated;
(b) exposed to a sintering cycle and subsequently heat-treated; and
(c) as cast.

The HIP cycles were performed at a temperature above the solutionization range normally
defined for this alloy [2,8]. The HIP temperature was chosen in order to dissolve matrix
and grain boundary precipitates remaining from the sintering cycle, while simultaneously
closing void defects produced by the dissolution. A secondary heat-treatment cycle was
performed in the homogenization temperature range to increase ductility and strength properties.

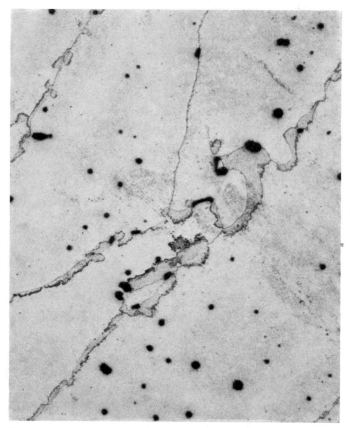

FIG. 2b—*Microstructure of the sintered and heat-treated condition, etched* (×100).

Specimen Preparation

All specimens were ground through a succession of silicon carbide papers to a 600-grit finish. Subsequent to a 6-μm diamond slurry polishing step, the specimens were given a final polish with 0.05-μm alumina. Immediately before being tested, all the specimens were ultrasonically cleaned in distilled water.

Electrochemical Techniques

Corrosion testing was performed using a Princeton Applied Research Model 173 potentiostat/galvanostat, with potentials referenced to a saturated calomel electrode. The polarization test cell configuration conformed to that previously described by Greene [9]. Three specimens of each type were used for anodic and cathodic polarization scans. Both methods were initiated only after a rest potential was reached, which was determined by no more than a 1-mV change in voltage over a 5-min period.

Potentiodynamic anodic polarization scans were performed at a rate of 10 mV/min into the transpassive region. This was followed by a reverse scan to detect the occurrence of a

FIG. 3a—*Microstructure of the sintered, HIP-treated, and heat-treated condition, in the as-polished state (×100).*

hysteresis, possibly indicating pitting. Cathodic polarization scans were performed at a rate of 5 mV/min.

All the testing took place in Ringer's injection solution, deaerated with nitrogen at least ½ h prior to, and during, testing. The electrolyte was adjusted to a pH of 7 ± 0.05 and maintained at a temperature of 37 ± 0.05°C.

Microscopic Examination

Standard metallurgical analyses were performed on representative specimens in order to evaluate various microstructural features. The specimens were etched in a 10% solution of chromic oxide (CrO_3) in water. The as-polished sections were analyzed prior to etching in order to detect porosity in the material. Surface chemical analyses were performed on the alloy in all conditions using a Cameca MBX electron microprobe, with wavelength-dispersive X-ray spectrometers. These analyses provided a chemical scan across the various microstructural features.

FIG. 3b—*Microstructure of the sintered, HIP-treated, and heat-treated condition, etched (×100).*

Results

Microstructural Evaluation

The effects of the various thermal cycles on the resultant microstructures are shown in Figs. 1a through 3b. The as-cast condition was found to exhibit a typical coarse grain structure, with discontinuous carbides occurring both at grain boundaries and within a highly cored matrix (see Figs. 1a and 1b). Evidence of interdendritic porosity was observed in the center of the bar and connected to the surface.

The material that had been sintered and heat-treated exhibited continuous grain-boundary carbide precipitates (see Figs. 2a and 2b). The regions where carbide dissolution occurred were found to contain porosity.

The effect of HIP, followed by a heat treatment cycle, is shown in Figs. 3a and 3b. It was found that the HIP cycle essentially eliminated porosity produced by the sintering treatment. Few carbides remained, and there was no evidence of grain boundary precipitation.

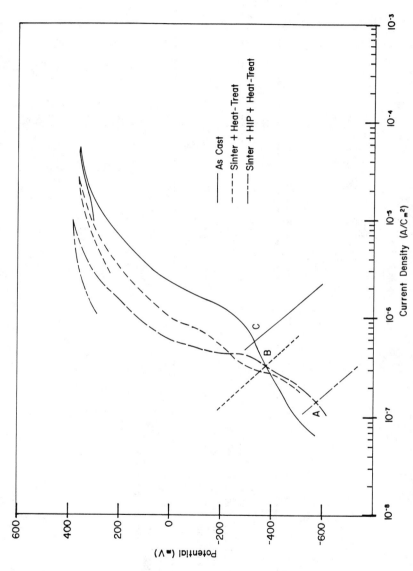

FIG. 4—*Anodic and cathodic polarization scans for the three microstructural conditions.*

TABLE 3—*Corrosion properties of the three material conditions.*[a]

Condition Treatment	E_{corr}, mV[a]	I_{corr}, A/cm^2	Breakdown Potential, mV	Corrosion Rate, μm/year
A sintering plus hot isostatic pressing plus heat treatment	−578	0.14	318	0.965
B sintering plus heat treatment	−350	0.33	290	2.261
C as cast	−335	0.58	308	3.962

[a]Key to symbols:
E_{corr} = resting potential.
I_{corr} = equilibrium current density.

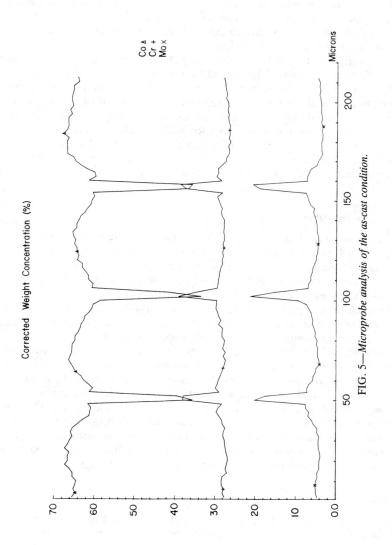

FIG. 5—*Microprobe analysis of the as-cast condition.*

Corrosion Testing

Anodic and cathodic polarization data are reported in Fig. 4 and Table 3. The breakdown potentials were found to be similar for the three conditions tested, being approximately 300 mV. Each condition exhibited a positive hysteresis when scanned in the reverse direction from the transpassive region, indicating a possible lack of susceptibility to pitting corrosion, under the conditions investigated. The sintered, HIP-treated, and heat-treated condition was found to exhibit the lowest current density in the passive region, followed by the sintered and heat-treated and the as-cast conditions.

The intersection of the anodic and cathodic polarization scans is shown in Fig. 4 at points A, B, and C for the sintered, HIP-treated, and heat-treated; the sintered and heat-treated; and the as-cast conditions, repectively. The sintered, HIP-treated, and heat-treated condition was found to exhibit the lowest equilibrium current density (0.14 A/cm^2), followed by the sintered and heat-treated (0.33 A/cm^2), and the as-cast (0.58 A/cm^2) conditions. In order to calculate the relative corrosion rates for each condition, Faraday's law was used in conjunction with these current densities (see the Appendix). The corresponding relative corrosion rates were found to be 0.965, 2.261, and 3.962 μm per year for the sintered, HIP-treated, and heat-treated; the sintered and heat-treated; and the as-cast conditions, respectively.

Surface Chemical Analyses

Surface chemical analyses are reported in Figs. 5, 6, and 7 for the as-cast; the sintered and heat-treated; and the sintered, HIP-treated, and heat-treated conditions, respectively. The data are plotted as a corrected weight percentage of cobalt, chromium, and molybdenum (discounting other elements) versus distance (scan across the surface). For the as-cast condition, the three peak regions occurring at approximately 50, 100, and 155 μm represent carbide regions. It is apparent that, for areas between carbides, there is both chromium and molybdenum depletion, as well as cobalt enrichment. These variations are caused by the elemental segregation inherent in the cored, as-cast condition.

The sintered and heat-treated condition exhibited a marked decrease in elemental segregation in comparison with the as-cast condition (see Fig. 6). The peaks occurring at approximately 310 μm represent a carbide remaining from the sintering process. The region scanned between 90 and 280 μm reveals some remaining areas of chromium depletion.

The sintered, HIP-treated, and heat-treated condition revealed a reduction in elemental segregation similar to that of the sintered and heat-treated condition (see Fig. 7). The peaks occurring in the 100-μm range correspond to one of the few carbides remaining from the as-cast condition. The regions distant to the carbide revealed a slightly greater reduction in segregation than was found in the sintered and heat-treated condition.

Discussion

The corrosion behavior of the Co-Cr-Mo alloy was found to be dependent on the microstructural condition. The as-cast condition revealed a carbide-containing, highly segregated microstructure. After sintering, most of the matrix carbides were solutionized, and a reduction in elemental segregation was observed. The liberation of chromium from carbide dissolution, as well as a more homogeneous microstructure, would be expected to result in a more stable, uniform oxide layer. These findings proved to be consistent with the observed corrosion behavior, which resulted in a lower equilibrium current density and a lower current density in the passive region for the sintered conditions than for the as-cast condition. Devine and Wulff [10] observed a similar reduction in chemical inhomogeneity in wrought and heat-

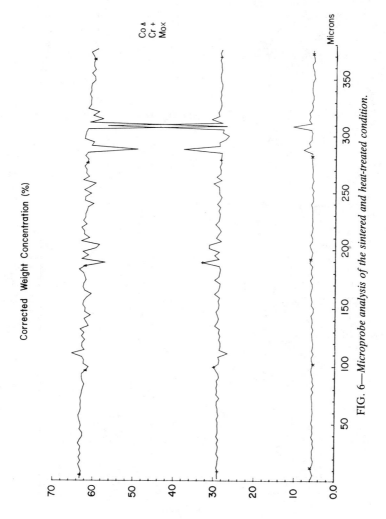

FIG. 6—*Microprobe analysis of the sintered and heat-treated condition.*

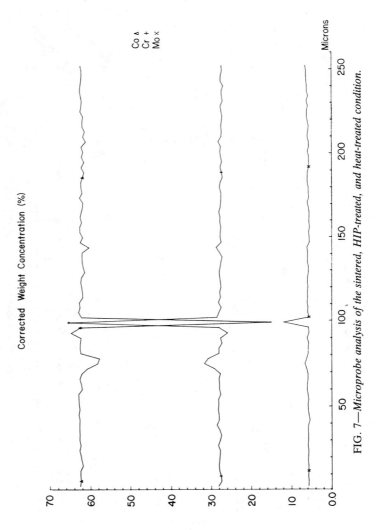

FIG. 7—*Microprobe analysis of the sintered, HIP-treated, and heat-treated condition.*

treated HS 21 alloy, relative to the cast and annealed condition. These authors noted an increased crevice corrosion resistance for the wrought condition, with the pitting potentials being equivalent for both conditions.

The slight improvement in the corrosion resistance of the HIP-treated material, over that of the sintered material which was not HIP-treated, is attributed to the further reduction in elemental segregation. Although there was a substantial amount of porosity exhibited by the sintered and heat-treated condition, it is not clear what effects this may have on the material's corrosion resistance. Microstructural evaluation did not reveal an increase in the extent of the existing porosity subsequent to corrosion testing. In addition, no susceptibility to pitting corrosion was observed, despite the presence of the pores generated from the sintering operation.

Conclusions

The thermal process commonly used to apply a porous Co-Cr-Mo coating to a substrate of the same alloy, can reduce elemental segregation inherent in the as-cast condition. The sintering treatment produces both porosity and grain boundary precipitates, resulting in reduced mechanical properties. Despite the amount of porosity induced by the sintering cycle, an improvement in corrosion resistance was observed. The effect of HIP sintered materials further reduces elemental segregation, resulting in a higher corrosion resistance and improved mechanical properties.

APPENDIX

Corrosion Rate Calculation

The calculations for the corrosion rate (in milliinches per year)[2] were made assuming a Cr_2O_3 oxide layer.

$$\text{Corrosion rate} = \frac{0.13\ I_{corr}\ (EW)}{d}$$

where

I_{corr} = equilibrium current density, A/cm^2,

EW = equivalent weight of the metal forming the oxide layer, g, and

d = density, g/cm^3.

Assuming the following values for the sintered, HIP-treated, and heat-treated condition:

I_{corr} = 0.14 A/cm^2,

EW = 17.332 g, and

d = 8.387 g/cm^3,

---

[2] 1 in. = 25.4 mm.

we arrive at the following result:

$$\text{Corrosion rate} = \frac{0.13 \ (0.14) \ (17.332)}{8.387}$$

$$= 0.038 \ \text{milliin./year}$$

$$= 0.965 \ \mu\text{m/year}$$

References

[1] Hodge, F. G. and Lee, T. S. III, "Effects of Processing on Performance of Cast Prosthesis Alloys," *Corrosion,* Vol. 31, No. 3, 1975, pp. 111–112.

[2] Dobbs, H. S. and Robertson, J. L. M., "Heat Treatment of Cast Co-Cr-Mo for Orthopaedic Implant Use," *Journal of Materials Science,* Vol. 18, 1983, pp. 391–401.

[3] Lorenz, M., Semlitsch, M., Panic, B., Weber, H., and Willert, H. G., "Fatigue Strength of Cobalt-Base Alloys with High Corrosion Resistance for Artificial Hip Joints," *Institution of Mechanical Engineers,* Vol. 7, 1978, pp. 241–250.

[4] Clemow, A. J. T. and Daniell, B. S., "Solution Treatment Behavior of Co-Cr-Mo Alloy," *Journal of Biomedical Materials Research,* Vol. 13, 1979, pp. 265–279.

[5] Georgette, F. S. and Davidson, J. A., "The Effect of HIP'ing on the Fatigue and Tensile Strength of a Cast, Porous-Coated Co-Cr-Mo Alloy," presented at the Eleventh Annual Meeting of the Society for Biomaterials, San Diego, 1985.

[6] Hollander, R. and Wulff, J., "New Technology for Mechanical Property Improvement of Cast Co-Cr-Mo-C Surgical Implants," *Institution of Mechanical Engineers,* Vol. 3, No. 4, 1974, pp. 8–9.

[7] Acharya, A., Freise, E., and Greener, E. H., "Open-Circuit Potentials and Microstructure of Some Binary Co-Cr Alloys," *Cobalt,* Vol. 47, 1970, pp. 75–80.

[8] Kilner, T., Pilliar, R. M., Weatherly, G. C., and Allibert, C., "Phase Identification and Incipient Melting in a Cast Co-Cr Surgical Implant," *Journal of Biomedical Materials Research,* Vol. 16, 1982, pp. 63–79.

[9] Greene, N. D., *Experimental Electrode Kinetics,* Rensselaer Polytechnic Institute, Troy, NY, 1965.

[10] Devine, T. M. and Wulff, J., "Cast vs Wrought Cobalt-Chromium Surgical Implant Alloys," *Journal of Biomedical Materials Research,* Vol. 9, 1975, pp. 151–167.

Walter P. Spires, Jr.,[1] *David C. Kelman,*[2] *and John A. Pafford*[3]

Mechanical Evaluation of ASTM F75 Alloy in Various Metallurgical Conditions

REFERENCE: Spires, W. P., Jr., Kelman, D. C., and Pafford, J. A., **"Mechanical Evaluation of ASTM F75 Alloy in Various Metallurgical Conditions,"** *Quantitative Characterization and Performance of Porous Implants for Hard Tissue Applications, ASTM STP 953,* J. E. Lemons, Ed., American Society for Testing and Materials, Philadelphia, 1987, pp. 47–59.

ABSTRACT: This study evaluated the mechanical properties and microstructure, in various metallurgical conditions, of an alloy meeting the requirements of the ASTM Specification for Cast Cobalt-Chromium-Molybdenum Alloy for Surgical Implant Applications (F 75-82). Standard mechanical property analyses such as tensile and yield strength, percentage of elongation, and rotating-beam fatigue tests were performed. The materials were tested in the as-cast, as-cast plus solution-annealed, as-cast plus hot-isostatic-pressed (HIP) plus solution-annealed, and as-cast plus HIP plus sintered conditions. Chemical analyses of the test specimens were reviewed to determine the extent of the correlation between mechanical properties and certain elements within the alloy, as has been suggested by other authors.

No effect on ultimate tensile strength was observed with a reduction of carbon content from 0.30 to 0.24%, but a reduction of the yield strength and a corresponding increase in ductility were observed. The as-sintered surface appeared to reduce the fatigue properties of the test specimens. A direct correlation between the fatigue properties, static mechanical properties, and microstructure was not established.

KEY WORDS: porous implants, sintering, fatigue, hot isostatic pressing, cobalt-chromium-molybdenum alloy, implant materials

With the advent in recent years of porous-coated materials for use in total joint replacements in humans, concern over maintaining the mechanical and metallurgical integrity of these implants has become increasingly important to manufacturers and surgeons. The improvements in mechanical properties achieved during the late 1970s by thermal and mechanical working of the cobalt-chromium alloys have, for the most part, been lost as a result of the sintering processes. After sintering, the fine-grained microstructures that result from thermal working are nearly indistinguishable from the microstructures of castings. There is also concern over incipient melting [1], as well as other grain-boundary phenomena that occur and phases that precipitate out at these grain boundaries during thermal treatment.

Since the functional end result of the prosthesis is of primary importance, the details of the sintering operation are not widely published—because of their proprietary nature and other business considerations. What is necessary is a careful examination of the resulting microstructures, alloy chemistries, static mechanical properties, and fatigue properties. If ASTM standards for surgical implants are to be further developed and implemented, uniform methods of testing and evaluation will have to be agreed upon between the various manufacturers and other interested parties.

[1] Director, Research and Commercialization, Dow Corning Wright, Arlington, TN 38002.
[2] Senior research engineer, DePuy, Warsaw, IN 46580.
[3] Supervisor, Custom and Specialty Products, Dow Corning Wright, Arlington, TN 38002.

TABLE 1—*Chemical analysis of the low- and high-carbon test materials.*

Carbon Content	Component Element, weight %										
	Carbon	Manganese	Silicon	Nickel	Chromium	Molybdenum	Phosphorus	Sulfur	Iron	Cobalt
Low carbon	0.24	0.43	0.81	0.14	28.55	6.12	0.014	0.005	0.29	balance
High carbon	0.30	0.56	0.78	0.17	28.73	5.75	0.011	0.006	0.22	balance

Materials and Methods

This study undertook an evaluation of the static and dynamic mechanical properties, microstructure, and carbon content of an alloy meeting the requirements of the ASTM Specification for Cast Cobalt-Chromium-Molybdenum Alloy for Surgical Implant Applications (F 75-82), which was subjected to various thermal processes. The test methods used were (1) tension testing on a closed-loop hydraulic Instron test machine and (2) a rotating-beam fatigue test performed on a Fatigue Dynamics Inc. rotating-beam machine. The tension specimen preparation and tests were conducted in accordance with the ASTM Methods of Tension Testing of Metallic Materials (E 8-85). The rotating-beam fatigue tests were performed in air at a frequency of 65 Hz. The fatigue test was terminated if (1) the specimen fractured, or (2) the specimen went through 10^7 cycles without fracturing. In the event that the specimen reached 10^7 cycles, the test was continued with the load increased. This continued until fracture occurred. The porous-coated fatigue specimens were coated with two layers of -18 to $+30$-mesh spherical powder conforming to the requirements of ASTM Specification F 75-82. The material evaluated was an ASTM F75 alloy with master-heat carbon content analyses of (1) 0.30% and (2) 0.24%. The balance of the composition remained virtually unchanged (See Table 1). The material was tested in the following conditions:

(a) as cast,
(b) solution annealed,
(c) cast plus hot isostatic pressed (HIP) plus solution annealed,
(d) HIP plus sintered plus solution annealed, and
(e) sintered plus HIP plus solution annealed.

The high-carbon (0.30%) material was evaluated in all five metallurgical conditions. The lower carbon material (0.24%) was evaluated only in the fifth condition (sintered plus HIP plus solution annealed).

TABLE 2—*Method of specimen preparation.*

Sequential Process Step	Method A	Method B	Method C
As cast	X	X	X
Thermally processed	X	NA[a]	NA
Rough machined	X	NA	NA
Low-stress ground	X	NA	NA
Polished	X	NA	NA
Rough machined	NA	X	X
Low-stress ground	NA	X	X
Polished	NA	X	X
Thermally processed	NA	X	X
Polished	NA	X	NA
Straightened	NA	X	X
Specimen size, mm	9.5 by 4.8	12.7 by 6.4	12.7 by 6.4
Surface finish, roughness average, μm	0.31	0.31	1.17

[a] NA = not applicable.

High-Cycle Fatigue Test Parameters

The high-cycle fatigue test parameters were the following:

(*a*) test specimen size: 9.5 by 4.8 mm versus 12.7 by 6.4 mm;
(*b*) test material source: test bars versus hip stems;
(*c*) test specimen preparation method: A, B, or C;
(*d*) thermal history;
(*e*) surface finish; and
(*f*) carbon content.

Specimen Preparation

Descriptions of the preparation methods are found in Table 2.

Results

Tension Test Results

The results of the mechanical tests, including those for ultimate tensile strength, yield strength, elongation, and reduction of area are included in Figs. 1 through 4, respectively. The sources of the tension test specimens were separately cast and machined specimens and distal portions of hip stems. Previous work indicated no difference between the two sources of test material.[3] A composite table showing the properties of the high-carbon versus low-carbon materials in the sintered plus HIP plus solution-annealed condition is shown in Fig. 5.

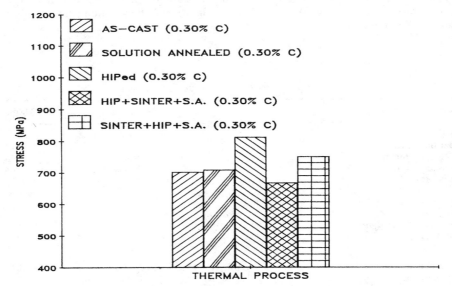

FIG. 1—*Ultimate tensile strength of ASTM F75 alloy.*

---

[3] "Dow Corning Laboratory Notebook," No. 6646, Dow Corning Wright, Arlington, TN, p. 19.

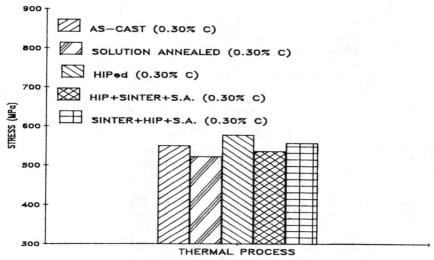

FIG. 2—*Yield strength of F75 alloy.*

High-Cycle Fatigue Test Results

The thermal histories and specimen preparation methods are listed in Table 3. The high-cycle fatigue limits versus the conditions given in Table 3 are shown in Fig. 6. The curves for stress versus number of cycles (S/N) for the various material conditions are shown in Figs. 7 through 12. It was not possible to establish a fatigue curve for the HIP plus solution-annealed condition from the current results because the results were all close to the 10-million-cycle location on the graph.

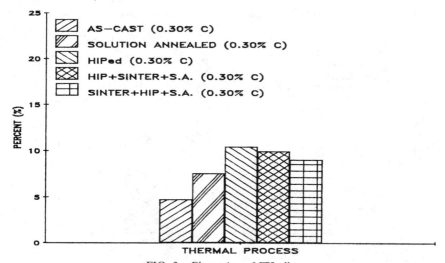

FIG. 3—*Elongation of F75 alloy.*

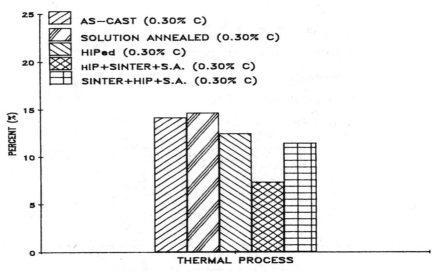

FIG. 4—*Reduction of area of F75 alloy.*

Discussion

Tension Test Results

The benefits of postcasting thermal processing of ASTM F75 alloy may be seen in the tension test results (Figs. 1 through 4). There was no statistical difference in the ultimate tensile strengths of the high-carbon (0.30%) and the low-carbon (0.24%) materials processed by sintering plus hot isostatic pressing plus solution annealing (Fig. 5). A statistically significant decrease of 10% in yield strength was observed for the low-carbon (0.24%) material (see Table 4). Corresponding to the decrease in yield strength, an increase in elongation and in reduction of area of 35 to 40% and 15%, respectively, was observed. Representative

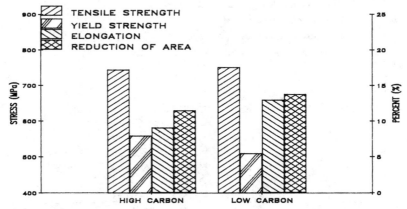

FIG. 5—*Effect of carbon content on sintered/hot-isostatic-pressed/solution-annealed F75 alloy.*

TABLE 3—*Thermal conditions of and preparation methods for the high-cycle fatigue test specimens.*

Specimen Code	Thermal Cycle	Carbon Content, Weight %	Test Method	Specimen Source	Porous Coated
1AT	as cast	0.30	A	test bar	no
2AT	solution annealed	0.30	A	test bar	no
3AH	HIP—sintered—solution annealed	0.30	A	hip stem	no
4BT	sintered—HIP—solution annealed	0.24	B	test bar	no
5CT	sintered—HIP—solution annealed	0.24	C	test bar	yes
6CT	sintered—HIP—solution annealed	0.24	C	test bar	no

TABLE 4—*Mechanical properties of low- and high-carbon materials after sintering plus hot isostatic pressing plus solution annealing.*

Carbon Content[a]	Tensile Strength		Yield Strength		Elongation		Reduction of Area	
	Mean, MPa	Standard Deviation	Mean, MPa	Standard Deviation	Mean, %	Standard Deviation	Mean, %	Standard Deviation
High carbon	746.7	54.5	554.7	9.7	9.1	1.4	11.1	1.5
Low carbon	748.8	61.4	510.9	30.3	13.0	1.52	13.8	3.3

[a] Based on 21 and 75 tension tests of the high- and low-carbon material, respectively.

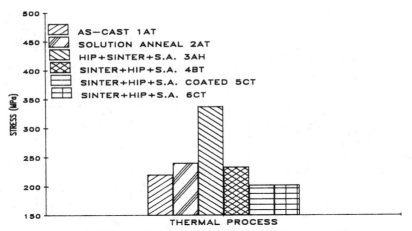

FIG. 6—*High-cycle fatigue limits of F75 alloy.*

microstructures of the high-carbon and low-carbon materials are shown in Figs. 13 and 14, respectively. As can be seen in Fig. 13, there is a significant amount of eutectic grain boundary carbide that was not resolutionized during the solution annealing process. Numerous attempts were made to refine these microstructures and, in most cases, these were unsuccessful. Microstructures for the lower carbon materials show a much finer dispersion of grain boundary carbides, and those within the matrix were resolutionized (see Fig. 14).

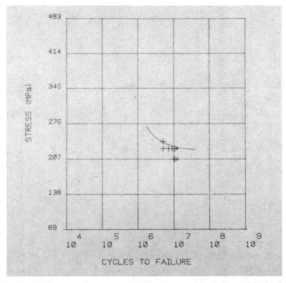

FIG. 7—*S/N curve for as-cast F75 alloy, smooth bar—Method A.*

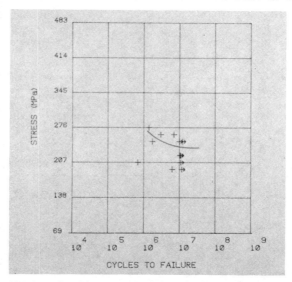

FIG. 8—S/N *curve for solution-annealed F75 alloy, smooth bar—Method A.*

Fatigue Test Results

The highest fatigue limits were obtained with the high-carbon material processed so that the hot isostatic pressing operation was performed prior to the sintering and solution annealing. The low ultimate tensile properties and the significant amount of eutectic carbide along the grain boundaries of these materials would lead one to predict lower fatigue

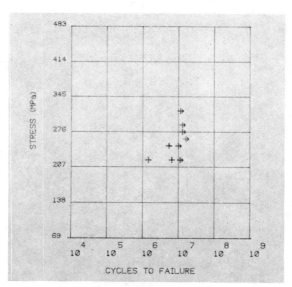

FIG. 9—S/N *curve for hot-isostatic-pressed (HIP) and solution-annealed F75 alloy, smooth bar—Method A.*

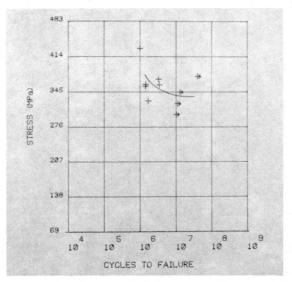

FIG. 10—S/N *curve for HIP/sintered/solution-annealed F75 alloy, smooth bar—Method A.*

properties. Additional testing is being conducted to determine if this result was an anomaly based on sample size or grain orientation within the test specimen.

The fatigue properties obtained from the limited number of specimens evaluated in specimen preparation Method C yielded similar results for both porous-coated and non-porous-coated test specimens. This indicates that the rough as-sintered surface may contribute to

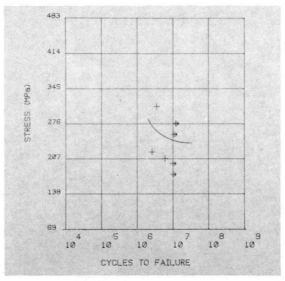

FIG. 11—S/N *curve for sintered/HIP/solution-annealed F75 alloy, smooth bar—Method A.*

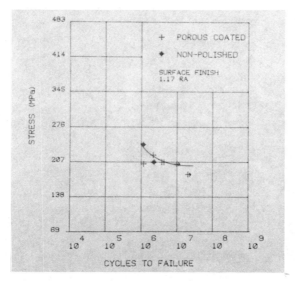

FIG. 12—*S/N curve for sintered/HIP/solution-annealed F75 alloy, porous coated and unpolished.*

a reduction of the fatigue properties. This may account for the reduction in fatigue properties observed by Georgette and Davidson [2] in a single layer of powder. As the number of layers increases, the increase in section modulus may override the surface effect.

It is very difficult to obtain high-cycle fatigue specimens that have been machined prior to the thermal processing and that maintain concentricity within 0.05 mm after processing.

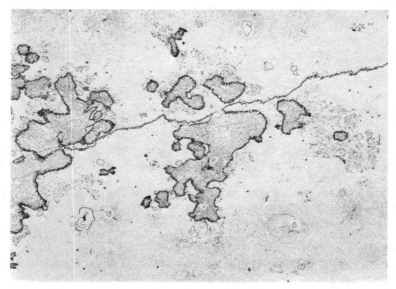

FIG. 13—*Microstructure of sintered/HIP/solution-annealed F75 alloy, at ×100 for 0.30% carbon content.*

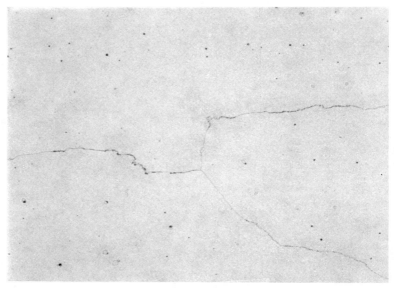

FIG. 14—*Microstructure of sintered/HIP/solution-annealed F75 alloy, at ×100 for 0.24% carbon content.*

Some authors have buried the specimens in ceramic powder, but this does not duplicate the thermal process utilized for actual implants [3]. Mechanical straightening of specimens should be avoided if at all possible to avoid damaging the specimens.

Conclusions

Based on these test results:

1. The as-sintered surface condition appears to reduce the fatigue properties of the test specimens. This is influenced by the presence of beads and the number of layers of coating applied.

2. After thermal processing, there was no apparent correlation between fatigue properties, static mechanical properties, and microstructure. Further tests are being run to evaluate some of the unexplained results.

3. Machining the high-cycle fatigue bars prior to thermal processing introduces additional variables. Care must be exercised to ensure that the process duplicates that utilized for the actual implants and produces acceptable test specimens.

4. The lower carbon content material test specimens had reduced yield strength and carbide content with an increase in ductility.

5. To avoid having only one or two grain boundaries at the reduced diameter and to obtain more consistent results, specimens with a minimum grain boundary diameter of 6.4 mm should be utilized.

References

[1] Kilner, T., Pilliar, R. M., and Weatherly, G. C., "Incipient Melting During the Heat Treatment of Cast-Co-Cr Surgical Implants," *Transactions,* Society of Biomaterials Meeting, Walt Disney World, FL, 1982.
[2] Georgette, F. S. and Davidson, J. A., "The Affect of HIPing on the Fatigue and Tensile Strength of a Cast, Porous Coated Co-Cr-Mo Alloy," *Transactions,* Society of Biomaterials Meeting, San Diego, CA, 1985.
[3] Vue, S., Pietrobon, T., Weatherly, G. C., and Pilliar, R. M., "The Fatigue Strength of Porous Surfaced Co-Cr-Mo Alloy," *Transactions,* Society of Biomaterials Meeting, San Diego, CA, 1985.

David J. Levine[1]

Metallurgical Relationships of Porous-Coated ASTM F75 Alloys

REFERENCE: Levine, D. J., **"Metallurgical Relationships of Porous-Coated ASTM F75 Alloys,"** *Quantitative Characterization and Performance of Porous Implants for Hard Tissue Applications, ASTM STP 953,* J. E. Lemons, Ed., American Society for Testing and Materials, Philadelphia, 1987, pp. 60–73.

ABSTRACT: Discontinuous carbide formation along the grain boundaries was found to be a desirable condition for obtaining ASTM-recommended strength and ductility levels in a sintered alloy containing 0.3% carbon by weight and meeting the requirements of the ASTM Specification for Cast Cobalt-Chromium-Molybdenum Alloy for Surgical Implant Applications (F 75-82). Excessive carbides in the grain boundaries of this alloy caused a severe reduction in ductility, while relatively carbide-free grain boundaries produced acceptable ductility but a marked reduction in the ultimate and 0.2% yield strengths. This behavior was also true for an ASTM F75 alloy with a lower carbon level (0.24% by weight). However, the latter alloy failed to meet the strength levels specified in ASTM Specification F 75-82. Subsequent aging of similar specimens led to carbide precipitation within the grains and dissolution of grain boundary particles, which caused an immediate loss of ductility. In addition, there was evidence for the precipitation of intermetallic, topologically close-packed (TCP) phases, which have been shown to have deleterious effects on the room-temperature ductilities of cobalt-base alloys. This phenomenon occurs in alloys with both high (0.3%) and low (0.24%) carbon contents. The precipitate population and the severity of ductility loss were time and temperature dependent.

KEY WORDS: porous implants, ASTM Specification F 75-82, cobalt-base alloys, porous coatings, sintering, tensile strength, ductility, carbide formation, sigma phase, carbon depletion

The process of sintering beads to a substrate, both of which meet the requirements of the ASTM Specification for Cast Cobalt-Chromium-Molybdenum Alloy for Surgical Implant Applications (F 75-82), necessitates thermal exposures at about 1300°C, which is well above the carbide eutectic temperature. Exposing the material for 1 to 3 h, which is necessary to form a strong bond between the beads and the substrate as well as bonds among the beads, causes microstructural changes that may have severe detrimental effects on the mechanical properties of the alloy. The severity of these effects varies with the amount of eutectic formed at the grain boundary, as well as with the morphology of the grain boundary phases. Pilliar [1] has pointed out that a continuous grain boundary carbide phase can result in mechanical property losses. The severity of these effects, however, is directly related to the carbon level in the ASTM F75 alloy. A lower carbon level in the alloy will produce a lesser amount of eutectic phases at the grain boundaries, thus alleviating the problem. Lower carbon levels, however, also tend to result in a reduction in strength and may result in failure to meet the requirements of ASTM Specification F 75-82. In addition, carbon-

---

[1] Technical director, Astro Met Associates, Inc., Cincinnati, OH 45215.

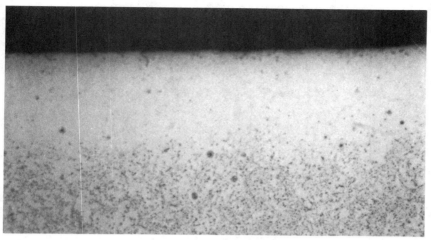

FIG. 1a—*Photomicrograph showing a carbon-depleted zone in a 0.3% carbon ASTM F75 alloy (×200 magnification).*

depleted surface areas showed a significant amount of sigma-type phase formation (see Figs. 1a and 1b).

The purpose of this work was to investigate workable methods for eliminating or reducing the continuous massive grain boundary carbide formation with post-porous-coating heat treatments, while maintaining relatively high carbon levels in the alloy for strength purposes.

Experimental Work and Results

The experimental work consisted of exposing tensile bars of two different carbon levels, both meeting the requirements of ASTM Specification F 75-82, to various thermal cycles.

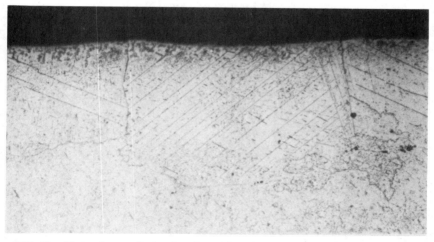

FIG. 1b—*Photomicrograph showing sigma phase formation in the carbon-depleted zone after 2 h at 1065°C in a 0.3% carbon F75 alloy (×200 magnification).*

TABLE 1—Chemical analyses of F75 alloys, in weight percent.

Lot No.	Co	Cr	Mo	C[a]	Fe	Mn	Si	N
21930	balance	27.66	6.34	0.295	0.24	0.46	0.82	0.063
22357	balance	27.29	5.45	0.238	0.24	0.48	0.82	0.045

[a] For convenience the carbon levels in the text are referred to as 0.3 and 0.24%, respectively.

TABLE 2—Mechanical properties of F75 alloys.

Specimen Condition	0.3% Carbon[a]				0.24% Carbon[a]				Microstructural Appearance
	Ultimate Tensile Strength MPa (ksi)	0.2% Yield Strength MPa (ksi)	Elongation, %	Reduction of Area, %	Ultimate Tensile Strength MPa (ksi)	0.2% Yield Strength MPa (ksi)	Elongation, %	Reduction of Area, %	
As cast	855 (124)	558 (81)	6	8	827 (120)	503 (73)	8	10	dentritic
Solution annealed	869 (126)	545 (79)	12	13	862 (125)	517 (75)	15	14	partially dendritic
Thermal Cycle A	710 (103)	469 (68)	9	11	724 (105)	427 (62)	14	15	discontinuous carbide at grain boundaries (0.3% carbon only)
Thermal Cycle B	634 (92)	450 (71)	4	6	614 (89)	359 (52)	16	13	heavy carbide at grain boundaries (0.3% carbon only)
Thermal Cycle C	606 (88)	414 (60)	12	11	641 (93)	400 (58)	13	15	no carbide at grain boundaries
Thermal Cycle C plus aging	738 (107)	552 (80)	3	3	524 (76)	427 (62)	2	3	precipitation

[a] The carbon levels are rounded off from the values given in Table 1.

The carbon levels selected were approximately 0.24 and 0.3% by weight. All other constituents in the two lots of material were similar and within the limits of ASTM Specification F 75-82 (Table 1).

All the tension tests were performed on an Instron tension testing machine, in duplicate, at room temperature, with strain rates of 0.005 mm/mm/min through 0.2% yield strength, and of 0.05 mm/mm/min to failure. One tested specimen of each condition was sectioned, mounted, and polished. In addition, selected specimens were analyzed by energy-dispersive X-ray analysis (EDAX) and scanning electron microscopy (SEM) techniques to identify the nature and composition of the constituents at the grain boundaries. The polished specimens were then examined metallographically, and the correlations between the microstructural characteristics and tensile properties were established. Microstructural and grain boundary observations, described in Table 2, together with the data obtained from the tension tests, suggested a definite and predictable relationship between the microstructure and the mechanical properties for these alloys.

Discussion

To create a porous coating on an ASTM F75 alloy prosthesis for better implant fixation, the beads must be sintered to the substrate at approximately 1300°C, which is well above the carbide eutectic temperature. Rapid cooling led to massive carbide concentration at the grain boundaries (see Fig. 2a), which resulted in a severe reduction in the ductility. This has been described by Kilner [2].

The extent of the carbide formation at the grain boundaries is largely dependent on the carbon level in the alloy (see Figs. 2a and 2b). The photomicrographs depict the microstructures of F75 alloy specimens with carbon levels of 0.3 and 0.24% by weight, respectively. Both were exposed to Thermal Cycle B, and, as shown in Table 2, the elongation of the 0.3% carbon specimen measured only 4%, while the 0.24% carbon specimen showed an elongation of 16% but a low yield strength. This is in line with data generated by others. Kilner [2] points out that, in order to eliminate the grain boundary carbide formation, the carbon levels must be drastically reduced (0.17%). However, this causes a serious reduction in the mechanical properties.

It has been found that careful control of the sintering parameters and cooling rates for F75 alloy substrates with a 0.3% carbon level can produce structures with desirable amounts of carbides in the grain boundaries, and with resulting strengths and ductilities within the limits of ASTM Specification F 75-82. A more acceptable distribution of carbides is shown in Figs. 3a and 3b. The data in Table 2 (Thermal Cycle A) show both strength and ductility within the limits of ASTM Specification F 75-82 for the 0.3% carbon specimens. The specimen with 0.24% carbon shows a 0.2% yield strength of 427 MPa (62 ksi), which is 21 MPa (3 ksi) below the ASTM limits, when processed at the same thermal cycle.

The photomicrographs shown in Figs. 4a and 4b depict the microstructures of the 0.3 and 0.24% carbon specimens, respectively, after exposure to Thermal Cycle C (Table 2). Note the complete absence of carbides at the grain boundaries. Thermal Cycle C caused all carbides to stay in solution, which is not necessarily desirable. The data in Table 2 show that the strength levels of these specimens fall below the minimum strength standards of ASTM Specification F 75-82, with 0.2% yield strengths of 414 and 400 MPa (60 and 58 ksi), respectively, for the 0.3% and 0.24% carbon levels.

Duplicate specimens of those shown in Figs. 4a and 4b, aged at 1065°C for 2 h, show secondary $M_{23}C_6$ carbide precipitation (Figs. 5a and 5b) and, as shown in Table 2, a marked decrease in ductility for specimens of both carbon levels. The specimen containing 0.3% carbon showed some increase in the 0.2% yield strength, but there was only minimal 0.2%

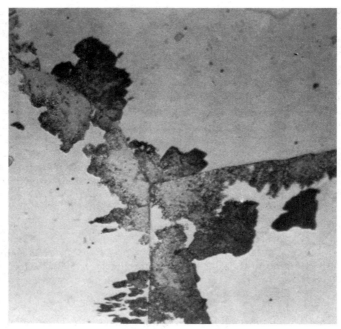

FIG. 2a—*Photomicrograph showing massive grain boundary carbides after Thermal Cycle B in a low-ductility, 0.3% carbon F75 alloy (×200 magnification).*

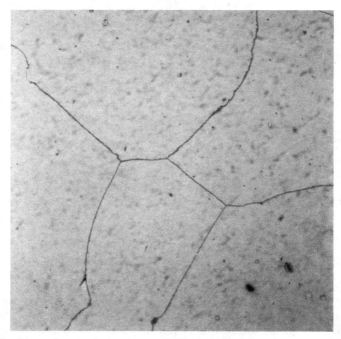

FIG. 2b—*Photomicrograph showing no visible grain boundary formation after Thermal Cycle B in a good-ductility, 0.24% carbon F75 alloy (×200 magnification).*

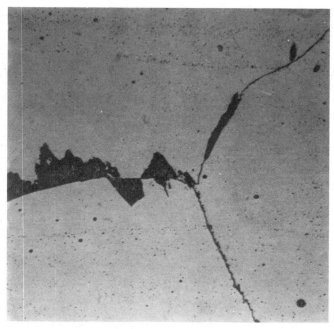

FIG. 3a—*Photomicrograph showing discontinuous grain boundary carbides after Thermal Cycle A in a good-ductility, good-strength, 0.3% carbon F75 alloy (×200 magnification).*

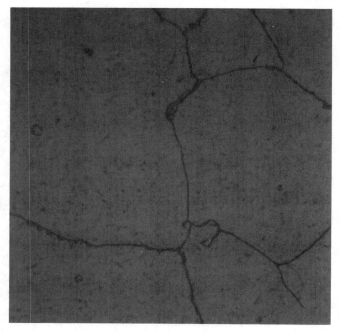

FIG. 3b—*Photomicrograph showing no visible grain boundary carbide formation after Thermal Cycle A in a good-ductility, low 0.2% yield strength, 0.24% carbon F75 alloy (×200 magnification).*

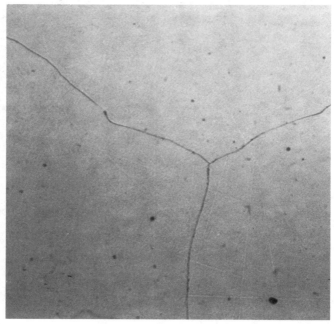

FIG. 4a—*Photomicrograph showing no visible carbides in grain boundaries after Thermal Cycle C in a good-ductility, low-strength, 0.3% carbon F75 alloy (×200 magnification).*

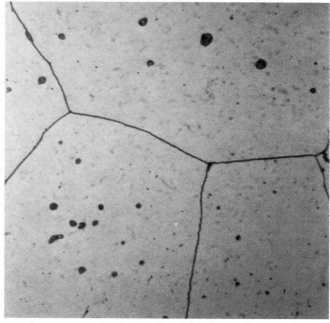

FIG. 4b—*Photomicrograph showing no visible carbides in grain boundaries after Thermal Cycle C in a good-ductility, low strength, 0.24% carbon F75 alloy (×200 magnification).*

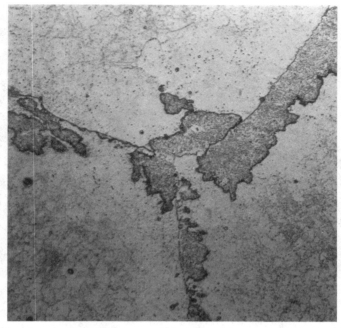

FIG. 5a—*Photomicrograph showing heavy precipitation of secondary carbides in a low-ductility, 0.3% carbon F75 alloy (×200 magnification).*

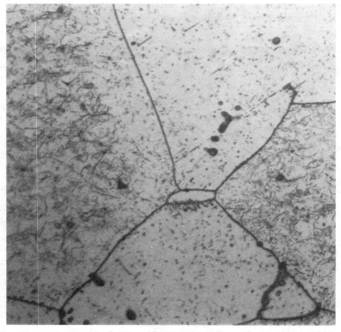

FIG. 5b—*Photomicrograph showing heavy precipitation of secondary carbides in a low-ductility, 0.24% carbon F75 alloy (×200 magnification).*

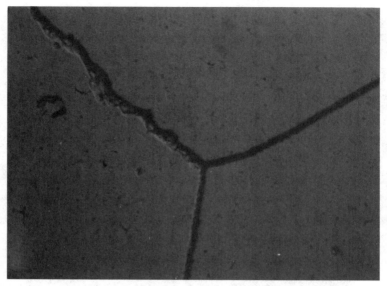

FIG. 6a—*Photomicrograph showing the grain boundary appearance of a ductile 0.24% carbon F75 alloy specimen (×1000 magnification).*

yield strength increase in the 0.24% carbon specimen. This again indicates the role of the carbon level in the F75 alloy system.

It is interesting to note that aging at 1065°C for 2 h causes carbides to precipitate within the grains rather than at the grain boundaries. At ×1000 magnification (Figs. 6a and 6b)

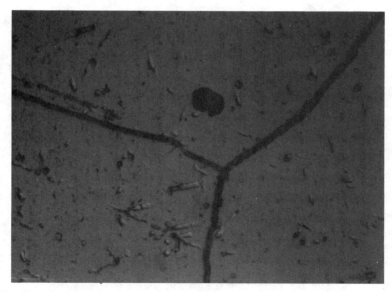

FIG. 6b—*Photomicrograph showing the grain boundary appearance of a brittle 0.24% carbon F75 alloy specimen (×1000 magnification).*

the grain boundaries of aged and unaged specimens appear to be similar. Nevertheless, the tension test failures were intergranular in both cases. The specimen depicted in Fig. 6a showed only 1% elongation, while the specimen in Fig. 6b had an elongation of 15%.

Scanning electron microscopy (SEM) studies of the grain boundaries (Fig. 7) did not reveal any visible differences between the ductile (*left*) and brittle (*right*) specimens. [The grain boundary in the ductile specimen is thicker than that in the brittle specimen, but statistical sampling of the grain boundaries of the whole specimens proved these differences not be be significant.]

Corresponding electron-dispersive spectra (EDS) analysis revealed that the material at the grain boundaries of both specimens consisted of chromium, cobalt, and molybdenum, indicating the presence of $M_{23}C_6$ carbides, which were rigidly fixed and most likely to appear as fine or blocky carbides [3]. The X-ray peaks shown in Fig. 8 also reveal a relationship between the composition of the carbide formers (M) in the $M_{23}C_6$ carbide and the ductility of the material. The profile of the ductile specimen (15% elongation) is shown in Fig. 8 (*left*) and the EDS profile for the brittle specimen (1% elongation) is shown in Fig. 8 (*right*). Note that the grain boundary composition of the ductile material shows a higher cobalt, but a lower molybdenum, X-ray peak than the EDS profile of the brittle specimen.

Additional SEM work performed on the facets of the fracture itself on the same specimens revealed the presence of small embedded particles (carbides) along the facets, that is, the grain boundaries. Figure 9 (*left*) shows the appearance of a ductile fracture surface, while a brittle fracture surface is depicted in Fig. 9 (*right*). The nature and the mechanism by which the particles function are not clearly understood, but it can be postulated that their presence at the grain boundary improves ductility by inhibiting grain boundary fracture. Similar trends have been shown in thoria-dispersed (TD) nickel, TD Nichrome, and doped tungsten.[2]

It seems, therefore, that aging at 1065°C for 2 h not only precipitates carbides within the grains but also dissolves the particles at the grain boundaries. This phenomenon was described by Sims and Hagal [4]. Because the failure mode of all specimens tested was intergranular, it seemed logical that precipitation within the grains would have little effect on the ductility. The loss of ductility in the specimens heat treated at 1065°C for 2 h could therefore be contributed to the dissolution of the particles at the grain boundaries.

Conclusions

This work suggests a definite and predictable relationship between the microstructural characteristics and the mechanical properties of F75 alloys. On this basis the following conclusions can be made:

1. Discontinuous carbide formation in the grain boundary (approximately 50/50) is a desirable condition for strength and ductility in F75 alloys.

2. Massive and continuous grain boundary carbides cause a reduction in ductility without an increase in strength.

3. Completely clean grain boundaries without any carbide formation are indicative of high ductility but low strength, at values possibly outside the limits of ASTM Specification F 75-82 (depending on the carbon level).

4. Total solutioning of the carbides followed by a 1065°C 2-h heat treatment causes immediate loss of ductility, as a result of changes in the grain boundary constituents. In addition, because F75 alloy is metastable, the appearance of secondary $M_{23}C_6$ carbide may

---

[2] Generally accepted theory of grain boundary pinning.

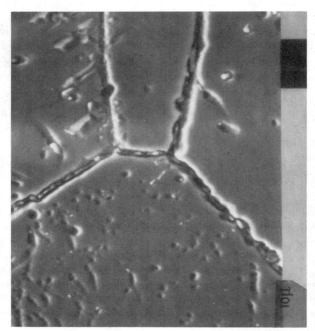

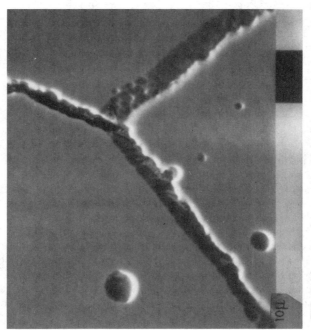

FIG. 7—Scanning electron micrograph of the same specimens as in Figs. 6a and 6b. Note the similar grain boundary appearance in the ductile (left) and brittle (right) specimens (×2000 magnification).

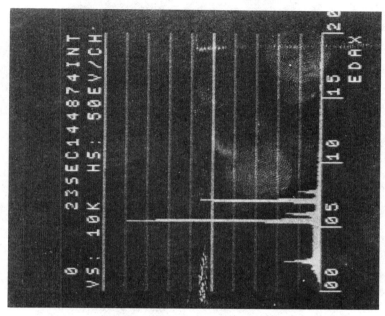

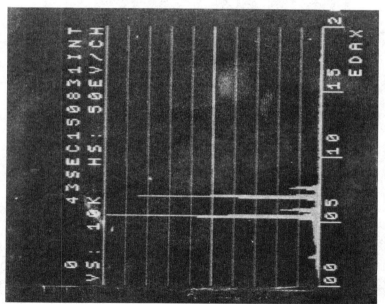

FIG. 8—*EDAX profiles of grain boundary constituents of the same specimens as in Figs. 6a and 6b. Note the higher cobalt-to-chromium ratio in the ductile specimen* (left).

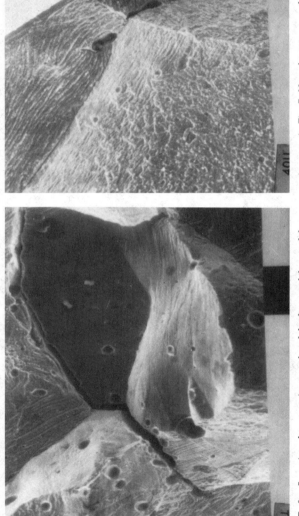

FIG. 9—Scanning electron micrographs of the fractured facets of the same specimens as in Fig. 7. Note the presence of particles in the ductile (left) specimen and the absence of particles in the brittle specimen (right) (×500 magnification).

lead to the formation of a topologically close-packed (TCP) intermetallic phase [4]. To produce an acceptable microstructure, the time- and temperature-dependent parameters must be carefully controlled.

5. In a 0.24% carbon F75 alloy material, the grain boundary phase composition of a ductile specimen has a higher cobalt-to-chromium ratio and a lower molybdenum level than is found in a brittle specimen.

Acknowledgments

The author wishes to recognize E. G. Clark of the University of Cincinnati, who performed the SEM analyses. Mechanical testing was performed by Metcut Research.

References

[1] Pilliar, R. M., "Powder Metal-Made Orthopedic Implants with Porous Surface for Fixation by Tissue Ingrowth," *Clinical Orthopaedics and Related Research,* June 1983.
[2] Kilner, T., "The Relationship of Microstructure to the Mechanical Properties of Cobalt-Chromium-Molybdenum Alloy Used for Prosthetic Devices," Ph.D. thesis, Centre for Biomaterials, University of Toronto, Toronto, Ontario, Canada, 1984.
[3] Sims, C. T., "A Contemporary View of Cobalt Base Alloys," *Journal of Metals,* December 1969.
[4] Sims, C. T. and Hagal, W. C., "The Superalloys," General Electric Co., April 1972.

Characterization of Pore Dimensions

Gary Hamman[1]

Comparison of Measurement Methods for Characterization of Porous Coatings

REFERENCE: Hamman, G., **"Comparison of Measurement Methods for Characterization of Porous Coatings,"** *Quantitative Characterization and Performance of Porous Implants for Hard Tissue Applications, ASTM STP 953*, J. E. Lemons, Ed., American Society for Testing and Materials, Philadelphia, 1987, pp. 77–91.

ABSTRACT: A comparison was conducted of measurement techniques used to quantify porous coatings applied to implant surfaces for biological fixation. Four types of coatings were measured, which included two bead size ranges, one wire pad, and one plasma flame-sprayed coating.

The purpose of the study was to determine the relationships between the measurement techniques and their resultant values when compared using two-dimensional sections. Basic stereological principles were used in the approach to this study. An automated image analysis system was utilized in making the determinations. Pore size measurement was conducted using two methods: (1) a line intercept method with serial measurements conducted at multiple levels parallel to and in plane with the substrate surface and (2) an interactive user-defined region measurement, in which the evaluator determined which features within the structure constituted individual pores. Comparison of data from both measurement methods was made contrasting the line intercept with the user-defined measurements and the planar-section with the cross-section measurements. Average values for the percentage of porosity in each coating were compared. The values obtained on cross sections were compared with values obtained on planar sections; the results indicated that the volume fraction of porosity observed in the cross sections was reproducibly 5% higher than that measured in the planar sections.

In the pore size measurements, two measurement methods and two section planes were compared. The line intercept method data showed a general tendency toward a smaller pore size when compared with the user-defined pore size data. This trend was observed in the pore size measurements in both cross sections and planar sections. The automated line intercept measurement did not exclude minimum values; therefore, its population of very small intercepts appears to skew the average pore size toward a lower average.

The choice of measurement method and the section plane selection appear to have significant influence on the average values obtained when measuring pore sizes on two-dimensional specimens.

KEY WORDS: porous implants, porous-coated implants, porosity, pore size, density

The application of a porous coating to orthopedic devices for enhancement of mechanical fixation is becoming increasingly more popular. The potential for improving mechanical interlocking of the implant with bone cement or ingrown tissue has resulted in a large variety of porous systems becoming available. Because of the complexity of pore size definition and the diversity of coating types, a definition of pore size and a standardized approach to its measurement appear to be warranted. Many clinical researchers report an average pore size [1] and coating thickness, while others report a range of porosity. This study compares methods by measuring four types of porous coatings for orthopedic implants. Two measurement methods were used to obtain data in two different section planes.

---

[1] Manager, Metallurgical Laboratory, Zimmer, Inc., Warsaw, IN 46580.

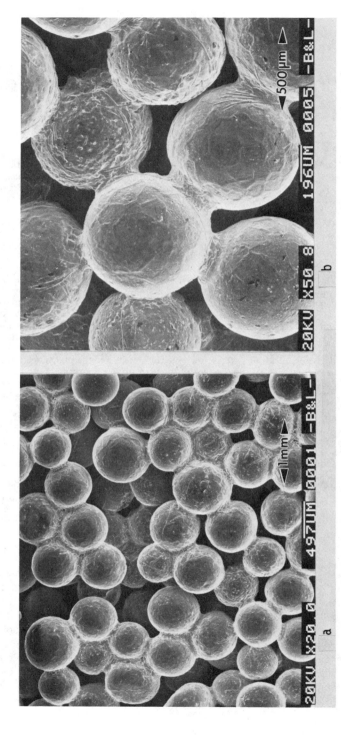

FIG. 1—*Photomicrographs of sintered Co-Cr-Mo alloy beads, 350 to 750 μm in diameter: (a) topographical view via scanning electron microscopy (SEM); (b) topographical SEM view with bead-to-bead bonding present.*

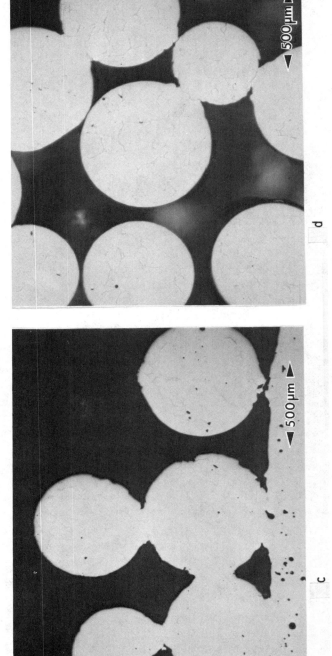

FIG. 1—Continued: (c) metallographic cross section, as polished; (d) metallographic planar section, as polished.

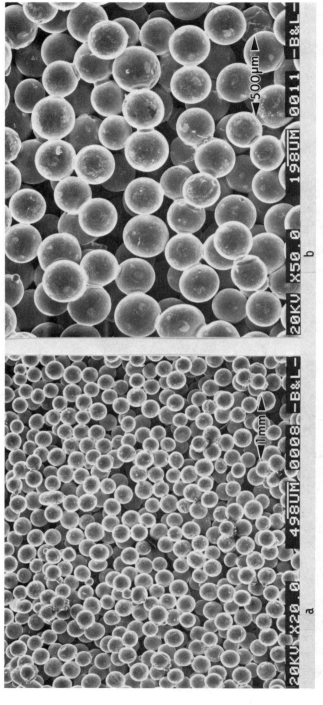

FIG. 2—*Photomicrographs of sintered Co-Cr-Mo alloy beads, 100 to 250 μm in diameter: (a) topographical view via SEM; (b) topographical SEM view showing interconnected beads.*

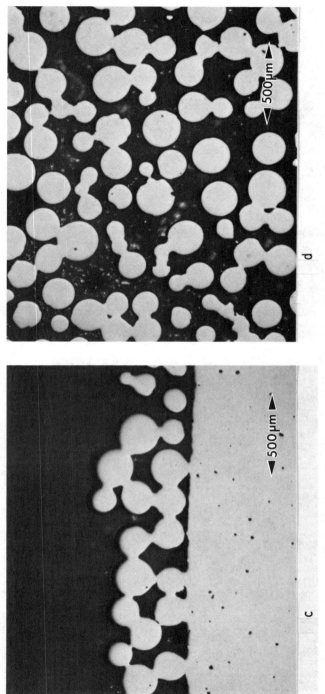

FIG. 2—Continued: (c) metallographic cross section, as polished; (d) metallographic planar section, as polished.

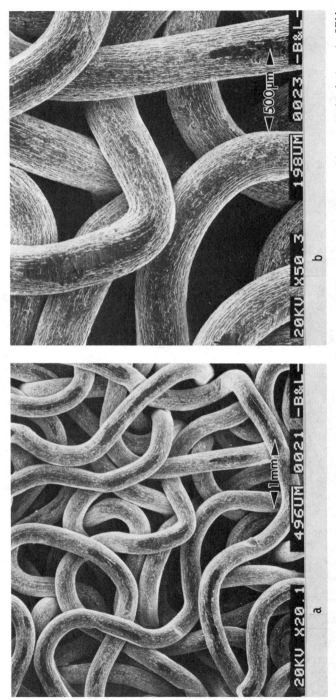

FIG. 3—Photomicrographs of diffusion-bonded titanium wire mesh (fiber metal), 275 μm in wire diameter: (a) topographical view via SEM; (b) topographical SEM view showing the interlaced wire.

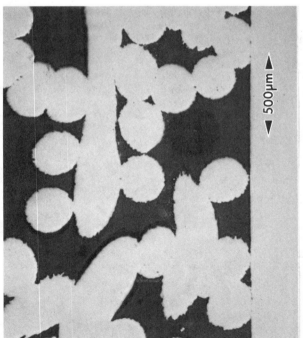

FIG. 3—*Continued: (c) metallographic cross section, as polished; (d) metallographic planar section, as polished.*

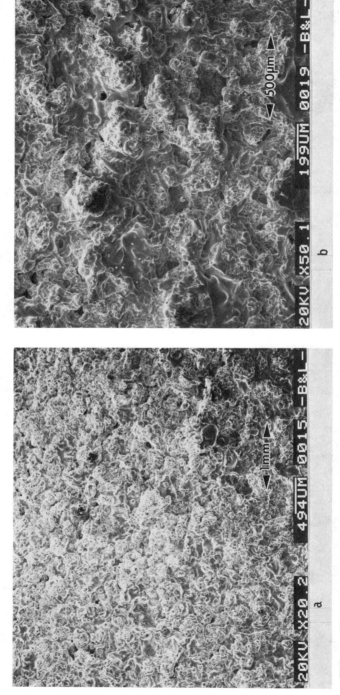

FIG. 4—*Photomicrographs of plasma flame-sprayed titanium alloy:* (a) *topographical view via SEM;* (b) *topographical SEM view illustrating the morphology of the deposited titanium alloy.*

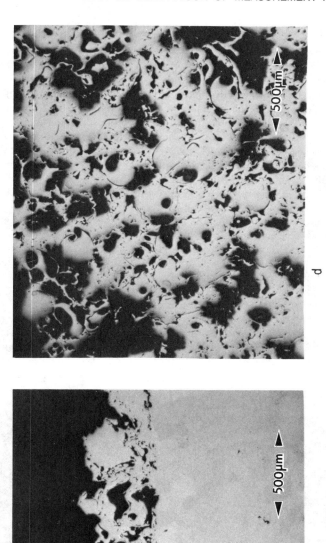

FIG. 4—*Continued*: (c) *metallographic cross section, as polished*; (d) *metallographic planar section, as polished.*

Materials and Methods

Four porous systems were chosen for evaluation, using two measurement methods for pore size: (1) a line intercept method with serial measurements conducted at multiple levels parallel to and in plane with the substrate surface, and (2) an interactive user-defined region measurement method, in which the evaluator determined which features within the structure constituted individual pores. The dimensional characteristics described were determined using scanning electron microscopy (SEM) of each coating surface.

The first porous system, shown in Fig. 1, consisted of a layer of cobalt-chromium-molybdenum beads [ASTM Specification for Cast Cobalt-Chromium-Molybdenum Alloy for Surgical Implant Applications (F 75-82)] on the same cast alloy substrate. The coating thickness was approximately 1.5 mm, and the cobalt alloy bead size ranged from 350 to 750 μm.

The second porous coating type (Fig. 2) also consisted of ASTM F75 cobalt-chromium-molybdenum alloy beads bonded to a cast cobalt alloy substrate with an average coating thickness of 0.6 mm. The bead size range of this cobalt alloy was 100 to 250 μm in diameter.

The third porous coating type (Fig. 3) consisted of an unalloyed titanium wire mesh [ASTM Specification for Unalloyed Titanium for Surgical Implant Applications (F 67-83)] diffusion bonded to a titanium alloy substrate [ASTM Specification for Wrought Titanium 6Al-4V ELI Alloy for Surgical Implant Applications (F 136-84)]. The unalloyed wire diameter was 275 μm.

A fourth porous coating type (Fig. 4) consisted of a plasma flame-sprayed titanium alloy (ASTM Specification F 136-84) deposited on an unalloyed titanium substrate (ASTM Specification F 67-83). The powder particle size prior to spraying was 170 to 300 μm in diameter.

All the porous-coated specimens were vacuum embedded in an epoxy cold mount medium prior to sectioning, and standard metallographic methods (see Appendix) were employed to prepare the specimens for examination in the planar and cross-sectional orientations

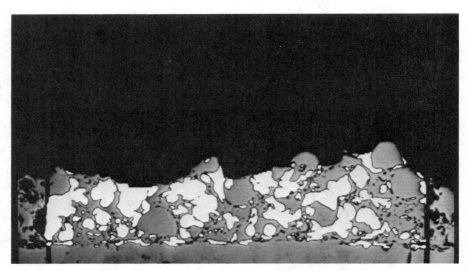

FIG. 5—*Metallographic cross section of a plasma flame-sprayed titanium alloy as measured for percentage of porosity. The video image is shown here with the void regions detected. These regions appear in white. The perimeter of the measurement zone has been defined using a light pen. The image analysis system calculates the percentage of area of the entire field identified as porous.*

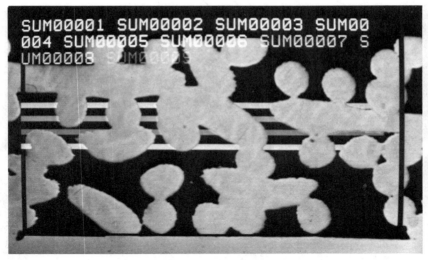

FIG. 6—*Metallographic cross section of a titanium wire mesh (fiber metal) as measured for pore size using the line intercept method. The white horizontal line is displayed at multiple levels of the section. The linear distance between wire sections is determined by the image analysis system. The numerical information indicates the tabulation and storage of data for each line.*

under reflected light. The specimen preparation method provided a two-dimensional image for characterization. The metallographic finish provided high contrast for distinction between the metallic alloy and void regions.

Automated image analysis was used to make the quantitative determinations on all specimens (see Appendix).

Percentage of Porosity

An image analysis system[2] was employed to determine the percentage of porosity, or volume fraction of pore space, in each type of porous-coated system. The method involved detection of the dark void region between the polished metallic bodies in the two-dimensional section planes (perpendicular and planar). The percentage of area occupied by this region was then measured (see Fig. 5).

Pore Size Measurement by Line Intercept Method

The pore size for each individual porous coating type was also measured in the planar and cross-sectional planes. The line intercept method employed a computer program to control the detection and measurement of the image analysis system. The program created an electronic frame for moving a horizontal line of detection across a specified region. This horizontal line of detection measured the spacing between metallic bodies (see the Appendix), and this spacing was interpreted as the pore size (see Fig. 6).

---

[2] Bausch and Lomb Omnicon 3500.

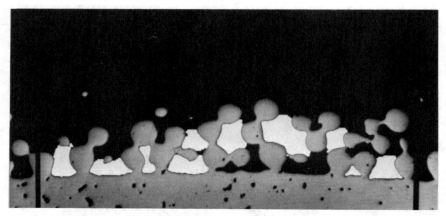

FIG. 7—*Metallographic cross section of a Co-Cr-Mo beaded coating illustrating pore size measurement using the user-defined method. The white regions are void areas identified by the image analysis instrument operator using a light pen. The maximum chord of each area is determined as a measurement of pore size.*

Pore Size Measurement by User Defined Pore Method

A second method for determining pore size was investigated, which involved interactive automated image analysis [2]. This measurement method was performed on planar and cross-sectional specimens of each of the four types of porous coatings examined. The technique involved identifying a porous region for measurement using a light pen technique, in which the boundary of the pore was defined by the instrument operator. A total of ten pore areas was identified per specimen field (see Appendix), and the instrument measured the maximum chord of the subjectively defined pore regions (see Fig. 7).

Results

When comparing the average values for porosity percentage in each of the four coatings evaluated, the percentage of porosity observed was 5% higher in the planar sections than in the cross sections. This orientation trend was observed in all four types of coating. Figure 8 illustrates the comparative averages in the porosity percentage measurement.

Figure 9 illustrates the comparative values obtained on the four coating types using the line intercept method of measurement. A comparison of the average pore size noted in the four systems indicates that, in general, the cross-sectional pore size was greater than the planar pore size. The line intercept method was performed in a manner that did not exclude minimum values determined by the horizontal measurement frame; therefore, small interstices between metallic bodies were tabulated, which may have skewed the average toward a lower value.

Figure 10 illustrates the results obtained using the user-defined measurement method. Here the average values were significantly higher than the line intercept measurement values. The interactive definition of pore area resulted in larger values, which appear to be the result of the exclusion of the small pores; in contrast, very small pores were included by the line intercept program. All four coating types indicated a larger average pore size in the planar section than in the cross section.

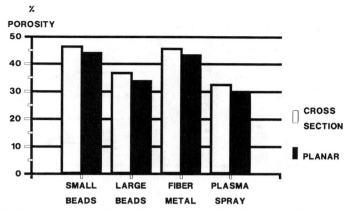

FIG. 8—*Comparative display of average porosity percentage values for the four porous coating types measured.*

Discussion

A comparison of measurement techniques used to quantify porous coatings applied to implant surfaces for biological fixation has indicated that the average values obtained vary significantly, depending on the section plane orientation and method employed to define and measure pore size. The diverse nature of porous coatings for orthopedic implants and the variability in the morphology encountered in the evaluation of these devices warrants standardization of terminology and a definition of what constitutes a pore. The pore size measurement methods used in this exercise are perhaps not the optimum techniques for comprehensive characterization of coatings; however, they have served to demonstrate the potential variability in mean values that can occur in quantitative evaluation of a porous system.

Conclusions

While only minor differences were observed between the cross sections and planar sections when the porosity percentage was measured, significant differences were noted when cross

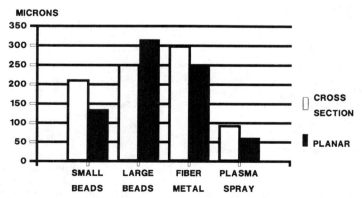

FIG. 9—*Line intercept method results for pore size for each of the four porous coatings measured.*

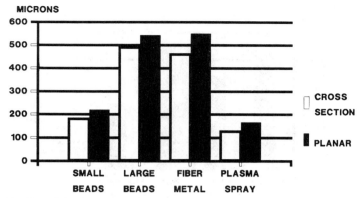

FIG. 10—*User-defined method results for pore size for the four porous systems measured.*

sections were compared with planar sections, using both the line intercept method and the user-defined pore size measurement method. The use of automated systems to perform line intercept measurement may skew the average values toward the lower side because of the indiscriminate inclusion of very small areas. The user-defined pore size measurement is a semiautomated approach and potentially more subjective in definition of pore size.

The choice of both section plane and measurement method appears to influence the average value obtained when measuring pore size. Therefore, the potential for high variability in the average value reported (as much as 200 μm) can exist if a standard method is not used in comparison. The addition of a pore size range and method of measurement description would be advisable when describing pore size.

Acknowledgment

The author wishes to thank S. L. Hartle, of Zimmer, Inc., for his valuable assistance in specimen preparation and measurement.

APPENDIX

Metallographic Sections

Metallographic sections were ground and polished using successively finer abrasives until a 0.05 γ alumina finish was acquired. Single sections in each orientation were considered for measurement. The cross sections were approximately 1.5 cm in length, while the planar sections were made at the midpoint thickness of the coating and had a rough area of 1.5 cm².

Image Analysis Measurement Parameters

The measurements described in the text employed standard techniques used in stereological determinations. The basic measurement routines are programmable menu options within the image analysis operating system. The line intercept routine, user-defined pore size, and porosity percentage determinations all consisted of process commands that consistently determine a specific quantitative value within a given area. The use of image analysis permits the acquisition of a large population of test data for a more statistically sound conclusion [3]. The speed at which the measurement is made, coupled with untiring precision, make this type of instrumentation invaluable for quantitative characterization of this nature. The instrument was calibrated using a standard micrometre slide at the magnifications used in making these measurements.

Percentage of Porosity

The relative coating density or porosity percentage measurement used a process command to measure the area percentage. The polished metallic specimens were measured using reflected light microscopy, which made all metallic areas in the image bright. The nonreflective epoxy embedding media, which filled the porous regions of the coating, appeared dark in the video image. The image analysis instrument's gray level threshold setting was adjusted to allow detection of all dark, nonmetallic regions. The measurement process routine was manual-interactive, and the area percentage measurement was made on a single field, subject to image editing. The percentage of porosity reported is actually the average percentage of area of the field image measured that is occupied by void regions. The planar sections filled the entire screen image, making direct, nonedited measurement possible. The cross-sectional specimens required image editing to define the upper boundary of the porous coating prior to the porosity percentage determination. An example of this method is shown in Fig. 5, in which the upper boundary follows the coating profile.

Line Intercept Pore Size

The line intercept measurement method for pore size determination utilized the automation features of the image analysis instrument. The attribute measured was the longest dimension of a detected image within a specified and variable location image frame. The programmable system permitted the establishment of a program routine to set a horizontal line of measurement across the field electronically. It was moved from top to bottom, during which process measurement of the horizontal void spacing or the distance between metallic features was accomplished. The cross sections were measured one field at a time, with the upper boundary being a straight line from the highest feature. The planar sections permitted the addition of autostage and focusing commands, which allowed the preset control and measurement of multiple fields for greater statistical information. The data for each line were collected and assimilated by the system, and the mean spacing, range, maximum, minimum, and standard deviation were provided.

User-Defined Pore Size Measurement

The measurement of pore size by a user-defined method used the image analysis operator's definition of a porous region for detection and quantification. This interactive mode of operation employed a light pen technique, which the operator used to trace the boundary of a single pore or void region. The technique required defining ten individual void regions per field examined and determining the maximum chord or maximum length of continuous horizontal intercept within each identified feature. This method introduced subjectivity because of the discrimination required from the operator. The criteria for pore region identification attempted to represent the range of sizes present from the smallest to the largest, but did not include all the pore areas present.

References

[1] Heck, D. A., Nakajima, I., Chao, E. Y., Kelly, P. J., "The Effect of Immobilization on Biologic Ingrowth into Porous Titanium Fiber Metal Prostheses," *Transactions,* 30th Annual Meeting of the Orthopaedic Research Society, Atlanta, GA, 6–9 Feb. 1984.
[2] Underwood, E. E., *Quantitative Stereology,* Addison-Wesley, Reading, MA, 1970.
[3] Underwood, E. E., *Quantitative Stereology,* Addison-Wesley, Reading, MA, 1970, p. 13.

Todd S. Smith[1]

Morphological Characterization of Porous Coatings

REFERENCE: Smith, T. S., **"Morphological Characterization of Porous Coatings,"** *Quantitative Characterization and Performance of Porous Implants for Hard Tissue Applications, ASTM STP 953,* J. E. Lemons, Ed., American Society for Testing and Materials, Philadelphia, 1987, pp. 92–102.

ABSTRACT: The morphological characteristics of porous coatings used on orthopedic joint replacement devices may have significant effects on their long-term performance. A method is presented for accurately quantifying coating morphologies. Ten coatings, six commercially available, were investigated. The characteristics measured were the average volume porosity, volume porosity gradient, average pore size, average pore size gradient, and pore size distribution. Computerized image analysis equipment is used to collect data from metallographically prepared sections through the coatings. The data, from which the morphological parameters are determined, are then corrected for sectioning effects.

KEY WORDS: porous implants, porous coatings, pore size, volume porosity, stereology, image analyzer, morphology

Bone cement has been implicated as a major contributor to implant loosening because of its brittle nature and its poor interfacial relationship with the metallic implant and the bone tissue [1–4]. As such, the use of porous-coated implants for biological fixation is receiving an enthusiastic response [5–7]. Several porous systems, both polymeric and metallic, are presently being clinically investigated.

While all these systems are porous, each is morphologically unique, as it functions as a dynamic interface between living tissue and the implant. Since the morphological features of the porous coating may have significant effects on the long-term performance of the arthroplasty, it becomes necessary to characterize these systems physically in as quantitative a manner as possible. In the past, researchers have made very cursory measurements from metallographic sections or scanning electron microscope (SEM) micrographs of coatings to determine the porosity characteristics. These methods are not quantitative in nature and can be highly subjective.

The porous systems studied can be broadly classified as sintered metal bead coatings. They may be described as random non-close-packed arrays of equal-sized or nonequal-sized metal spheres forming a three-dimensional network of interconnected pores.

The morphological characteristics of interest include the average volume porosity, volume porosity gradient, average pore size, pore size gradient, and pore size distribution. Of these parameters, the average pore size is the most difficult to develop. Past researchers [8,9] have limited themselves to reporting the average pore size only. While this parameter may be the single most significant attribute of a porous coating, it is my belief that the importance of the pore size and volume porosity gradients should not be underestimated, as it is the

[1] Director, Materials Engineering, DePuy, Inc., Warsaw, IN 46580.

distributed, or gradual, transfer of stress from implant to bone that tends to favor the viability of surrounding osseous tissues. These gradients are thought by this author to be fundamental to the efficient transfer of stress from the implant to the surrounding bone tissues. We know from experience with bone cement that an abrupt interface between dissimilar materials is prone to failure [10].

Since the direct measurement of porosity parameters in three dimensions is not feasible, the principles of stereology form the basis of this measurement technique [11–13]. Briefly stated, two-dimensional measurements obtained from metallographically prepared sections through the porous coating *must* be converted to the true three-dimensional parameters [14,15]. Past methods stop short of this step, giving rise to inaccurate results. A path or algorithm is needed for this conversion, which necessitates certain simplifying assumptions. The first is the pore shape. Any unit of pore shape could be used, provided the relationship between the dimensions of the average random planar section through the pore shape and the corresponding three-dimensional dimensions is known. These relationships have been determined for several simple shapes [11]. For the purposes of this investigation a pore is defined as a sphere formed by the interstice created between any three or more sintered beads.

Considerable discussion can be had on possible inaccuracies arising from the assumption of a spherical unit pore shape. In a completely analogous situation, that of the determination of true three-dimensional grain size from the analysis of two-dimensional metallographic sections, the literature reveals that only a marginal change in grain size results from a variety of different grain shape assumptions [16,17]. Although pore shapes are not necessarily spherical, the issue is only a simplifying means for achieving the conversion to three dimensions. Analyses using a variety of unit shapes show that large differences in assumed pore shape will result in only minor changes in the result [11,18]. Figure 1 shows the relationship between the average grain diameter and the average linear transverse across a grain for three different grain shapes [16]. It becomes even less important when it is the comparative result between systems that is sought. The second assumption is that the real distribution of pore size within the coating is continuous [19]. In previous investigations by this author (unpublished), serial sectioning of these coatings has substantiated the validity of this assumption.

Given these two assumptions, mathematical methods have been developed by which the three-dimensional distributions of pores can be determined from two-dimensional distributions obtained directly from parametric analyses of random cross sections [20,21,24]. The reason that pore size distribution data collected from planar sections do not represent the true spacial size distributions has to do with sampling bias with respect to feature size and truncations [14,18]. The feature size sampling effect occurs because the probability of intersecting a pore on planar sectioning *increases* with the pore diameter. Small pores are sampled less often than their actual frequency, as illustrated in Fig. 2. If uncorrected for this effect, the pore size distribution would be biased toward a larger pore size.

$$D = \begin{cases} 1.50\ \bar{L} & \text{SPHERE} \\ 1.62\ \bar{L} & \text{PENTAGON DODECAHEDRON} \\ 1.68\ \bar{L} & \text{TETRAKAIDECAHEDRON} \end{cases}$$

FIG. 1—*Pore shape assumption. The equalities show the relationship between the volume equivalent pore diameter and the average lineal intercept length,* $\bar{L}$, *across the pore.*

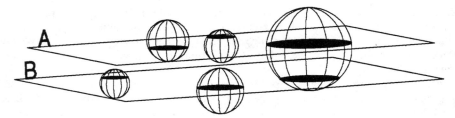

FIG. 2—*Sampling effect. A spacial distribution of five unequal-sized pores is shown inter-sected by two sectioning planes, A and B. The distribution analysis based on Plane A shows two small pores to one large. The analysis of Plane B yields the same result. Yet the true spacial distribution is* four *small pores to* one *large pore.*

The truncation effect occurs because the sectioning plane cuts segments from the pores which are smaller than the pore diameter. This is illustrated in Fig. 3. This effect would tend to decrease the pore size distribution and average pore size values. The net result of both effects on an average pore size depends on the characteristics of the coating being analyzed. If uncorrected data are used, the pore size frequency distribution will definitely be incorrect.

The objective of this paper is to disclose a technique for the rapid and precise determination of several morphological characteristics of porous coatings in three dimensions from two-dimensional parameters and, using this technique, to compare the morphological charac-teristics between several commercially available and laboratory-produced porous coatings.

Methods and Materials

The method presented here is able to characterize several porous coatings, using the principles of stereology in conjunction with the vast data-gathering capabilities of the image-analyzing computer. The morphological parameters considered effective in characterizing and differentiating coatings include the average volume porosity, volume porosity gradient, average pore size, pore size gradient, and pore size frequency distribution. Ten different porous coatings were characterized, six of which are commercially available on implants designed for biological fixation. The remaining four were produced in our laboratories (see Table 1). The techniques used for their fabrication were identical to those used for the Porocoat porous coating available on the AML hip implant system. The only difference between these coatings was that various different powder size fractions were employed.

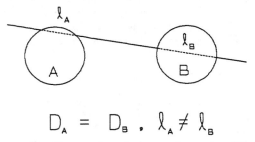

$$D_A = D_B , \quad \ell_A \neq \ell_B$$

FIG. 3—*Truncation effect. Two equal-sized pores are shown, A and B. A common sectioning plane intersects each at a different location, creating sections of different dimensions, ℓ_A and ℓ_B. The analysis of this section would make Pore B larger than Pore A, yet the pores are equal in size.*

TABLE 1—*Relationship between the volume equivalent diameter pore size and the volume porosity.*

Porous System	Mean Pore Size, μm	Mean Volume Porosity, %
60/80 DePuy	280	46.7
50/70 DePuy	397	49.0
50/60 DePuy	345	44.6
45/60 HTCC[a]	456	51.8
45/60 DePuy	619	52.2
40/50 DePuy	473	50.0
30/40 DePuy	622	51.4
Richards[b]	696	48.0
PCA[c]	670	48.8
DCW[d]	390	43.7

[a] Howmet Turbine Components Corp., Dover, NJ. The samples were taken from porous-coated buttons supplied by HTCC.
[b] Richards Medical Co., Memphis, TN. The samples were taken from Tricon-M tibial components.
[c] Howmedica, Inc., Rutherford, NJ. The samples were taken from PCA total knee tibial components.
[d] Dow Corning Wright, Memphis, TN. The samples were taken from Whiteside Ortholoc total knee tibial components.

Numerous samples of each coating were metallographically mounted, using an opaque glass-filled epoxy mounting medium so that all the pore space was fully impregnated. The samples were then sectioned at a 15° angle to the plane of the coating substrate. This has the advantage of artificially spreading out the coating to facilitate the image analysis. However, any sectioning angle can be used without change in the result. Low-angle sectioning allows visualization of the porosity from the direction of bony invasion without resorting to

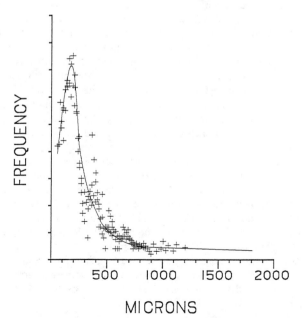

FIG. 4—*Raw data of the pore size frequency distribution for the −50+70 mesh porous coating. The computer-generated fitted curve is superimposed on the raw data.*

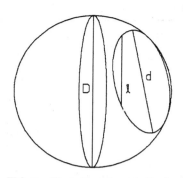

D = TRUE DIAMETER OF PORE

d = DIAMETER OF RANDOM SECTION

ℓ = RANDOM LINEAR INTERCEPT

$\bar{L}$ = MEAN INTERCEPT LENGTH

$\bar{d}$ = MEAN SECTION DIAMETER

D = 3/2 $\bar{L}$ (SPHERE)

$\bar{d}$ = 4/π $\bar{L}$

FIG. 5—*Illustration of an idealized pore showing the relationships between the true diameter, the average section diameter, and the average section intercept length.*

the tremendously time-consuming parallel serial sectioning techniques [22,23]. The specimens were carefully ground and polished through 0.3-μm alumina using standard metallographic practice. Much care was exercised to eliminate the occurrence of bead pullout and edge rounding during grinding and polishing.

The analysis was performed on a Quantimet image analyzer used in conjunction with a reflecting light microscope. The analyzer scans the field of the light image and separates the bright reflective phase from the dark pore space. Each pore space can be classified by its area, length and width, and perimeter. Several thousands of pores can be identified, counted, measured, and classified by the computer analyzer. Magnification was adjusted to obtain optimal differentiation between the polished metal phase and the pore space.

Volume porosity is easily determined. The area percentage of pore space is measured and divided by the total area scanned. The area percentage of pore space is numerically equivalent to the volume pore fraction. The volume porosity gradient is determined by restricting the field of analysis to specific distances from the solid substrate. The local volume porosity is determined and is displayed graphically by plotting it versus the substrate distance.

The coating characteristics based on pore dimensions are more difficult to determine. Data acquisition again is facilitated by the image analyzer. For this determination it is programed to measure the distribution of linear intercept chord lengths or pore area equivalent diameters. The analyzer scans the sample, identifying the pore spaces, and categorizes them according to size class, frequency of occurrence, and location. These data yield the uncorrected pore size distribution. It is plotted for further analysis in Fig. 4. Although these data can be used directly for further analysis, curve smoothing can be employed, which will greatly facilitate the derivation of the average pore size. The smoothed curve for these data is shown superimposed on the raw data. The curve smoothing was accomplished using a software package with the computer graphic system used to plot the curves in this investigation.[2] Curve smoothing is not necessary if enough data are gathered from the image analysis phase of this technique. These data must now be corrected for sectioning effects.

The data are first modified to correct for the truncation effect. Figure 5 shows the relationships between the possible measured quantities from the image analyzer and the actual size of the pore. Essentially, each pore size range is upgraded by a factor of 1.5 when the distribution of intercept chord lengths is used and by a factor of 1.27 if the average pore section diameter data are used.

[2] Advanced Surface Design Software, Computervision Corp., Bedford, MA.

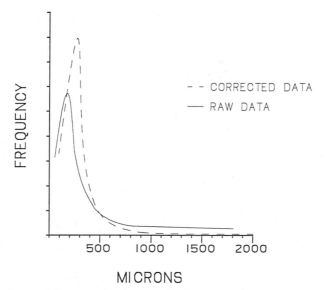

FIG. 6—*Pore size frequency distributions, illustrating the difference between the smoothed curve for the raw data and that for the corrected data.*

It has been shown that the measured distribution from planar sections is related to the actual distribution by an integral equation [20]. An iterative solution to this relationship was developed by Lord and Willis [19] and can be employed successfully to adjust the distribution from the two-dimensional analysis to the true three-dimensional distribution. The details of this procedure are described in the sited reference, a discussion of which is not within the scope of this paper. The corrected data can then be plotted to show graphically the true spacial pore size frequency distribution (see Fig. 6).

Determination of the average pore size is now a straightforward procedure. The mean pore volume (MPV) is calculated by multiplying the volume of each pore size class, V, by its frequency of occurrence, f. The summation of the products is then divided by the sum, Σ, of the frequencies.

$$MPV = \frac{\Sigma(f \times V)}{\Sigma f}$$

The average pore size volume equivalent diameter [22], the mean pore size, D, is then calculated.

$$D = 2 \times \left(\frac{3MPV}{4\pi}\right)^{1/3}$$

Results and Discussion

The results of the mean volume porosity measurements appear in the right-hand column of Table 1. The values vary between 43.7% (Dow Corning Wright) and 52.2% (45/60 DePuy

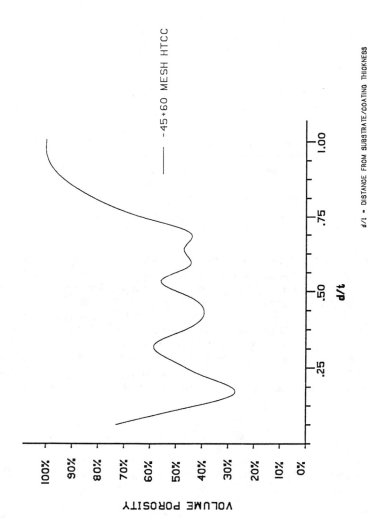

FIG. 7—*Graph of the average volume porosity gradient for the* −45+60 *porous coating from Howmet Turbine Components Corp.*

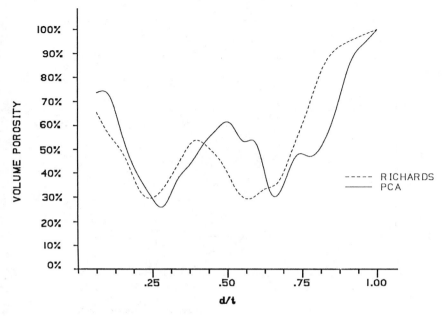

d/t = DISTANCE FROM SUBSTRATE/COATING THICKNESS

FIG. 8—*Graph of the average volume porosity gradient for the Howmedica, Inc., PCA porous coating and that from the Richards Medical Co.*

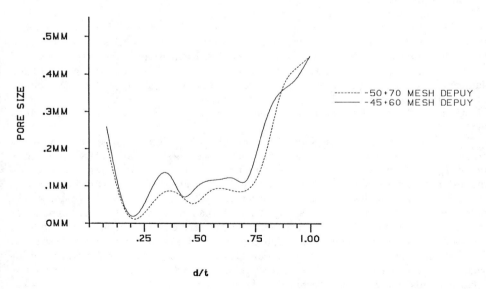

d/t = DISTANCE FROM SUBSTRATE/COATING THICKNESS

FIG. 9—*Graph of the average pore size gradient for the* −50+70 *and* −45+60 *DePuy laboratory fabricated coatings.*

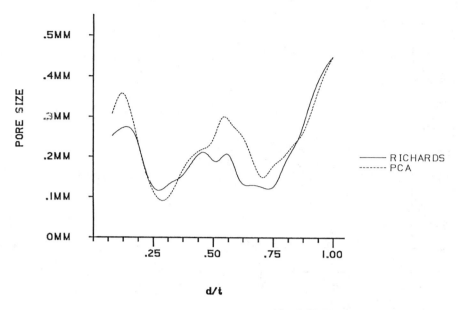

d/t = DISTANCE FROM SUBSTRATE/COATING THICKNESS

FIG. 10—*Graph of the average pore size gradient for the Howmedica, Inc., PCA porous coating and that from the Richards Medical Co.*

laboratory sample). Interestingly, there was no particular dependence of the mean volume porosity on the bead size comprising the coating.

The average volume porosity gradients for all samples analyzed show a dampened sinusoidal increasing curve. A representative curve, that of the 45/60-mesh Howmet sample, is shown in Fig. 7. An initial high value at and just above the substrate is due to the space below the radius of an almost fully close-packed first layer of beads. Successive layers of beads become increasingly random in order and less densely packed, giving rise to the dampened feature of these curves. As the outer boundary is approached the local volume porosity rapidly approaches 100%. Coatings comprised of larger beads show fewer cycles of the sinusoidal shape as the relative number of bead layers is less. This is shown in Fig. 8.

The average pore size gradient shows a trend similar to that of the average volume porosity gradient, a dampened sinusoidal increasing curve (see Fig. 9). The average pore size is relatively large between the substrate and the first layer of beads. It decreases to a low as the distance from the substrate approaches the bead radius, where most of the beads are in contact. It sinusoidally increases through several cycles, depending on the coating-thickness-to-bead-diameter ratio, then rapidly increases to a maximum as the outer boundary of the coating is approached. The vertical positioning of each curve is roughly proportional to the average pore size of the coating. Compare Fig. 9 with Fig. 10 (the larger bead size and pore size coatings). The data for these curves were not corrected for the truncation or sampling effects since the absolute values for this characteristic are not important. The relative trends would not be affected. It is fairly evident that the larger pores occupy the outer regions of the coating.

Figure 4 shows the frequency distribution of data directly obtained from the image analyzer. For comparison purposes both the uncorrected and corrected distributions for the 50/

70-mesh sample appear in Fig. 6. The corrected curve shows a peak at a larger pore size, owing to the truncation effect, while the corrected distribution of the larger pores shows a decrease due to the correction for the sampling effect.

The average pore size determinations appear in the middle column of Table 1. The value is expressed as an average diameter measurement based on an average pore volume equivalent diameter determination. It is interesting to note that the volume fraction occupied by larger pores overcomes the much higher frequencies of the smaller pore size classes and serves to drive up the average pore size value. This phenomenon serves to remind us of the three-dimensional nature of the pore, and that the average pore size is not the arithmetic average of the pore diameter-frequency product from two-dimensional measurements. After analyzing these results further, there is another anomaly. The 50/60-mesh laboratory sample and the coating of Dow Corning Wright sample show a much lower average pore size than one would expect based on the powder size fraction comprising these coatings. The explanation appears to be related to a lower volume porosity (higher density) particular to these two coatings.

Conclusions

The image analyzing computer, used in conjunction with a reflecting light microscope, is an extremely useful tool in facilitating the analysis of porous coatings used on prosthetic implants. The science of stereology can be used effectively to characterize accurately specific morphological parameters of porous coatings. Stereology allows the determination of the true three-dimensional characteristics to be determined from analyses of two-dimensional metallographic sections. The metal bead coatings investigated all show volume porosity and pore size gradients through the thickness. Larger pores tend to occupy the outer regions of the coatings. The average pore size tends to be dependent on the powder mesh size as well as the volume porosity.

References

[1] Amstutz, H. C., Markolf, K. L., McNeice, G. M., and Gruen, T. A., "Loosening of Total Hip Components: Cause and Prevention," *The Hip: Proceedings of the Fourth Open Scientific Meeting of The Hip Society,* C. V. Mosby, St. Louis, 1976, pp. 102–116.

[2] Gruen, T. A., McNeice, G. M., and Amstutz, H. D., "Modes of Failure: Cemented Stem-Type Femoral Components—A Radiographic Analysis of Loosening," *Clinical Orthopaedics,* Vol. 141, 1979, pp. 17–27.

[3] Willert, H. G., Ludwig, J., and Semlitsch, M., "Reaction of Bone to Methacrylate after Hip Arthroplasty: A Long-Term Gross, Light Microscopic and Scanning Electron Microscopic Study," *Journal of Bone and Joint Surgery,* Vol. 56A, 1974, p. 1368.

[4] Willert, M. D., "Tissue Reactions Around Joint Implants and Bone Cement," *Symposium on Arthroplasty of the Hip,* G. Chapcal, Ed., Thieme, Stuttgart, 1973, pp. 11–21.

[5] Smith, T. S., "Rationale for Biological Fixation of Prosthetic Devices," *SAMPE Journal,* Vol. 21, No. 3, May/June 1985.

[6] Engh, C. A. and Bobyn, J. D., "Biologic Fixation of Hip Prosthesis: A Review of the Clinical Status and Current Concepts," *Advances in Orthopaedic Surgery,* Vol. 18, 1984, pp. 136–149.

[7] Collier, J. P., Kennedy, F. E., Mayor, M. B., and Townley, C. O., Stress Distribution in the Human Femur: The Role of Femoral Prosthesis Geometry and the Mechanics of Fixation," *Transactions,* 30th Annual Meeting of the Orthopaedic Research Society, Atlanta, GA, February 1984.

[8] Walsh, K. S., Cook, S., and Haddad, R., "Evaluation of Parameters Affecting Porous Alloy Implants Retention," *Transactions,* 30th Annual Meeting of the Orthopaedic Research Society, Atlanta, GA, February 1984.

[9] Bobyn, J. D., Pilliar, R., and Cameron, H., "The Optimum Pore Size for the Fixation of Porous-Surfaced Metal Implants by the Ingrowth of Bone," *Clinical Orthopaedics and Related Research,* No. 150, July/August 1980.

[10] Huiskes, R., "Properties of the Stem-Cement Interface and Artificial Hip Joint Failure: A Work-shop on the Bone-Implant Interface," *Proceedings*, Workshop on the Bone-Implant Interface, American Academy of Orthopaedic Surgeons, Chicago, IL, 13–16 Sept. 1983.

[11] Aaron, H. B., Smith, R. D., and Underwood, E. E., "Spatial Grain-Size Distribution from Two-Dimensional Measurements," *Proceedings*, First International Congress on Stereology, Vienna, Austria, 18–20 April 1963, pp. 16-1–16-8.

[12] Underwood, E. E., "Practical Solutions to Stereological Problems," *Practical Applications of Quantitative Metallography, ASTM STP 839*, J. L. McCall and J. H. Steel, Jr., Eds., American Society for Testing and Materials, Philadelphia, 1984, pp. 160–179.

[13] Saltykov, S. A., "The Determination of the Size Distribution of Particles in an Opaque Material from Measurement of the Size Distribution of their Sections," *Stereology,* Springer-Verlag, New York, 1967, pp. 163–173.

[14] Exner, H. E., *International Metallurgical Reviews*, Vol. 17, 1972, pp. 25–42.

[15] Durand, M. C. and Warren, R., "Computer Synthesis of Microstructures in Stereological Analy-sis," *Proceedings*, Swedish Symposium on Non-Metallic Inclusions in Steel, Sodertalje, Sweden, April 1981.

[16] Bodyako, M. N., Kasichev, V. P., and Naumovich, N. V., *Practical Metallography*, Vol. 17, May 1980, pp. 232–237.

[17] Thompson, A. W., *Metallography*, Vol. 5, 1972, pp. 366–369.

[18] Vander Voort, G. F., "Grain Size Measurement," *Practical Applications of Quantitative Metal-lography, ASTM STP 839*, J. L. McCall and J. H. Steel, Jr., Eds., American Society for Testing and Materials, Philadelphia, 1984, pp. 85–131.

[19] Lord, G. W. and Willis, T. F., "Calculation of Air Bubble Size Distribution from Results of a Rosiwal Transverse of Aerated Concrete," *ASTM Bulletin*, October 1951, pp. 56–61.

[20] Chan, J. W. and R. L. Fullman, *Transactions of the A.I.M.E.*, Vol. 206, May 1956, pp. 610–612.

[21] Fullman, R. L., *Transactions of the A.I.M.E.*, Vol. 197, March 1953.

[22] Rhines, F. N., Craig, K. R., and Rouse, D. A., *Metallurgical Transactions*, Vol. 7A, November 1976, pp. 1729–1734.

[23] DeHoff, R. T. and Aigeltinger, E. H., "Experimental Quantitative Microscopy with Special Ap-plications to Sintering," R. T. DeHoff and F. N. Rhines, Eds., *Quantitative Microscopy*, McGraw-Hill, New York, 1968.

[24] Schuckher, F., "Grain Size," *Quantitative Microscopy*, McGraw-Hill, New York, pp. 201–265.

Biodegradation and Biological Analyses

Raymond A. Buchanan,[1] *E. Douglas Rigney, Jr.,*[2] *and Charles D. Griffin*[3]

Biocorrosion Studies of Ultralow-Temperature-Isotropic Carbon-Coated Porous Titanium

REFERENCE: Buchanan, R. A., Rigney, E. D., Jr., and Griffin, C. D., **"Biocorrosion Studies of Ultralow-Temperature-Isotropic Carbon-Coated Porous Titanium,"** *Quantitative Characterization and Performance of Porous Implants for Hard Tissue Applications, ASTM STP 953,* J. E. Lemons, Ed., American Society for Testing and Materials, Philadelphia, 1987, pp. 105–114.

ABSTRACT: Static, *in vitro* electrochemical and gravimetric corrosion studies were conducted to evaluate the effectiveness of 0.5-μm ultralow-temperature-isotropic (ULTI) carbon coatings in reducing the metal-ion release rates of porous titanium and solid Ti-6Al-4V alloy. The results indicated significant reductions for both materials, with the corrosion rate of carbon-coated porous titanium being reduced to that of solid uncoated Ti-6Al-4V. Analyses of the results indicate a minimum expected carbon-film lifetime of approximately 38 years.

KEY WORDS: ultralow-temperature-isotropic (ULTI) carbon, porous titanium, corrosion, carbon coatings, biocorrosion, titanium alloy, porous implants

Ultralow-temperature-isotropic (ULTI) carbon coatings applied to metallic surgical implant devices offer the potential of substantially reducing the rate of metal-ion release to surrounding tissues, while presenting to the tissues a highly biocompatible material [1,2]. This potential benefit becomes especially important in view of the introduction of porous metal surface layers on orthopedic implants for the purpose of enhanced bone fixation. The porous metal layers have increased surface areas, which necessarily produce higher levels of corrosion products. Because of concerns over the possible deleterious biological effects of metal ions released during the corrosion process [3], studies are now being made of surface modifications intended to ameliorate the effect of increased surface area.

The purpose of the present work was to quantitatively determine the effectiveness of ULTI carbon coatings in reducing the corrosion of porous commercially pure titanium. Also, ULTI carbon-coated solid Ti-6Al-4V alloy was studied for comparison.

Static, *in vitro* electrochemical and gravimetric corrosion tests were performed in three different electrolytes with varying degrees of aggressiveness. These procedures allowed not only an evaluation of the effectiveness of the carbon coating in reducing the titanium corrosion rate, but also an estimation of the lifetime of the carbon coating in simulated tissue

[1] Professor, Department of Materials Science and Engineering, University of Tennessee, Knoxville, TN 37996.
[2] Graduate research assistant, Department of Biomedical Engineering, University of Alabama at Birmingham, Birmingham, AL 35294.
[3] Head of engineering, Carbomedics, Inc., Austin, TX 78752.

fluids. The electrochemical corrosion tests were conducted in a solution composed of 90% saline and 10% serum—an electrolyte that has been employed by a number of investigators to simulate tissue fluids. The gravimetric or weight-loss corrosion tests were conducted in a 95% saline/5% hydrochloric acid (HCl) solution and an 80% saline/20% HCl solution. The hydrochloric acid solutions were selected because of two considerations. In order to generate titanium weight losses sufficiently large to be accurately measured, a more aggressive electrolyte than the saline/serum solution had to be selected. In addition to other acids, hot hydrochloric acid is known to attack titanium readily. The advantage to hydrochloric acid lies in its relevance to the wound-healing processes at the site of a surgical implant. During wound healing the environment is known to become more acidic [4,5]. Furthermore, tissue fluids are characterized by concentrations of chloride ions. The constituents of hydrochloric acid, therefore, have appropriate biological relevance and also provide a rapid attack of titanium.

Experimental Procedures

All the specimens were fabricated by Carbomedics, Inc., of Austin, Texas. The ULTI carbon coatings were applied with vacuum-vapor-deposition procedures previously described in the literature [6,7]. The coatings thus produced were approximately 0.50 μm thick. The test specimens, shown in Fig. 1, consisted of the following: uncoated and ULTI carbon-coated specimens consisting of void-metal-composite porous commercially pure (CP) titanium surface layers [Grade 2, ASTM Specification for Unalloyed Titanium for Surgical Implant Applications (F67-83)] on solid Ti-6Al-4V substrates; uncoated and ULTI carbon-coated solid Ti-6Al-4V specimens [ASTM Specification for Wrought Titanium 6Al-4V ELI Alloy for Surgical Implant Applications (F 136-84)]; and low-temperature-isotropic (LTI) carbon-coated graphite specimens. The non-carbon-coated specimens were passivated in a 30% nitric acid solution at 54 ± 6°C for 30 min. All the specimens were sterilized in a steam autoclave at 121°C for 30 min before testing.

The electrochemical testing consisted of the generation of anodic polarization curves in an electrolyte of 10% calf serum (Gibco 200-6170) and 90% isotonic saline (0.9% NaCl) at a temperature of 37.0 ± 0.5°C and a pH of 7.00 ± 0.05. Hereafter, this electrolyte will be referred to as Electrolyte 1. The polarization tests were conducted in the following manner. The test specimen was first placed in the polarization cell and allowed to stabilize over a period of 1 h. The potential was then potentiostatically stepped in the noble direction in 50-mV increments every 5 min. At the end of each 5-min period, the anodic current was recorded, then divided by the specimen's *nominal* surface area to produce the current density. Anodic polarization curves were produced by plotting the log of current density as a function of potential.

The two electrolytes employed for the gravimetric or weight-loss measurements were the following: a 95% saline/5% HCl solution at 90°C (Electrolyte 2) and a 80% saline/20% HCl solution at 90°C (Electrolyte 3). The HCl acid employed was a 37.5% concentration. Each specimen was carefully weighed, placed in a polytetrafluoroethylene (PTFE) mesh basket, and immersed in a given electrolyte for a specified time period. The specimen was then withdrawn, dried in a 80°C oven for 15 min, and reweighed. The analytical balance employed for the weight measurements had a resolution of 1 mg. During each weight determination, triplicate measurements were made and then averaged.

Results and Discussion

The anodic polarization curves produced in the saline/serum electrolyte at 37°C, Electrolyte 1, are presented in Fig. 2. The error bars represent the total ranges of values

FIG. 1—*Corrosion test specimens:* (a) *porous titanium,* (b) *ULTI carbon-coated porous titanium,* (c) *solid Ti-6Al-4V,* (d) *ULTI carbon-coated Ti-6Al-4V or LTI carbon-coated graphite.*

determined in either two or three independent evaluations. These curves give the anodic dissolution rates, in terms of current density, from potentials approximately 50 mV above the open-circuit corrosion potentials to +500 mV versus the saturated calomel electrode (SCE). Note that the current densities were based on the nominal surface areas. Whereas

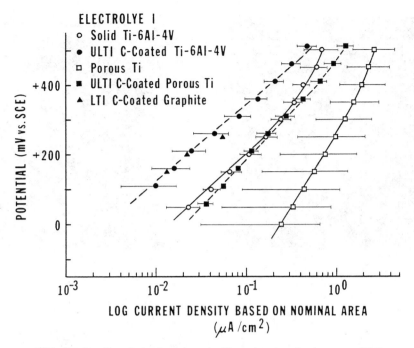

FIG. 2—*Anodic polarization curves in Electrolyte 1, saline/serum at 37°C.*

the nominal and true surface areas are equivalent for solid, smooth-surfaced specimens, the nominal area is less than the true area for porous specimens. The data were analyzed based on the nominal areas so that Fig. 2 would allow direct comparisons, for 1 cm² of nominal surface area, of the amount of corrosion acceleration in going from a solid to a porous surface, and of the amount of corrosion reduction achieved by ULTI carbon coatings on both solid and porous surfaces.

As shown in Fig. 2, the anodic current densities for the porous titanium were appreciably larger than those for the solid Ti-6Al-4V. The primary reason for this effect is believed to be the larger true surface area associated with the porous titanium. It is also apparent that the ULTI carbon coatings reduced the anodic current densities for both the porous titanium specimens and the solid Ti-6Al-4V specimens. Indeed, the ULTI carbon coatings reduced the current densities for the porous titanium to levels comparable to those produced by the solid uncoated Ti-6Al-4V. After running anodic polarization curves on the carbon-coated specimens, the authors noted that the surfaces were no longer as black, bright, and shiny in appearance; that is, they no longer looked as highly polished. This type of behavior is normally caused by the corrosion process producing an etched surface.

The results of the gravimetric studies in the more aggressive saline/5% HCl electrolyte at 90°C, Electrolyte 2, are presented in Fig. 3. These results represent two independent evaluations, shown as separate data points. The total weight of the ULTI carbon film on a solid Ti-6Al-4V specimen was calculated, based on the specimen size and film thickness, to be only 0.52 mg. Since the resolution of the measurement was 1 mg, the weight losses shown in Fig. 3 were due to titanium or titanium alloy dissolution, not to carbon film dissolution. Within the resolution of the weight-loss measurements, and over a time period of 4 h, it can be observed that Electrolyte 2 did not attack the LTI carbon-coated graphite or the

ULTI carbon-coated solid Ti-6Al-4V specimens but readily attacked the uncoated porous titanium specimens. The weight-loss rates for the ULTI carbon-coated porous titanium and uncoated solid Ti-6Al-4V specimens were low and comparable. These results indicate significant corrosion protection by ULTI carbon coatings on both porous titanium and solid Ti-6Al-4V surfaces.

The results of the gravimetric studies in the most aggressive saline/20% HCl electrolyte at 90°C, Electrolyte 3, are presented in Fig. 4. Within the resolution of the weight loss measurements, it is clear that Electrolyte 3 did not attack the LTI carbon-coated graphite specimens, and did not attack the ULTI carbon-coated solid Ti-6Al-4V specimens until after a delay time of 15 h. The uncoated solid Ti-6Al-4V was readily attacked, and the uncoated porous titanium, even more so. The ULTI carbon-coated porous titanium specimens were attacked by the solution, but were not rapidly attacked until after a delay of approximately 1 hr. At the 1-h point in time, the corrosion rates of the carbon-coated porous titanium specimens and the uncoated solid Ti-6Al-4V specimens were comparable. In these gravimetric studies, it was observed that, after the delay times, the carbon coatings appeared to undergo a flaking-off process. This action would indicate that the electrolyte was penetrating the carbon film through surface flaws and attacking the titanium or Ti-6Al-4V substrates at the interfaces.

In summarizing the gravimetric test results in the aggresssive saline/HCl solutions, it was observed that *until carbon-film breakdown occurred* (15 h for the carbon-coated solid Ti-6Al-4V in Electrolyte 3, 1 h and >4 h for the carbon-coated porous titanium in Electrolytes

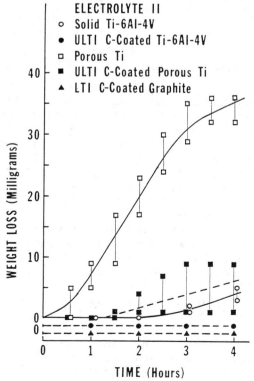

FIG. 3—*Gravimetric results in Electrolyte 2, saline/5% HCl at 90°C.*

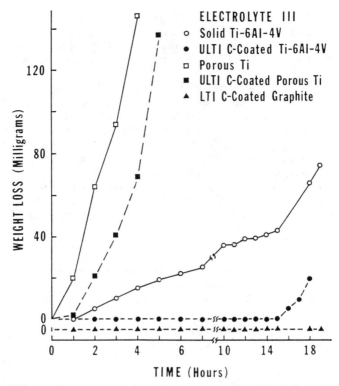

FIG. 4—*Gravimetric results in Electrolyte 3, saline/20% HCl at 90°C.*

3 and 2 respectively) the following phenomena occurred: (1) the Ti-6Al-4V was not attacked at all on the carbon-coated solid Ti-6Al-4V, indicating *total* protection by the carbon film, and (2) the titanium was attacked on the carbon-coated porous titanium specimens, but at a reduced rate comparable to that on the uncoated solid Ti-6Al-4V specimens, indicating *partial* protection by the carbon film.

Efforts were made to understand the mechanisms associated with the behavior of the ULTI carbon coatings in the three electrolytes employed in this study. These evaluations resulted in estimated minimum lifetimes for the carbon coatings in the saline/serum simulated-tissue-fluid electrolyte, Electrolyte 1.

It was first necessary to determine the corrosiveness of the three electrolytes toward both titanium (or Ti-6Al-4V) and carbon. To this end, the Tafel extrapolation method was employed to evaluate the corrosion current densities [8]. For the carbon evaluations, LTI carbon-coated graphite specimens were used to preclude any influence of titanium-substrate corrosion on the results. The measured corrosion rates, in terms of corrosion current densities (i_c values), are presented in Table 1. Also shown in Table 1 are the calculated corrosion penetration rates and relative aggressiveness factors, both of which were based on the i_c values.

Two possible hypotheses involving the mechanisms of carbon-film breakdown were evaluated. The first hypothesis was that of carbon film breakdown by the mechanism of carbon dissolution, schematically illustrated in Fig. 5a. In accordance with this model, the carbon dissolution behavior would be that represented by the anodic polarization curve of the ULTI

TABLE 1—*Corrosion rate parameters.*

	Ti-6Al-4V			Carbon		
	Corrosion Rate i_c, µA/cm²	Penetration Rate, µm/year	Relative Aggressiveness Toward Ti-6Al-4V[a]	Corrosion Rate i_c, µA/cm²	Penetration Rate, µm/year	Relative Aggressiveness Toward Carbon[a]
Electrolyte 1 (saline/ 10% serum at 37°C)	0.008	0.069	1	0.003	0.013	0.38
Electrolyte 2 (saline/ 5% HCl at 90°C)	1 100	9 560	137 500	...	...	...
Electrolyte 3 (saline/ 20% HCl at 90°C)	3 900	33 800	487 500	1.0	4.36	125

[a] In relation to the aggressiveness of Electrolyte 1 on Ti-6Al-4V.

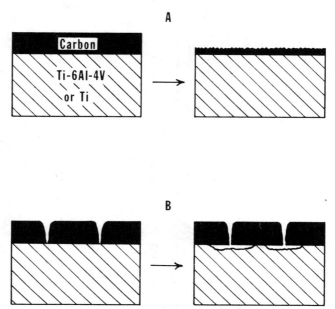

FIG. 5—*Schematics illustrating* (a) *carbon-film breakdown by the carbon dissolution mechanism and* (b) *carbon-film breakdown by the flaking-off process involving interfacial corrosion.*

carbon-coated solid Ti-6Al-4V in Fig. 2, which closely matched that of the LTI carbon-coated graphite in the lower potential range close to the open-circuit corrosion potentials. One could argue that these two polarization curves were caused by species in the electrolyte undergoing oxidation, and not by the dissolution of carbon. However, the etched carbon surfaces produced by the polarization tests appear to indicate that some carbon corrosion was occurring. Furthermore, the purpose of this hypothesis was to predict a minimum carbon film lifetime based on the worst possible situation. By utilizing the respective penetration rates shown in Table 1 and the known thickness of the ULTI carbon coating, 0.50 μm, the carbon film lifetime was predicted for each electrolyte. The predicted lifetime in the saline/serum electrolyte was approximately 38 years; in comparison, the predicted lifetime in the saline/20% HCl electrolyte was approximately 0.115 year or 1000 h. However, with reference to Fig. 4, the actual carbon-film lifetime in Electrolyte 3 was only 15 h. Consequently, the authors concluded that the mechanism of carbon-film breakdown in Electrolyte 3 was not consistent with the first hypothesis, but was more consistent with the second.

The second hypothesis refers to the breakdown of the carbon-film by the flaking-off process caused by dissolution of titanium at the carbon/titanium interface, schematically illustrated in Fig. 5b. Access to the interface by the electrolyte would have to be gained through flaws in the carbon film. As seen in Table 1, Electrolyte 3 was 487 500 times more aggressive toward titanium than Electrolyte 1. Note that Electrolyte 3, in comparison with Electrolyte 1, was much more aggressive toward titanium than toward carbon. The actual lifetime of the carbon film on the solid Ti-6Al-4V material in Electrolyte 3 was 15 h, and the breakdown mechanism appeared to be the flaking-off process. By utilization of the known corrosion rates of titanium in the two electrolytes, the expected lifetime in Electrolyte 1 was calculated to be approximately 835 years, that is, 487 500 × 15 h. Similarly, the actual lifetime of the carbon film on the porous titanium material in Electrolyte 3 was approximately 1 h, and the calculated lifetime in Electrolyte 1 was approximately 56 years. When the Electrolyte 2 results were employed (with a delay time of greater than 4 h and a relative aggressiveness of 137 500), the calculated lifetime for the carbon film on the porous titanium material in Electrolyte 1 was 63 years.

As a consequence of these analyses, the authors believe that in the most aggressive electrolyte, Electrolyte 3, the mechanism of ULTI carbon-film breakdown was that corresponding to the second hypothesis. This is consistent with visual observations of the behavior of specimens during the gravimetric tests—that is, flaking-off of the carbon film—and was a direct consequence of the fact that the saline/HCl solutions were much more aggressive toward titanium than toward carbon. However, for Electrolyte 1, which was the electrolyte selected to simulate tissue-fluid conditions, the authors believe that, although both mechanisms would be operative, the carbon-film breakdown would *first* occur by the carbon-film dissolution mechanism corresponding to the first hypothesis, which yielded a predicted lifetime of 38 years, in comparison with the second-hypothesis predictions of 835 years for the solid material and 56 years for the porous material. The 38-year predicted lifetime for the ULTI carbon film probably represents a minimum value, since it does not take into account the normal decrease in corrosion rate as a function of time.

In view of these evaluations, and with reference to the electrochemical results presented in Fig. 2 for the saline/serum electrolyte at 37°C, the authors believe that the curve for the ULTI carbon-coated Ti-6Al-4V represents to a large extent the corrosion of ULTI carbon. This conclusion is consistent with the etched appearance of the carbon surface after each polarization test. It is also indirectly confirmed by the gravimetric results; that is, there was no measurable Ti-6Al-4V corrosion until after carbon-film breakdown. This conclusion necessarily means that the rate of metal-ion release from the carbon-coated solid Ti-6Al-4V in the saline/serum electrolyte was much lower than was indicated by the anodic current densities in Fig. 2. The curve for the ULTI carbon-coated porous titanium was shifted to higher current densities for two reasons: (1) the true carbon surface area was considerably higher, and (2) the carbon coating probably did not totally cover the area associated with the porosity. The second reason is consistent with the gravimetric results in Figs. 3 and 4; that is, small titanium weight losses were measured for the carbon-coated porous titanium specimens even during the >4 h and 1 h delay times before breakdown.

Additional experiments could potentially aid in elucidating the relative roles of the two carbon-film breakdown mechanisms in the saline/serum electrolyte. Specifically, the authors recommend that scanning electron microscopy or scanning Auger electron microscopy, or both, be performed on ULTI carbon-coated samples that have been subjected to accelerated corrosion by electrochemical polarization at constant anodic potential for different time periods. These surface analytical techniques should be especially sensitive to the second-hypothesis mechanism.

It should be emphasized that all of the previous results, analyses, and conclusions (especially those related to predicted carbon-film lifetimes) were based on studies conducted under static conditions; that is, no mechanical stresses, wear, abrasion, fretting, or similar conditions were involved.

Summary of Results and Conclusions

Under the static, *in vitro* conditions of this investigation, the following results were obtained relative to the effectiveness and stability of ULTI carbon films on solid Ti-6Al-4V and porous titanium:

1. For solid Ti-6Al-4V, the ULTI carbon films significantly reduced the anodic dissolution rates in the saline/serum solution at 37°C and the weight-loss rates in the saline/HCl solutions at 90°C.

2. For porous titanium, and for a given nominal surface area, the ULTI carbon films reduced the anodic dissolution rates in the saline/serum solution and the weight-loss rates in the saline/HCl solutions to values comparable to those of solid uncoated Ti-6Al-4V.

3. Analyses of two mechanisms of carbon-film breakdown, and extrapolation of these data to the saline/serum simulated-tissue-fluid solution, yielded a minimum carbon-film lifetime of approximately 38 years.

The electrochemical and gravimetric results of ths *in vitro* study have indicated that ULTI carbon coatings on porous titanium and solid Ti-6Al-4V provide significant protection from metallic corrosion under static conditions. Furthermore, the lifetime of the coatings, estimated to be in excess of 38 years, is sufficiently long to ensure clinical relevance.

References

[1] Haubold, A. D., Shim, H. S., and Bokros, J. C., "Carbon in Medical Devices," *Biocompatibility of Clinical Implant Materials, CRC Series in Biocompatibility,* Vol. 2, D. F. Williams, Ed., Chemical Rubber Company, Cleveland, OH, 1981, pp. 3–42.

[2] Bokros, J. C., "Carbon in Medical Devices," *Ceramics in Surgery,* P. Vincenzini, Ed., Elsevier, Amsterdam, 1983, pp. 199–214.

[3] "Metal Ion Release from Orthopaedic Implants," *Technology in Orthopaedics,* Zimmer, Warsaw, IN, 1984.

[4] Sutow, E. J., Pollack, S. R., and Korostoff, E., *Journal of Biomedical Materials Research,* Vol. 10, 1976, pp. 671–693.

[5] Guyton, A. C., *Textbook of Medical Physiology,* 4th ed., Saunders, Philadelphia, 1971, p. 492.

[6] Marinkovic, Z. and Roy, R., *Carbon,* Vol. 14, 1976, pp. 329–331.

[7] Agarwal, N. K., Shim, H. S., and Haubold, A. D., *Extended Abstracts,* 13th Biennial Conference on Carbon, American Carbon Society, Irvine, CA, 1977, p. 338.

[8] Fontana, M. G. and Greene, N. D., *Corrosion Engineering,* 2nd ed., McGraw-Hill, New York, 1978, pp. 342–344.

Brian J. Edwards[1] and Paul Higham[1]

Anodic Polarization of Porous-Coated Vitallium Alloy—Effect of Passivation

REFERENCE: Edwards, B. J. and Higham, P., **"Anodic Polarization of Porous-Coated Vitallium Alloy—Effect of Passivation,"** *Quantitative Characterization and Performance of Porous Implants for Hard Tissue Applications, ASTM STP 953*, J. E. Lemons, Ed., American Society for Testing and Materials, Philadelphia, 1987, pp. 115–123.

ABSTRACT: Passivation has the same effect on porous-coated Vitallium as it has on the smooth alloy. It decreased the current during anodic polarization by as much as two orders of magnitude. A potential scan reversal at $+700$ mV for all specimens resulted in a hysteresis for passivated specimens (both smooth and porous) but not for nonpassivated specimens. The protection potential was $+296 \pm 68$ and $+345 \pm 69$ mV versus the saturated calomel electrode (SCE) for porous and smooth passivated specimens, respectively.

There were two objectives of this work. The first was to see if a simple anodic polarization test could detect any basic difference in corrosion behavior between porous and smooth Vitallium. The second was to see if a standard passivation treatment affected these two geometries differently, again as detected by anodic polarization. The authors found that this porous configuration responds to polarization in a manner similar to the smooth surface in either the passivated or nonpassivated condition. There is an increase in current associated with the increased surface area. However, the increase in current appeared to be less than one would predict, assuming a linear proportionality between the current and surface area.

All tests were performed on Vitallium in the form of modified patella buttons. The specimens were manufactured by Howmedica, Inc. The porous surface was identical to the P.C.A. configuration. All specimens, both smooth and porous-coated, underwent the same production heat cycle.

KEY WORDS: porous implants, cobalt-chromium alloy, anodic polarization, passivation

Metallic porous surfaces used for surgical implants pose some interesting questions as to how this configuration will affect the corrosion process. The simplest consideration is that there will be some increase in corrosion current which may be directly proportional to the increase in surface area. Other possible important factors include crevices formed at the bonding points of these surfaces and any restrictions the porous configuration may impose on the movement of species (corrosion products, oxygen). The degree to which these factors affect corrosion behavior will probably be different for the different porous systems because of their variety of pore sizes and configurations (for example, beads as opposed to wire mesh).

---

[1] Research engineer and manager, respectively, Research and Development, Orthopaedics Division, Howmedica, Inc., Rutherford, NJ 07070.

[2] A cobalt-based alloy whose typical composition is as follows (all values are in weight percent): chromium 27.9, molybdenum 5.9, nickel 0.07, silicon 0.72, manganese 0.67, iron 0.28, carbon 0.22, and the balance cobalt. This is within the requirements of the ASTM Specification for Cast Cobalt-Chromium-Molybdenum Alloy for Surgical Implant Applications (F 75-82).

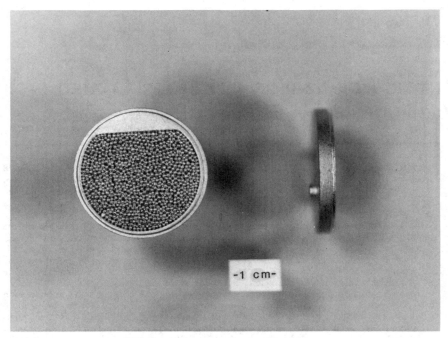

FIG. 1—*Porous-coated corrosion specimen.*

Experiments described in this paper seek to identify any effect the P.C.A. porous surface may have on the passive film breakdown and repair process of Vitallium.[2] The response of porous versus smooth Vitallium to the passivation process has also been investigated.

Experimental Procedure

Anodic polarization tests were conducted on smooth and porous-coated Vitallium corrosion specimens in deaerated 0.9% sodium chloride (NaCl) solution at 37°C. The specimens were modified patella buttons, that is, flat Vitallium disks 25.4 mm (1 in.) in diameter with a concave section on one side. For the smooth specimens the concave section was left uncoated. For the porous-coated specimens the concave section contained the porous coating, which consisted of two layers of 595 to 840-μm (20 to 30-mesh) Vitallium spheres sintered in place (Fig. 1). The concave side of the disk was exposed to the test solution using a specially designed holder (Fig. 2).

Mounted specimens were placed in the solution, which had been deaerated with nitrogen for 20 min. The purge was continued throughout the test. The specimen potential versus the saturated calomel electrode (SCE) after 1 h in solution was recorded as the rest potential. Anodic polarization was begun at the rest potential at a scan rate of 10 mV/min. Anodic polarization data from smooth specimens that had undergone the same heat treatment cycle as the sintered specimens was collected for comparison with the data for the porous surface. The specimens designated passivated were treated in accordance with the ASTM Recommended Practice for Surface Preparation and Marking of Metallic Surgical Implants (F 86-84) prior to testing.

Results

Average curves from at least three runs in each condition were established. Each curve begins at the lowest potential (at 100-mV intervals) where all runs had a current value. Comparisons are made for passivated porous versus passivated smooth surfaces (Fig. 3), nonpassivated porous versus nonpassivated smooth surfaces (Fig. 4), passivated porous versus nonpassivated porous surfaces (Fig. 5) and passivated smooth versus nonpassivated smooth surfaces (Fig. 6).

The 1-h rest potentials for the passivated and nonpassivated porous specimens were -34 ± 70 and -428 ± 42 mV, respectively. The range of data represents 95% confidence intervals calculated by Student's t test. The rest potentials for the passivated and nonpassivated smooth specimens were $+86 \pm 157$ and -404 ± 17 mV, respectively. The passivated porous specimens had lower currents than the nonpassivated porous specimens by as much as two orders of magnitude at the same potential. This difference narrowed with increasing scan potential (see Fig. 5).

The passivated porous surfaces developed a hysteresis with potential scan reversal, whereas the nonpassivated surfaces did not. The smooth specimens revealed this same behavior. The potential at which the reverse scan intersects the forward scan is the protection potential E_p. Generally, damage to the material's passive film can propagate at potentials greater than E_p [1]. E_p was $+296 \pm 68$ mV for the passivated porous surface and $+345 \pm 69$ mV for the passivated smooth surface.

The area of the smooth specimens exposed to solution was 5.8 cm^2, calculated by addition of the surface areas from simple geometric shapes comprising the total surface. The porous-coated surface area was obtained as follows. The number of spheres required for the porous

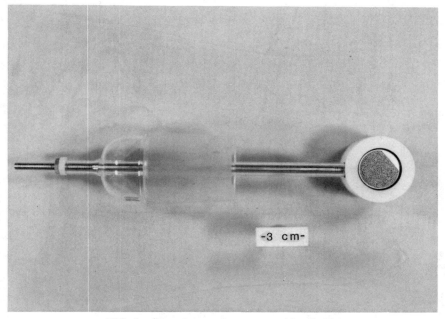

FIG. 2—*Corrosion specimen mounted in a holder.*

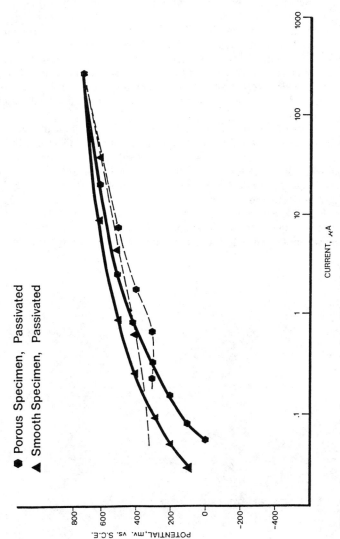

FIG. 3—*Average polarization curves (at least three runs) for porous-coated and smooth passivated Vitallium. Broken lines indicate the reverse scans.*

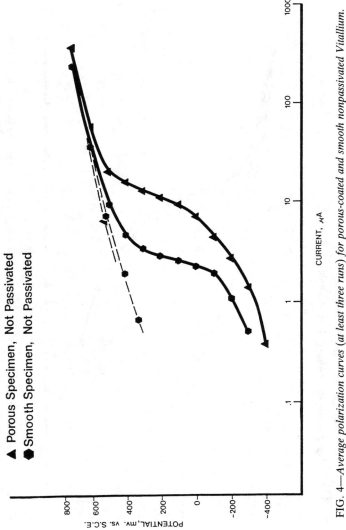

▲ Porous Specimen, Not Passivated
◆ Smooth Specimen, Not Passivated

POTENTIAL, mv. vs. S.C.E.

CURRENT, ₃A

FIG. 4—*Average polarization curves (at least three runs) for porous-coated and smooth nonpassivated Vitallium.*

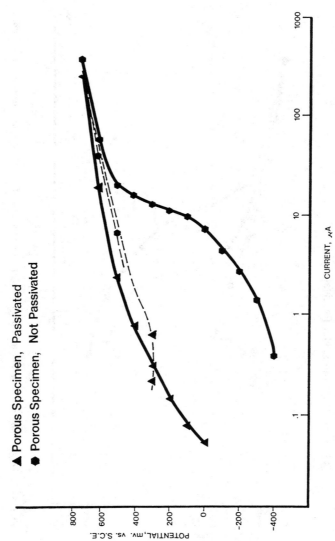

FIG. 5—*Average polarization curves (at least three runs) for passivated and nonpassivated porous-coated Vitallium.*

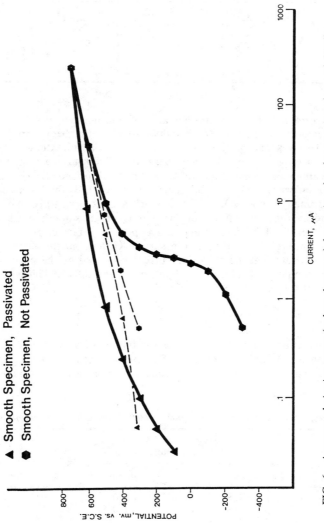

FIG. 6—*Average polarization curves (at least three runs) for passivated and nonpassivated smooth Vitallium.*

TABLE 1—*Average currents for smooth and porous-coated Vitallium specimens in the passivated condition.*

Potential, mV	Average Current ± 95% Confidence Interval, μA		Ratio of Porous to Smooth Surface Currents
	Porous Surface	Smooth Surface	
+100	0.08 ± 0.03	0.03	2.7
+200	0.15 ± 0.05	0.05 ± 0.04	3.0
+300	0.32 ± 0.14	0.10 ± 0.05	3.2
+400	0.78 ± 0.30	0.25 ± 0.10	3.1
+500	2.4 ± 1.0	0.84 ± 0.34	2.8
+600	20 ± 6.4	8.6 ± 6.5	2.3

surface was determined. Then the surface area of that number of spheres was calculated using an average diameter ($d = 0.0689 ± 0.006$ cm) which had been empirically determined by Fricker [2]. This was taken as the porous surface area. Consideration of bonding geometry was neglected since this would only decrease the calculated surface area. The "worst case" porous surface area thus obtained was 16.2 cm². The total surface area of a porous specimen exposed to solution was therefore 22 cm² (that is, 16.2 + 5.8). By this "worst case" calculation the porous specimens exposed 3.8 times as much surface area to solution as the smooth specimens. The average current values for porous versus smooth specimens at 100-mV intervals are summarized in Tables 1 and 2. The porous surface currents in Tables 1 and 2 are approximately three times the smooth surface currents at the same potential. These results tend to indicate that the corrosion current produced by the porous geometry may be less than one would predict based on simple area-ratio effects. This would be consistent with the results of Lucas et al. [3].

Discussion

The most interesting result of this work is the finding of a hysteresis for passivated but not for nonpassivated Vitallium. This effect has also been reported for a different specimen geometry of a cobalt-chromium alloy very similar to Vitallium [4], which meets the require-

TABLE 2—*Average currents for smooth and porous-coated specimens in the nonpassivated condition.*

Potential, mV	Average Current ± 95% Confidence Interval, μA		Ratio of Porous to Smooth Surface Currents
	Porous Surface	Smooth Surface	
−300	1.4 ± 0.5	0.5 ± 0.2	2.8
−200	2.7 ± 0.8	1.1 ± 0.2	2.4
−100	4.4 ± 0.3	1.9 ± 0.8	2.3
0	7.2 ± 0.4	2.3 ± 0.4	3.1
+100	9.5 ± 2.1	2.6 ± 0.3	3.6
+200	11 ± 4	2.9 ± 0.3	3.8
+300	13 ± 4	3.4 ± 0.1	3.8
+400	16 ± 5	4.7 ± 0.1	3.4
+500	20 ± 10	9.6 ± 1.0	2.1
+600	58 ± 7.2	37 ± 15	1.6

ments of the ASTM Specification for Cast Cobalt-Chromium-Molybdenum Alloy for Surgical Implant Applications (F 75-82).

The reverse scans collected in both studies are fairly independent of the specimen's initial surface condition. Figure 5, for instance, shows that for the passivated and nonpassivated porous-coated alloy the reverse scan is similar. A hysteresis indicates that polarization has damaged the passive film so that it cannot repassivate or recover to its initial level of protection at the higher polarization. The reverse scan forms a hysteresis with passivated specimens because their forward scan current is low, that is, their initial passive film is effective. It does not form a hysteresis with nonpassivated specimens, whose forward scan current is much higher, because they have a less effective initial passive film. In the case of this alloy, a hysteresis does not indicate a susceptibility to pitting corrosion [4]. A more extensive discussion of the cause of a hysteresis in the passivated surgical cobalt-chromium alloy is presented by Lucas et al. [4].

Conclusions

The geometry of the porous surface does not change this alloy's corrosion behavior as measured by anodic polarization. The increased alloy surface area of the porous surface does shift the polarization curve to higher currents, but this is an area effect. Passivation has the same beneficial effect on the porous alloy as it has on the smooth alloy.

References

[1] Syrett, B. and Wing, S., *Corrosion,* Vol. 34, No. 4, April 1978, pp. 138–145.
[2] Fricker, D., "Porous Coated Metal," internal report, Howmedica, Inc., Limerick, Ireland, 1983.
[3] Lucas, L., Lemons, J., Lee, J., and Dale, P., *Transactions,* Eleventh Annual Meeting of the Society for Biomaterials, Vol. 8, 1985, p. 80.
[4] Lucas, L., Buchanan, R., Lemons, J., and Griffin, C., *Journal of Biomedical Materials Research,* Vol. 16, 1982, pp. 799–810.

Linda C. Lucas,[1] Jack E. Lemons,[2] James Lee,[1] and Paul Dale[1]

In Vitro Corrosion of Porous Alloys

REFERENCE: Lucas, L. C., Lemons, J. E., Lee, J., and Dale, P., **"*In Vitro* Corrosion of Porous Alloys,"** *Quantitative Characterization and Performance of Porous Implants for Hard Tissue Applications, ASTM STP 953,* J. E. Lemons, Ed., American Society for Testing and Materials, Philadelphia, 1987, pp. 124–136.

ABSTRACT: Porous metallic surfaces are currently being applied to both orthopedic and dental implant devices. The increased surface areas provided by these systems have caused many in the health sciences to be concerned over an accelerated release of corrosion products into the adjacent and systemic tissues. The objective of this study was to evaluate the corrosion characteristics of six available porous alloy systems by utilizing electrochemical corrosion analyses. The surface area provided by the selected porous structures increased 1.2 to 7.2 times over that of the solid forms of the alloys. The corrosion rate for the porous alloys was in a range of 1.2 to 5.2 times greater than that for the solid alloys. In general, the increase in corrosion rate at the corrosion potential was not of the same magnitude as the increase in surface area. At potentials greater than the corrosion potential, the corrosion rates for the porous alloys were normally greater than the rates experienced by the solid alloys. In conclusion, the porous alloy systems investigated demonstrated increased corrosion magnitudes. Since the long-term effects of the released metallic constituents have not been evaluated at this time, surface treatments aimed at reducing the corrosion rates should continue to be evaluated.

KEY WORDS: porous implants, porous alloys, corrosion, Ti-6Al-4V alloy, Co-Cr-Mo alloys, implant alloys, electrochemical corrosion analyses

Porous alloys continue to find expanded utilization as biologic ingrowth or polymethyl methacrylate fixation surfaces for surgical implant devices. Selected alloys of titanium and cobalt have been the leading substrates for these applications. Questions concerning the relative corrosion susceptibilities of porous alloys manufactured by different industrial firms have resulted in continued laboratory *in vitro* research and development programs. The objective of this study was to evaluate the corrosion characteristics of six porous cobalt-base and titanium-base alloy systems.

Materials and Methods

Materials

The porous alloy systems chosen for evaluation included (1) commercially pure (CP) titanium wire on a Ti-6Al-4V substrate—Fiber metal system (Zimmer, USA, Inc.), (2) sintered Ti-6Al-4V alloy spheres on a Ti-6Al-4V substrate (Johnson and Johnson Products, Inc.), (3) sintered Co-Cr-Mo alloy spheres on a Co-Cr-Mo substrate (Johnson and Johnson Products, Inc.), (4) a plasma-sprayed Co-Cr-Mo coating on a Co-Cr-Mo substrate (Ortho-

---

[1] Associate professor and graduate students, respectively, Department of Biomedical Engineering, University of Alabama at Birmingham, Birmingham, AL 35294.

[2] Professor, Department of Biomaterials, University of Alabama at Birmingham, Birmingham, AL 35294.

FIG. 1—*Fiber metal system—CP titanium wire on a Ti-6Al-4V solid substrate.*

FIG. 2—*Sintered spheres on a solid substrate.*

paedic Products Division/3M), (5) solid Co-Cr-Mo with only 0.16% of the overall surface area coated with a porous Ti-6Al-4V plasma coating (Biomet, Inc.), and (6) solid Ti-6Al-4V with only 0.16% of the overall surface area coated with a porous Ti-6Al-4V plasma coating (Biomet, Inc.).

The fiber metal system was composed of commercially pure titanium wire diffusion-bonded on a Ti-6Al-4V solid substrate. This fabrication method resulted in an irregular interconnected porosity, as shown in Fig. 1. The porous specimens fabricated by sintering either Co-Cr-Mo or Ti-6Al-4V spheres on a solid substrate demonstrated a regular porosity, as shown in Fig. 2. The Co-Cr-Mo plasma-coated solid Co-Cr-Mo specimens provided a dense irregular surface (Fig. 3) and not an interconnected porosity, as provided by the previously described surface modifications. The remaining two porous alloy systems investigated involved applying a Ti-6Al-4V plasma coating on the flat surface of cylindrically shaped corrosion specimens (Fig. 4). For these porous systems, two sets of corrosion specimens were fabricated. One set contained a solid Ti-6Al-4V cylinder with a Ti-6Al-4V plasma-coated end. The second set contained a solid Co-Cr-Mo cylinder with a Ti-6Al-4V plasma-coated end. All of the porous alloys listed in Table 1 were provided by their manufacturers in a clean and passivated condition. The surface areas for each alloy system were obtained by the manufacturers utilizing quantitative microscopy techniques. The increases in surface area provided by the porous surfaces are given in Table 1. A complementary series of solid control specimens was also evaluated. The solid specimens were provided in a clean and passivated state by the manufacturers.

Methods

Triplicate anodic and cathodic polarization curves were generated for the porous alloys and also for a complementary series of solid control specimens [2]. A schematic drawing of the potentiostatic polarization test system employed in this study is shown in Fig. 5. This

FIG. 3—*Plasma-sprayed Co-Cr-Mo on a Co-Cr-Mo solid substrate.*

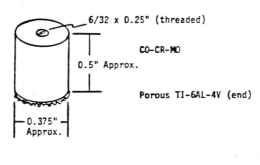

6/32 x 0.25" (threaded)

CO-CR-MO

0.5" Approx.

Porous TI-6AL-4V (end)

0.375" Approx.

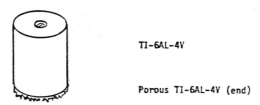

TI-6AL-4V

Porous TI-6AL-4V (end)

FIG. 4—*Schematic drawings of porous-coated corrosion specimens.*

TABLE 1—Corrosion characteristics of the six solid and porous titanium-base and cobalt-base alloys.

Alloy	Measured Corrosion Potential E_c, mV versus SCE[a]	Surface Area Increase, ×	Corrosion Rate Increase at E_c, ×
CP titanium (solid)	−14		
CP titanium wire/Ti-6Al-4V (porous)	−10	6.0	3.4
Ti-6Al-4V (solid)	−50		
Ti-6Al-4V spheres/Ti-6Al-4V (porous)	−75	7.0	5.2
Co-Cr-Mo (solid)	−10		
Co-Cr-Mo spheres/Co-Cr-Mo (porous)	−35	7.2	2.5
Co-Cr-Mo (solid)	−96		
Co-Cr-Mo plasma coating/Co-Cr-Mo (porous)	−103	2.0	1.2
Co-Cr-Mo (solid)	−10		
Ti-6Al-4V plasma coating/Co-Cr-Mo (porous)	−72	1.3	1.5
Ti-6Al-4V (solid)	−50		
Ti-6Al-4V plasma coating/Ti-6Al-4V (porous)	−41	1.2	1.8

[a] SCE = saturated calomel electrode value.

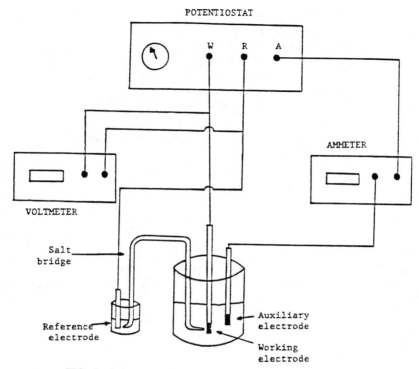

POTENTIOSTAT

VOLTMETER

AMMETER

Salt
bridge

Reference
electrode

Auxiliary
electrode

Working
electrode

FIG. 5—*Schematic of the potentiostatic polarization test system.*

system was used for the production of anodic and cathodic polarization curves. The electrolyte was 0.9% sodium chloride solution adjusted to pH 7.00 ± 0.05 with sodium bicarbonate. The temperature was maintained at 37 ± 1°C. A salt bridge connected a standard calomel electrode (SCE) to the polarization cell. The electrochemical studies were conducted with the electrolyte in the aerated condition. Aeration was achieved by diffusing oxygen through the electrolyte for 15 min prior to experimentation and throughout the test. The aerated condition was studied because oxygen acts as part of the cathodic process and also plays a role in the repassivation, that is, the growth, of the passive film.

The corrosion characteristics of both the solid and porous alloy specimens were evaluated using specimens that had been previously passivated by their manufacturers. As a final step, all the specimens were steam-sterilized for 20 min at 121°C. After sterilization, the individual test specimens were placed into the 0.9% sodium chloride electrolyte (pH 7.00 ± 0.05, temperature at 37 ± 1°C). The alloys were equilibrated in the test solution for 1 h prior to generating either the anodic or cathodic polarization curves. Tests were not initiated if the potential changed more than 1 to 2 mV over a 5-min period. The potential measured after the 1-h equilibrium time was the corrosion potential E_c. This value was recorded prior to conducting the anodic and cathodic polarization curves.

To develop the anodic polarization curves, the potential was set at 25 mV more noble than the corrosion potential (that is, the open-circuit potential). After a 5-min interval, the current was recorded and the current density was determined by dividing the current by the surface area of the test specimen. The surface area used for the current density calculations was the area of the test specimen and not the total surface area provided by the porous

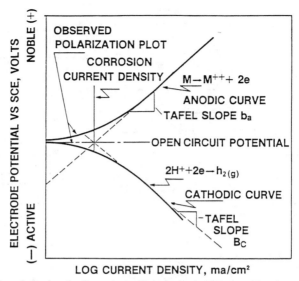

FIG. 6—*Hypothetical cathodic and anodic polarization diagram for determining the corrosion rate.*

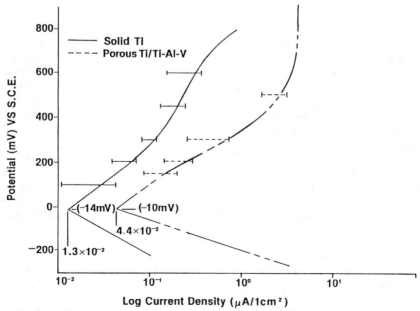

FIG. 7—*Anodic and cathodic polarization curves for solid CP titanium versus porous fiber metal (CP titanium/Ti-6Al-4V) alloys (mean ± standard deviation).*

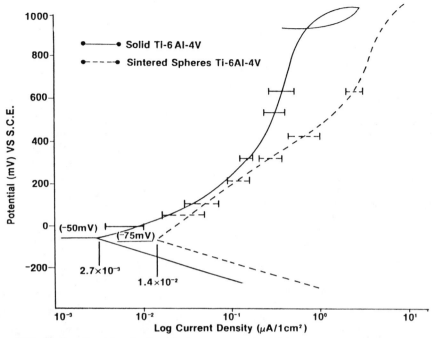

FIG. 8—*Anodic and cathodic polarization curves for solid Ti-6Al-4V versus porous alloys having Ti-6Al-4V spheres sintered on a Ti-6Al-4V substrate (mean ± standard deviation).*

surfaces. The potential was increased in steps of 50 mV, and the current density was recorded at each step after a 5-min period. This procedure was continued until the transpassive region was reached. When the current density exceeded 10^2 $\mu A/cm^2$, cyclic anodic polarization curves were produced by reversing the direction of potential stepping.

To develop the cathodic polarization curve, the potentiostat was set at 25 mV more active than the corrosion potential. The current was recorded at 25-mV steps with 3-min intervals. The steps of potential change were continued until the plot of the potential versus the log current density was linear. The linear portion of the cathodic polarization curve was extrapolated to the measured corrosion potential. The intersection of the cathodic curve at the corrosion potential determined the corrosion rate of the alloy. The anodic polarization curve then was extrapolated to the corrosion potential. A diagram exhibiting the method used for determining the corrosion rate of each alloy is shown in Fig. 6 [ASTM Practice For Standard Reference Method for Making Potentiostatic and Potentiodynamic Anodic Polarization Measurements (G 5-82)] [*3,4*].

Results

The results of the electrochemical corrosion analyses are shown in Figs. 7 through 12. The averaged anodic and cathodic polarization curves are displayed for each alloy. The intersection shown in each figure occurred at the measured corrosion potential E_c. The Tafel extrapolation of the cathodic polarization curve to the measured corrosion potential predicted the corrosion current rates i_c for each alloy. A summary of the corrosion data for

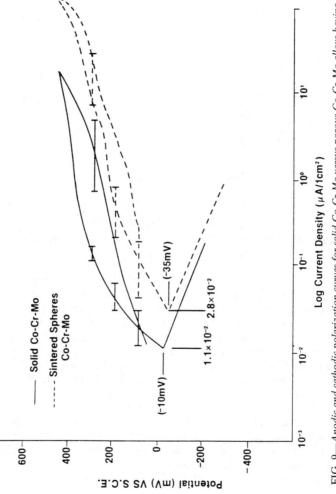

FIG. 9—*Anodic and cathodic polarization curves for solid Co-Cr-Mo versus porous Co-Cr-Mo alloys having Co-Cr-Mo spheres sintered on a Co-Cr-Mo substrate (mean ± standard deviation).*

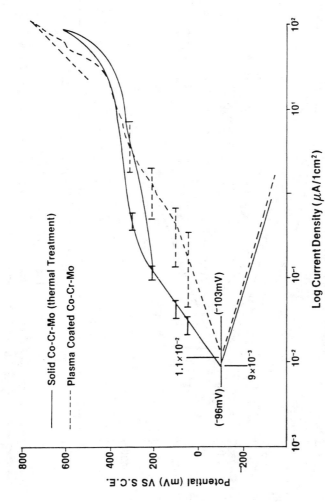

FIG. 10—*Anodic and cathodic polarization curves for solid Co-Cr-Mo versus Co-Cr-Mo plasma-coated Co-Cr-Mo substrates.*

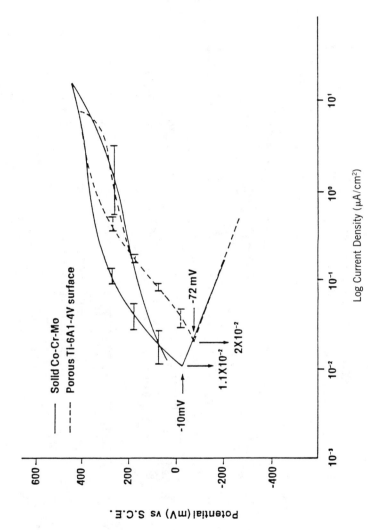

FIG. 11—*Anodic and cathodic polarization curves for solid Co-Cr-Mo versus solid Co-Cr-Mo alloys with 0.16% of the surface area containing a Ti-6Al-4V plasma-coated surface.*

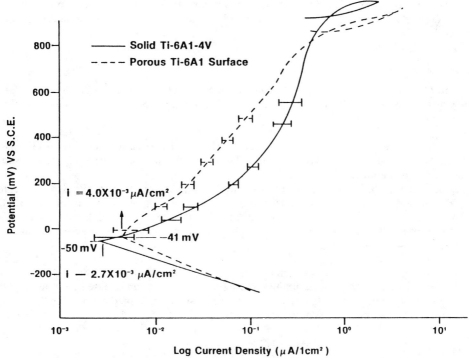

FIG. 12—*Anodic and cathodic polarization curves for solid Ti-6Al-4V versus solid Ti-6Al-4V alloys with 0.16% of the surface area containing a Ti-6Al-4V plasma-coated surface.*

each alloy is provided in Table 1. For each of the six alloy systems the measured corrosion potentials were not significantly different for the solid and porous forms of the alloys. The most significant difference in corrosion potentials was exhibited between the solid Co-Cr-Mo specimens and the Ti-6Al-4V plasma-coated end/Co-Cr-Mo specimens, which showed the respective values of -10 mV versus SCE and -72 mV versus SCE.

The increase in surface area provided by each of the porous alloys is given in Table 1. Surface area increases of 1.2 to 7.2 times were determined for the alloys in this study. The corrosion rate predicted for each alloy at the corrosion potential is shown on the individual graphs (Figs. 8 through 12) and also in Table 1. Corrosion rate increases of 1.2 to 5.2 times were determined for the alloys from the corrosion data. The porous alloys with the greatest surface areas (Figs. 7, 8, and 9) exhibited higher increases in the corrosion rates at the corrosion potential. However, these increases were not of the magnitude indicated by the surface area of the alloy. The porous alloy systems with the lowest surface area increases (Figs. 10, 11, and 12) did not exhibit a statistically significant corrosion rate increase at the corrosion potential. At potentials greater than the corrosion potentials, the increased surface area provided by the surface porosity resulted in higher current density values for the alloys shown in Figs. 7 through 11.

The results obtained from the corrosion evaluations of the six porous alloy systems revealed that the increased surface area provided by the porous alloys could result in increased corrosion rates. The question that must be addressed is, "What is an acceptable quantity of material to be released during long-term implantation?" Since there is no answer to this question at the present time, efforts aimed at reducing any increases should be investigated [5].

Conclusions

The objective of this study was to evaluate the corrosion characteristics of six available porous alloy systems utilizing electrochemical corrosion analyses. The following conclusions were made:

1. The increase in surface area provided by the porous structures resulted in increased corrosion rates at the corrosion potential. However, the increases were not of the same magnitude as the increase in surface area.
2. In general, the corrosion current densities exhibited by the porous alloys were greater than those exhibited by the solid alloys at potentials greater than the corrosion potential.
3. The measured corrosion potentials for the solid alloys were very similar to the measured corrosion potentials of the porous alloys when the composition of the substrate and the surface porosity were similar.

In summary, the increase in surface area provided by the porous alloys can result in increased corrosion rates, indicating that these porous systems can release accelerated quantities of metallic ions to the tissues. Thus, further investigations aimed at reducing the overall corrosion rates of the alloy systems are indicated.

References

[1] Spector, M. in *Biocompatibility of Orthopaedic Implants,* CRC Press, Boca Raton, FL, 1982, Chapter 5, pp. 89–128.
[2] Fontana, M. G. and Greene, N. D., *Corrosion Engineering,* McGraw-Hill, New York, 1978, Chapters 9–10.
[3] Lucas, L. C., Buchanan, R. A., and Lemons, J. E., *Journal of Biomedical Materials Research,* Vol. 15, 1981, p. 731.
[4] Lucas, L. C., Buchanan, R. A., Lemons, J. E., and Griffin, C. D., *Journal of Biomedical Materials Research,* Vol. 16, 1982, p. 799.
[5] Lemons, J., Compton, R., Buchanan, R., and Lucas, L., "Surface Modifications and Corrosion of Porous Fiber Titanium Alloy Systems," *Proceedings,* Second World Congress on Biomaterials, Washington, DC, April 1984.

DISCUSSION

E. P. Lautenschlager[1] (written discussion)—You conducted your tests in an aerated solution, while the tests of the previous paper given (B. Edwards and P. Higham), used a deaerated electrolyte. Why did you choose the aerated solution?

L. C. Lucas, J. E. Lemons, J. Lee, and P. Dale (authors' closure)—The oxygenated condition was studied because oxygen acts as part of the cathodic process and also plays a role in the repassivation of the passive oxide film. In addition, oxygen is a component of the biological environment.

[1] Department of Biological Materials, Northwestern University, Chicago, IL 60611.

Kirk J. Bundy[1] and Richard E. Luedemann[1]

Characterization of the Corrosion Behavior of Porous Biomaterials by A-C Impedance Techniques

REFERENCE: Bundy, K. J. and Luedemann, R. E., "**Characterization of the Corrosion Behavior of Porous Biomaterials by A-C Impedance Techniques,**" *Quantitative Characterization and Performance of Porous Implants for Hard Tissue Applications, ASTM STP 953,* J. E. Lemons, Ed., American Society for Testing and Materials, Philadelphia, 1987, pp. 137–150.

ABSTRACT: In this research, the corrosion behavior of Ti-6Al-4V extralow interstitial (ELI) and Co-Cr-Mo alloys with polished, grit-blasted, and porous-coated surfaces has been investigated. The A-C impedance technique and a variety of D-C methods were used for this purpose. Both highly stressed and nonstressed specimens were tested. Stress-enhanced release of corrosion products was demonstrated to occur in some cases. The A-C impedance method, which is capable of measuring both corrosion current and true surface area, showed that this effect was mainly due to an area increase caused by plastic deformation. The advantages and characteristics of the A-C impedance method for corrosion measurements of implant materials are also described.

KEY WORDS: porous implants, biomaterials, surgical implants, corrosion, corrosion rate, A-C impedance, Ti-6Al-4V ELI alloy, Co-Cr-Mo alloys, polarization, implant materials, porous electrodes

The use of porous metallic implants is becoming more frequent because of their greater efficiency of fixation to hard tissue. Since porous-coated devices have larger surface areas, more ions are released into the *in vivo* environment as a result of electrochemical dissolution processes. The clinical significance of the increased corrosion product release is presently under debate. A more informed debate will be possible if improved data regarding the corrosion rates experienced by surgical implants become available.

The primary objective of the present research is to investigate the utility of the A-C impedance method (which offers potential advantages over D-C methods) for measurement of the corrosion rates of implant materials, particularly porous materials. A secondary objective is to study the effect of applied static stresses on the corrosion rates of surgical implant materials. A third objective is to obtain a quantitative comparison of corrosion rates obtained by different experimental methods.

The A-C Impedance Technique

Although it has been used in electrochemistry for a considerable period of time, the A-C impedance method has only relatively recently been applied to corrosion measurements

[1] Associate professor and graduate student, respectively, Biomedical Engineering Department, Tulane University, New Orleans, LA 70118.

[1,2]. Like the D-C linear polarization (polarization resistance) method, it is nondestructive to the surface tested because only a small amplitude potential perturbation ($\leq \sim 20$ mV) is applied. In either case, the corrosion current density i_c is determined from the Stern-Geary equation [3]

$$i_c = \frac{1}{2.3 R_p A} \cdot \frac{\beta_a \beta_c}{(\beta_a + \beta_c)} \tag{1}$$

where β_a and β_c are the Tafel slopes for the anodic and cathodic reactions, respectively; R_p is the polarization resistance; and A is the electrode surface area.

Both methods determine the polarization resistance. Additional information can be obtained from the impedance measurement, however. In this method an overpotential η is applied to the test electrode

$$\eta = \eta_0 \sin \omega t \tag{2}$$

where $\omega = 2\pi f$, where f is the frequency of the perturbation and t is time. The resultant current I is phase shifted by an amount ϕ

$$I = I_0 \sin (\omega t + \phi) \tag{3}$$

which varies with ω. An A-C impedance measurement is, in effect, a measurement of ϕ versus ω. One way to accomplish this is to measure η and I and plot them orthogonally. This generates a Lissajous figure with an elliptical shape. An example is shown in Fig. 1.

If the electrode impedance is Z

$$Z = |Z| e^{-j\phi} = R - Xj \tag{4}$$

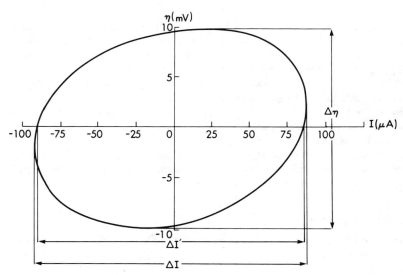

FIG. 1—*Experimentally determined Lissajous figure with parameters used in an A-C impedance measurement.*

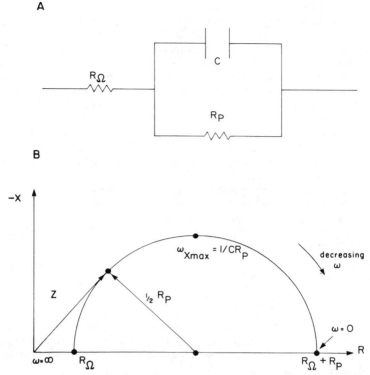

FIG. 2—*Electrical circuit and Nyquist plot for an electrode under charge transfer control:* (a) *electrical circuit corresponding to the electrode,* (b) *Nyquist plot for the electrode.*

where e = the base of natural logarithms, and $j = \sqrt{-1}$. The quantities which appear in this equation can be found from the Lissajous figure

$$|Z| = \frac{\Delta\eta}{\Delta I} \tag{5}$$

$$\sin\phi = \frac{\Delta I'}{\Delta I} \tag{6}$$

$$R = |Z|\cos\phi \tag{7}$$

$$-X = |Z|\sin\phi \tag{8}$$

There are various ways to display the data from A-C impedance measurements. In one method, termed a Nyquist plot, $-X$ versus R is plotted for the various values used in the measurement. Several pieces of information useful in describing the electrochemical behavior of the test material can be extracted from the Nyquist plot.

Every metal/electrolyte combination can be modeled in terms of an electrical circuit analog. To illustrate the information available from a Nyquist plot, consider a smooth electrode under charge transfer control. The circuit model for this case is shown in Fig. 2a.

The corresponding Nyquist plot is given in Fig. 2b. From the basic shape of the curve, knowledge of the dominant type of electrochemical process can be obtained. For example, a semicircle, such as that shown in Fig. 2b, indicates a charge-transfer controlled process [4]. A straight line at a 45° angle indicates a diffusion-controlled process.

From the high-frequency intercept on the real axis, the ohmic resistance R_Ω, can be determined. R_Ω is the resistance R involved in IR drop (which is the apparent overvoltage that occurs as a result of ohmic solution resistance when the net current I flows through the electrode). The straightforward determination of R_Ω is an advantage of the A-C impedance method in comparison with D-C methods. In certain situations of biomedical interest, such as *in vivo* corrosion measurement, IR drop could be substantial. With a D-C linear polarization measurement, IR drop can go unrecognized. The measured polarization resistance would be $R_p + R_\Omega$, and an underestimated i_c would be obtained from the Stern-Geary equation. The presence of IR drop is clear from a D-C potentiodynamic polarization measurement at high values of overpotential, but accurate compensation for it requires a separate determination of R_Ω (using the current-interruption technique).

From the low-frequency intercept, R_p can be obtained if R_Ω is known. By measuring the ω value where the magnitude of X is a maximum, electrode capacitance C can be found because it has been shown that [1,4]

$$\omega_{X\text{max}} = \frac{1}{R_p C} \tag{9}$$

Determination of the electrode capacitance can be of great use because it allows determination of the surface area of irregular objects. Once the specific double-layer capacitance C_{DL} has been determined by capacitance measurements on specimens of known area, the true area of a specimen can be determined from its capacitance

$$C = C_{DL} A \tag{10}$$

Unlike R_p or corrosion rates, C_{DL} only varies within relatively narrow limits for a wide range of metal and electrolyte combinations. A typical range of 10 to 40 μfarad/cm² has been reported [4]. Hence, it is not likely that microstructural differences due to differing heat treatments or surface preparation procedures in the same metal in a given electrolyte will have more than a minor effect on C_{DL}. This would mean that, by measuring C_{DL} in smooth specimens, the true surface area of grit-blasted, textured, flame-sprayed, fiber mesh, cancellous-bone-like, or porous-coated metallic materials could be found from A-C impedance measurements.

The same principles as those described for the smooth electrode under charge transfer control can be applied to porous materials. However, the theory for the electrochemical response of a porous electrode is somewhat more complicated. De Levie has presented the theory for a single pore [5], and Kramer and Tomkiewicz have described the A-C response of a random network of pores [6]. A porous network made of sintered spheres does not completely correspond to either of these conditions, but the basic ideas may still be applicable. In the theories just mentioned, a pore is analogous to a transmission line. Unlike a nonporous, smooth electrode, the impedance of a single pore Z_p is not identical to the interfacial impedance Z, but is given, rather, by

$$Z_p = (RZ)^{1/2} \coth (\rho L) \tag{11}$$

where R is the resistance of the electrolyte in the pore per unit length, L is the pore length, ρ is $(R/Z)^{1/2}$, and Z is the impedance of the metal/electrolyte interface in the pore per unit length. When the ohmic resistance to radial current flow in the pore is negligible, at the D-C limit

$$Z = \frac{R_p' A^*}{L} \tag{12}$$

where A^* is the surface area of the pore, and R_p' is its polarization resistance. Since i_c is independent of area, $R_p' A^* = R_p A$.

The measured impedance Z^m of n pores in parallel would be

$$Z^m = \frac{Z_p}{n} \tag{13}$$

For materials with passive films of high Z (and therefore a small value of ρL), Eq 11 reduces to

$$Z_p = \frac{Z}{L} \tag{14}$$

De Levie has shown for porous electrodes [5] that

$$\beta_a^t = 0.5\beta_a^m \tag{15}$$
$$\beta_c^t = 0.5\beta_c^m$$

where superscripts t and m stand for the true and measured values, respectively. The Stern-Geary equation thus becomes

$$i_c = \frac{1}{4.6R_p A} \cdot \frac{\beta_a^m \beta_c^m}{(\beta_a^m + \beta_c^m)} \tag{16}$$

The validity of the use of the porous electrode theory to analyze measured data can be ascertained by comparing the measured spectrum with that predicted by Eq 11 and by comparing Tafel slope data with Eq 15.

Materials and Methods

Smooth and porous-coated specimens of Co-Cr-Mo and Ti-6Al-4V extralow interstitial (ELI) alloys were employed in this investigation. Grit-blasted specimens of Ti-6Al-4V ELI alloy were also tested. The properties of the materials met the requirements set forth in the ASTM Specifications for Cast Cobalt-Chromium-Molybdenum Alloy for Surgical Implant Applications (F 75-82) and for Wrought Titanium Ti-6Al-4V ELI Alloy for Surgical Implant Applications (F 136-84). The materials were passivated by the supplier of the materials according to the procedure described in the ASTM Recommended Practice for Surface Preparation and Marking of Metallic Surgical Implants (F 86-84). In the authors' laboratory the surfaces were ultrasonically cleaned in detergent for 5 min, rinsed in distilled water, ultrasonically cleaned in ethanol for 5 min, rinsed in tap water, ultrasonically cleaned in distilled water for 5 min, and air dried. Although this procedure of not passivating the materials in our own laboratory probably induces more scatter in the data, we feel that this provides a more realistic simulation of the type of metal surface that is actually present on medical devices implanted in patients.

The test specimens of smooth and grit-blasted material were in the form of cylindrical rods 11.43 cm long and 0.635 cm in diameter. The smooth materials were polished with diamond and alumina abrasive pastes down to a 0.05-μm finish. The grit-blasted specimens were prepared by glass-bead blasting with 50 to 70 mesh beads. The porous-coated specimens were prepared by sintering 0.203-mm-diameter spheres onto 0.635-cm-diameter cylindrical substrates. The procedures used for the sintering were those used in the manufacture of porous-coated surgical implants. The thickness of the sintered layer is 0.866 mm.

The corrosion behavior of some of the specimens was determined under conditions of static loading. For the experiments with stressed specimens, a holder was used which loaded the specimens in three-point bending. The holder has been described in more detail elsewhere [7,8]. Basically, the surface of the holder, the loading points, and the area of the test specimen (except for a 1.5-cm length at the middle portion of the rod) are insulated from the environment. The corrosion behavior of the nonloaded controls is also tested while the specimens are in this holder.

Using a beam theory analysis for a material undergoing both elastic and plastic deformation, the deflection of a smooth beam at its midpoint is calculated; this deflection corresponds to a maximum stress level in the beam equal to the true stress σ_{uts}, which is present at the point of maximum load in a uniaxial tension test. The porous-coated specimens were subjected to the same deflection. A composite beam continuum theory analysis, which accounts for the presence of the porous layer, showed that for a porous-coated beam the maximum stress level is $0.81\sigma_{uts}$. The actual stresses present at some points on the surfaces of the sintered spheres could be larger than this because of the presence of microscopic stress concentrations neglected in the analysis. For the sake of simplicity of representation, in the tables that follow, the stress level in the porous layer is denoted σ_{uts}. It should be understood, however, that this actually represents the stress distribution described earlier.

Table 1 describes the set of specimens used in this study. Information regarding the number of specimens tested for the different material, surface condition, and maximum stress level combinations is given.

The electrolyte employed for these tests was a 37°C Ringer's solution. The corrosion cell was open to the atmosphere. The dissolved oxygen concentration was measured with a Yellow Springs Instrument Co. (YSI) dissolved oxygen meter and probe and was found to be in the 6 to 10-ppm range. The solution was not stirred in order to simulate *in vivo* conditions more closely.

Corrosion testing was conducted using commercial equipment available from EG & G Princeton Applied Research (PAR). The system can be operated in either a manual or a

TABLE 1—*Number of specimens employed in corrosion tests for different combinations of material, surface condition, and maximum stress level.*

| | Number by Material and Stress Level | | | |
| | Co-Cr-Mo | | Ti-6Al-4V ELI | |
Surface Condition	0	σ_{uts}	0	σ_{uts}
Polished	5	2	4	4
Grit-blasted	...	...	3	3
Porous	4	4	4	2

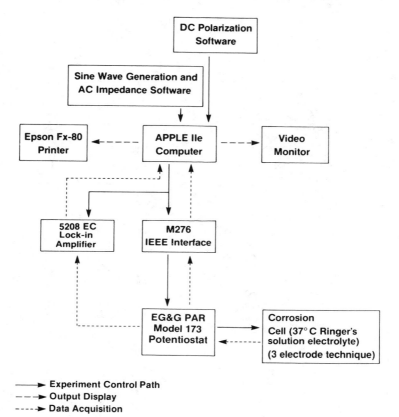

FIG. 3—*Block diagram of apparatus used for A-C impedance measurements and D-C corrosion tests.*

computer-controlled mode. A block diagram of the instrumentation is given in Fig. 3 for the computer-controlled case. When the system is operated in the computerized mode with the lock-in amplifier and the Model 368 A-C impedance software package, at frequencies above 1 Hz the lock-in amplifier enhances the signal-to-noise ratio and improves measurement accuracy. Below 10 Hz, a fast Fourier transform algorithm increases the speed of data acquisition at some expense to accuracy in comparison with a discrete sine wave approach.

Since a comparison of different corrosion rate determination methods was desired, various tests were run with each specimen. In the majority of cases, a specimen was subjected to a linear polarization test, an A-C impedance measurement, and a potentiodynamic polarization curve determination. This usually enabled three or four estimates of i_c to be made. The specimens were immersed in the electrolyte for a time sufficient to achieve a stabilized value of the corrosion potential E_c. Then, a D-C polarization resistance measurement was made following procedures similar to those outlined in the ASTM Practice for Conducting Potentiodynamic Polarization Resistance Measurements [G 59-78(1984)]. A 0.1-mV/s rate was used for these tests. The range of potential used was ±25 mV about the corrosion potential.

Next, A-C impedance measurements were conducted in the 0.01 to 65-Hz frequency range. Typically, measurements were made at about 30 different frequencies. The peak-to-peak

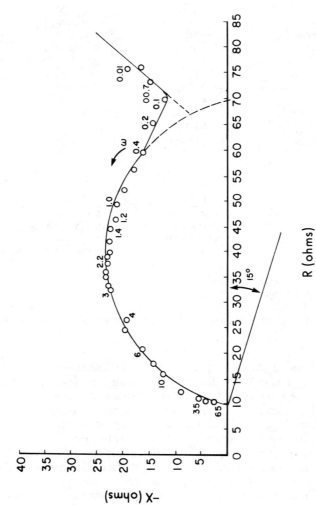

FIG. 4—Nyquist plot for a porous-coated Ti-6Al-4V ELI specimen in 37°C Ringer's Solution. The numbers given are frequencies in Hertz units.

value of the perturbation from the corrosion potential was 20 mV. It has been demonstrated, particularly in electrochemical systems showing inductive behavior, that the most accurate way to estimate R_p from A-C impedance tests is to use the true low-frequency D-C limit [2]. When such effects are absent, it is possible to identify $R_\Omega + R_p$ as the low-frequency intercept even in circumstances in which the Nyquist plot shows a depressed semicircle [9].

Finally, after these nondestructive tests were conducted, potentiodynamic polarization curves were measured, following procedures similar to those outlined in the ASTM Practice for Standard Reference Method for Making Potentiostatic and Potentiodynamic Anodic Polarization Measurements (G 5-82). The potential range typically employed for these tests was from 250 mV below E_c (to obtain β_c) up to a value that exceeded the breakdown potential. A 0.1-mV/s scan rate was employed for tests with Co-Cr-Mo, and a 1-mV/s rate was used for the tests with the titanium alloy. The potentiodynamic polarization tests were conducted to determine β_a and β_c for use in the Stern-Geary equation and also to obtain a further estimate of i_c using the Tafel extrapolation technique.

Results and Discussion

A Nyquist plot for a porous-coated Ti-6Al-4V ELI alloy specimen is shown in Fig. 4 to illustrate the type of experimental data obtained from an A-C impedance measurement. The 45° straight line at low frequencies characteristic of diffusion control is clearly evident. In the semicircular charge transfer control region, the frequency where the imaginary component has maximum magnitude ω_{Xmax}, can be identified, and the value of what would have been the low-frequency limit of the real component (in the absence of diffusion control) is also clearly apparent. The charge transfer control region shows what is termed a "depressed" semicircle. In this case the depression angle is 15°. This angle is measured between the real axis and the diameter passing through the high ω limit.

This type of Nyquist plot with a depressed semicircle can still be analyzed as a transmission line, although this requires some modification of the material presented previously. The complex impedance that appears in Eq 11 [6] is

$$Z = R_s + \frac{\dfrac{R'_p A^*}{L}}{1 + \omega^2\tau^2} - j\frac{\dfrac{\omega\tau R'_p A^*}{L}}{1 + \omega^2\tau^2} \qquad (17)$$

for the case in which the depression angle for Z is 0°. In Eqs 17 and 18, $\tau = R'_p C_{DL}(A^*/L)^2$, and R_s is the ohmic component for radial current flow in the pore.

For the case in which the depression angle for Z is $\alpha\pi/2$, the appropriate expression [10] is

$$Z = R_s + \frac{\dfrac{R'_p A^*}{L}\left\{1 + (\omega\tau)^{1-\alpha}\cos\left[(1 - \alpha)\dfrac{\pi}{2}\right]\right\}}{1 + 2(\omega\tau)^{1-\alpha}\cos\left[(1 - \alpha)\dfrac{\pi}{2}\right] + (\omega\tau)^{2(1-\alpha)}}$$
$$- j\frac{\dfrac{R'_p A^*}{L}(\omega\tau)^{1-\alpha}\sin\left[(1 - \alpha)\dfrac{\pi}{2}\right]}{1 + 2(\omega\tau)^{1-\alpha}\cos\left[(1 - \alpha)\dfrac{\pi}{2}\right] + (\omega\tau)^{2(1-\alpha)}} \qquad (18)$$

TABLE 2—*Anodic and cathodic Tafel slopes for various materials, surface conditions, and stress levels.*

		Tafel Slopes, mV/decade, by Stress Level			
		β_a		β_c	
Material	Surface Condition	0	σ_{uts}	0	σ_{uts}
Co-Cr-Mo	polished	176	295	97.5	77.5
	porous	365	275	95	200
Ti-6Al-4V ELI	polished	860.7	1050	50	90
	grit-blasted	1259.7	1528	101.3	79.7
	porous	1155	783	103	70

Using an expression of the form given in Eq 18 and employing an iteration procedure to find the values of the circuit parameters of Z for the data of Fig. 4, we found good agreement of the measured spectrum with the transmission line analogy for the semicircular region when Z had a 32° depression angle. Also, the diffusional tail was found to be consistent with an extension of the transmission line model which deals with radial diffusion in the pores [11]. For this more complicated circuit model, use of Eq 9 to calculate C is not valid quantitatively. The appropriate value of C must be determined from the iteration procedure.

Table 2 contains the mean values of the measured Tafel slopes for the zero stress controls and those specimens for which the maximum stress level is σ_{uts}. For the porous materials (except for the stressed porous Ti-6Al-4V ELI) at least one of the Tafel slope values is elevated above the "true" value for smooth, polished material by an amount qualitatively similar to that predicted by De Levie (see Eq 15). Similar findings have been previously reported [12]. As is often the case with electrochemical measurements, significant scatter was observed in the Tafel slope data. For the measurements given in Table 2, the standard deviation divided by the mean was 39.8% on the average.

Table 3 displays the average of the corrosion current density i_c. Corrosion currents were determined according to the methods indicated previously. Some explanation is necessary regarding the surface area values employed to obtain the current densities. For the polished specimens, A was taken to be the nominal geometrical area.

From the A-C impedance measurement with polished specimens, C_{DL} was found to be 13.40 ± 2.86 μfarad/cm² for the Co-Cr-Mo alloy and 13.82 ± 2.10 μfarad/cm² for the Ti-6Al-4V ELI. To find the surface area for the grit-blasted and porous materials, the assumption was made that the C_{DL} values just given applied to these surface conditions also. The surface areas of the grit-blasted and porous materials were found from the A-C impedance capacitance measurements and Eq 10. It was also assumed that C_{DL} does not change with stress. Conclusive establishment of the validity of these assumptions would require independent measurement of the actual surface area using, for example, the Brunauer, Emmett, and Teller (BET) technique, photomicrographic image analysis, or other appropriate methods.

For calculations with porous surfaces, the pore length L, which appears in Eqs 11, 12, and 14, was taken to be the thickness of the porous layer. The number of pores, which appears in Eq 13, was estimated by assuming that the sintered spheres were placed in a simple cubic array.

As was the case for the Tafel slope data, significant scatter was observed in the measurements of corrosion current density. The median of the standard deviation divided by the mean of i_c was 56.3%.

Inspection of Table 3 reveals that the estimates of i_c from the various electrochemical tests with a given material, surface condition, and stress level agree generally to within one order of magnitude (and often to within much closer limits). It cannot be said that one method will generally give a systematically higher or lower estimate of i_c than another method under all conditions. In themselves, these electrochemical tests cannot answer the question of which method is most accurate. To do this, comparison with direct measurements of the solution concentration (by means of atomic absorption spectroscopy or polarography, for example) would be necessary.

The corrosion current density i_c is apparently less for porous materials than for smooth materials at the same stress level. This is indicated without exception by all four corrosion rate determination methods for both materials. For unstressed materials, similar findings have been previously reported using Tafel extrapolation [12,13] and A-C impedance [12]. Although the current densities are lower for porous materials, the total current can still be substantially higher, as reported here, because of the increased surface area.

For polished Ti-6Al-4V ELI alloy, the application of stress that causes plastic deformation seems to increase i_c substantially (by a factor of 1.55 to 25.5, depending on the estimation method). Identification of a stress-enhanced ion release (SEIR) effect for polished Co-Cr-Mo is less certain. Two i_c estimation methods indicate that this is the case, but a third one shows the opposite. For grit-blasted Ti-6Al-4V ELI alloy the SEIR effect apparently disappears altogether, and for the porous material it is much less marked.

An explanation for these findings cannot be found when using information obtained from the D-C electrochemical tests. However, the underlying phenomena are clearly identified from the A-C impedance tests and are based on surface area changes. Table 4 displays the ratios of the true surface areas (found by A-C impedance determination of the electrode capacitance) to the nominal surface areas. Application of stresses that cause plastic deformation increase the total surface area exposed to the electrolyte by about 50 to 80%. Although the necessary experiments have not been performed, presumably a similar effect

TABLE 3—*Corrosion current densities based on surface areas, determined from capacitance measurements.*

Surface Condition	Stress Level	Current Density, nA/cm², by Method[a]			
		ATE	CTE	SG	ACI
Co-Cr-Mo					
Polished[b]	0	14.8	27.8	21.8	29.7
	σ_{uts}	36.6	14.7	74.0	...
Porous	0	3.0	...	...	6.14
	σ_{uts}	18.4	8.6	...	35.9
Ti-6Al-4V ELI					
Polished[b]	0	62.1	5.69	79.4	41.2
	σ_{uts}	347	145	123	...
Grit-blasted	0	105	72.4	14.4	103
	σ_{uts}	67.9	28.7	2.6	45.0
Porous	0	27.6	...	...	7.16
	σ_{uts}	44.4	18.7	...	16.9

[a] Key to abbreviations for method:
  ATE = anodic Tafel extrapolation.
  CTE = cathodic Tafel extrapolation.
  SG = Stern-Geary method.
  ACI = A-C impedance method.
[b] Based on the geometrical area.

TABLE 4—*Ratio of the true surface area determined by A-C impedance tests to the nominal surface area for various surface conditions, stress states, and implant materials.*

Material	Surface Condition	Ratio by Stress Level	
		0	σ_{uts}
Co-Cr-Mo	porous	6.25	9.56
Ti-6Al-4V ELI	porous	7.96	12.97
	grit-blasted	3.04	5.58

occurs with smooth specimens. This would mean that the stress-enhanced ion release effect found with smooth material (in terms of the current density based on the geometrical area) would not be as marked if the true area of stressed material were taken into account. Although the increased surface area due to plastic deformation is a plausible cause from the metallurgical point of view, because of the emergence of slip steps, the A-C impedance method provides a way of easily quantifying the surface area increase without a separate measurement.

To summarize, the A-C impedance method shows that the application of stress which causes plastic deformation can increase corrosion currents. This is accomplished mainly by an increase in the total surface area rather than by an increase in i_c. Although D-C electrochemical measurement methods can detect the increase in overall corrosion current, they cannot clarify the origin of the increase.

The amount by which applied stress or increased surface area (due to roughness or porosity), or both, leads to an increased body burden, resulting from a greater release of ions from an implant, can best be visualized in terms of the corrosion current per nominal unit area. This information is found by multiplying the true current density values (whose averages are given in Table 3) by the area ratios (whose averages are given in Table 4).

TABLE 5—*Corrosion current densities based on nominal area.*

Surface Condition	Stress Level	Current Density, nA/cm², by Method[a]			
		ATE	CTE	SG	ACI
Co-Cr-Mo					
Polished	0	14.8	27.8	21.8	29.7
Porous	0	18.8	...	...	37.9
Polished	σ_{uts}	36.6	14.7	74.0	...
Porous	σ_{uts}	176.6	82.2	...	344.8
Ti-6Al-4V ELI					
Polished	0	62.1	5.7	79.4	41.2
Grit-blasted	0	319.2	220.1	43.8	301.3
Porous	0	219.7	...	...	47.8
Polished	σ_{uts}	347	145	123	...
Grit-blasted	σ_{uts}	378.9	160.1	14.7	236.5
Porous	σ_{uts}	575.9	242.5	...	226.3

[a] Key to abbreviations for method:
    ATE = anodic Tafel extrapolation.
    CTE = cathodic Tafel extrapolation.
    SG = Stern-Geary method.
    ACI = A-C impedance method.

Table 5 displays the average i_c values based on the nominal area. As expected, the data indicate that, in most cases, rougher surfaces will release more corrosion products per unit of nominal area under equivalent material and stress conditions. The applied stress appears to enhance ion release from polished Ti-6Al-4V ELI and from both alloys in the porous-coated form. These effects are not clear for polished Co-Cr-Mo or grit-blasted Ti-6Al-4V ELI. In the latter case, it is possible, though not proven, that residual compressive stresses caused by grit blasting could mitigate SEIR caused by applied tensile loads [8].

The maximum increase in body burden due to both surface roughness and applied stress was observed with the cathodic Tafel extrapolation method. The stressed porous Ti-6Al-4V ELI, on the average, released 42.5 times the amount of ions that are released from a polished unstressed surface. Thus, besides the obvious mechanical damage that results, plastic deformation and high stress levels should be avoided in implants because of the possibly enhanced release of ions into the *in vivo* fluids and, therefore, the potentially heightened biocompatibility problems.

Conclusions

These studies have shown that the A-C impedance technique can be a very useful method for measuring the corrosion behavior of surgical implant materials. For *in vivo* measurements, the nondestructive nature of the method and the ease with which *IR* compensation can be achieved offer significant advantages.

In general, the ability of A-C impedance methods to measure the true surface area from electrode capacitance determinations is also an important characteristic of this technique. In this work, grit-blasted and porous-coated materials were investigated, but this attribute of the method could be of particular use for coatings of very irregular porosity, such as flame-sprayed layers.

In the specific research presented here, it was demonstrated that the A-C impedance test could show not only that stress-enhanced ion release can occur from implant surfaces, but also that the reason it occurs is mainly traceable to an increase in surface area caused by plastic deformation at the surface.

Although the question of the biocompatibility of porous-coated materials will ultimately be resolved by specialists in that area, corrosion studies such as the ones described here are useful in identifying the amounts of increased release of corrosion products from stressed porous surfaces. Reliable knowledge about the amount of material released is needed to determine the challenge posed to the body by surgical implant corrosion. In the worst case observed here, a 42.5-times increase in released material occurs.

Acknowledgments

The authors would like to express their appreciation to the Whitaker Foundation for the research grant that provided the financial support for this investigation. C. J. Williams and M. Kolakowski provided assistance with some of the experimental measurements. We are also grateful to De Puy, Inc., for providing materials for the testing program.

References

[1] McDonald, D. D., *Journal of the Electrochemical Society,* Vol. 125, No. 9, September 1978, pp. 1443–1449.
[2] Mansfeld, F., *Corrosion,* Vol. 37, No. 5, May 1981, pp. 301–307.
[3] Stern, M. and Geary, A., *Journal of the Electrochemical Society,* Vol. 105, 1958, p. 638.

[4] Bard A. and Faulkner, L., *Electrochemical Methods—Fundamentals and Applications,* Wiley, New York, 1980.

[5] De Levie, R. in *Advances in Electrochemistry and Electrochemical Engineering,* Vol. 6, P. Delahay and C. W. Tobias, Eds., Wiley/Interscience, New York, 1967, pp. 329–397.

[6] Kramer, M. and Tomkiewicz, M., *Journal of the Electrochemical Society,* Vol. 131, No. 6, June 1984, pp. 1283–1288.

[7] Bundy, K. J., Marek, M., and Hochman, R. F., *Journal of Biomedical Materials Research,* Vol. 17, 1983, pp. 467–487.

[8] Williams, C. J., "The Influence of Static Stress on the Corrosion Behavior of Surgical Implant Alloys," M.S. thesis, Tulane University, New Orleans, LA, April 1985.

[9] Mansfeld, F., Kendig, M. W., and Tsai, S., *Corrosion,* Vol. 38, No. 11, November 1982, pp. 570–580.

[10] Wang, D. Y. and Nowick, A. S., *Journal of the Electrochemical Society,* Vol. 126, No. 7, July 1979, pp. 1166–1172.

[11] De Levie, R., *Electrochimica Acta,* Vol. 8, 1963, pp. 751–780.

[12] Bundy, K. J. and Kolakowski, M., *Transactions,* 17th International Biomaterials Symposium, 25–28 April 1985, Vol. 8, p. 79.

[13] Lucas, L. C., Lemons, J., Lee, J., and Dale, P., *Transactions,* 17th International Biomaterials Symposium, 25–28 April 1985, Vol. 8, p. 80.

Lynne C. Jones,[1] David S. Hungerford,[2] Robert V. Kenna,[3] Guy Braem,[4] and Virginia Grant[5]

Urinary Excretion Levels of Metal Ions in Patients Undergoing Total Hip Replacement with a Porous-Coated Prosthesis: Preliminary Results

REFERENCE: Jones, L. C., Hungerford, D. S., Kenna, R. V., Braem, G., and Grant, V., "Urinary Excretion Levels of Metal Ions in Patients Undergoing Total Hip Replacement with a Porous-Coated Prosthesis: Preliminary Results," *Quantitative Characterization and Performance of Porous Implants for Hard Tissue Applications, ASTM STP 953,* J. E. Lemons, Ed., American Society for Testing and Materials, Philadelphia, 1987, pp. 151–162.

ABSTRACT: Porous-coated prostheses implanted without bone cement are currently being evaluated for use in patients undergoing total joint replacement (TJR). One parameter under study is the potential release of metal ions from these prostheses. In order to determine if there is a systemic increase in cobalt, chromium, or nickel levels within the body subsequent to total joint replacement with a porous-coated prosthesis, 24-h urine specimens were collected from patients prior to and subsequent to TJR with a PCA total hip prosthesis. Metal ion analysis was achieved using flameless atomic absorption spectroscopy. Increases in urinary cobalt and nickel excretion were detected in several patients at six months and in most patients at one year after surgery. However, these differences were not statistically significant. No differences between the preoperative and postoperative time periods (one week, six months, and twelve months) were detected for urinary levels of chromium. Although the metal ion levels for all of the patients studied appear to be in the range handled by the body's systemic compensatory mechanisms, which adjust levels of trace elements, continued follow-up is needed to determine the patterns and the long-term significance of metal ion release.

KEY WORDS: porous implants, biocompatibility, corrosion, total hip replacement, porous-coated prostheses, flameless atomic absorption, metal ion release

The issue of biocompatibility of orthopedic implants, although not new, has come recently to the fore in orthopedic surgery with the introduction of porous-coated prostheses, which can be used in the cementless application of total joint arthroplasty. Although all metallic implants are subjected to corrosive forces within the body, these forces may have a larger impact on porous-surfaced prostheses. Porous-coated implants have an increased total sur-

[1] Research associate, Orthopaedic Surgery, Johns Hopkins University School of Medicine, Baltimore, MD 21239.

[2] Professor, Orthopaedic Surgery, Johns Hopkins University School of Medicine, and chief, Division of Arthritis Surgery, Good Samaritan Hospital, Baltimore, MD 21239.

[3] Research associate, Orthopaedic Surgery, Good Samaritan Hospital, Baltimore, MD 21239.

[4] Research technician, Johns Hopkins University School of Medicine, Baltimore, MD 21239.

[5] Staff chemist, Chesapeake Bay Institute of the Johns Hopkins University School of Medicine, Baltimore, MD 21239.

face area, as well as intimate contact with bone. It has been suggested that these two factors may contribute to an accumulation of corrosion products retained in the body, which, if given sufficient quantities and time, may lead to potential immunological, toxicological, or carcinogenic effects [1–6]. Due to the limited clinical experience with porous-coated prostheses, there is a paucity of information available pertaining to metal ion release from these implants in patients.

The principal objective of our investigation was to determine if there were increased levels of the metal ions which constitute the metallic components retained by patients undergoing total hip arthroplasty. Analysis of metal ion levels in urine samples has been shown to give an accurate assessment of the total body load [7–13]. A long-term study was instituted in which 24-h urine samples from patients undergoing total hip replacement (THR) with a PCA total hip prosthesis were collected at various time intervals prior to and subsequent to surgery and were analyzed for changes in total metal ion content. Levels of the metals that comprise this prosthesis—cobalt, chromium, and nickel—were determined using flameless atomic absorption spectroscopy. The following is a report of our preliminary findings for up to twelve months after surgery.

Materials and Methods

The study population included 30 patients who were scheduled for cementless total hip replacement with a PCA total hip system (Howmedica, Inc., Rutherford, NJ). Participation in this clinical study, approved by the Institution Review Board, was on a volunteer basis; consent was obtained only after the study had been fully explained to each individual. At the time of this writing, information was available for various time periods up to one year for 17 patients. The following descriptive information was collected from each patient: occupation, location of residence, alcohol consumption, and level of tobacco smoking. All patients underwent cementless implantation of a PCA total hip prosthesis, an implant with an ultrahigh-molecular-weight polyethylene (UHMWPE) surface articulating with a cobalt-chromium alloy surface.

Twenty-four hour urine samples were collected at the following time periods: preoperatively, and postoperatively at one week, six months, and twelve months. [Urine specimens are continuing to be collected for each of these patients on an annual basis.] Urine collection was achieved using sterile Medi-Flex specimen containers (Tri-State Hospital Supply Corp., Howell, MI) and Medi 24-h urine containers (Medi, Inc., Holbrook, MA). The patients were asked not to ingest any vitamins (particularly B-complex vitamins) 48 h prior to collection of urine. They were also asked not to contaminate the collection vessels by rinsing or wiping the surfaces. At the time of receipt, the total 24-h urine output was measured and recorded. The samples were then agitated, and two 50-mL aliquots were drawn into sterile Corning centrifuge tubes (Corning, Corning, NY). These tubes were stored in a Revco ultralow-temperature freezer (West Columbia, SC) at −70°C until the time of analysis. An additional 10-mL sample was taken and used for analysis of specific gravity.

Calibration standards were evaluated in order to determine the accuracy of the methodology for each of the ions under analysis. Each of the standards was prepared from reference solutions purchased from the Fisher Scientific Co. (Fairlawn, NJ). Solutions ranging from 5 to 50 parts per billion (ppb) were analyzed. The diluent used was ultrapure deionized water (Millipore Milli-Q System, Bedford, MA). In addition, Standard Reference Material (SRM) No. 2670 (National Bureau of Standards, Washington, DC) was analyzed for normal and elevated chromium and nickel levels in urine. Both calibration evaluations yielded reproducible, valid results.

At the time of analysis, all the urine samples were removed from the freezer and placed in a warm-water bath until thawed. The tubes were then centrifuged at 200 rpm for 10 min. If a precipitate had formed, determinations of the volume of the supernatant and the precipitate were made. A concentration of approximately 60-μL of ultrapure nitric acid (Ultrex grade, J. T. Baker Chemical Co., Philipsburg, NJ) for each 10 mL of supernatant was added to the pellet only. The dissolved pellet was then returned to the supernatant.

Analytical determinations of the cobalt, chromium, and nickel content of the urine samples were made using the "method of additions" [14]. This involves analysis of three aliquots per specimen under study: these aliquots were (1) undiluted, (2) spiked with 2 ppb cobalt and 2 ppb chromium, and (3) spiked with 10 ppb nickel. All the specimens were analyzed using a Perkin-Elmer Model 400G atomic absorption spectrometer with a Model HGA 500 electrothermal source (Perkin-Elmer, Norwalk, CT). This technique included the use of pyrolyzed graphite furnace tubes and deuterium background correction. The analytic program was optimized for each phase (drying, ashing, and atomization) and for each metal (wavelengths and slit widths) [14,15].

Care was taken to minimize all possible sources of contamination. This included subjecting all containers, laboratory glassware, and equipment to acid washing with 10% nitric acid. In addition, all triplicate values were reviewed to detect possible decreases in analytic efficiency, as well as periodic sampling error or contamination. If the three values were inconsistent, the specimen was reanalyzed.

Statistical analysis of the results was achieved using analysis of variance techniques for related measurements, and the comparisons were made using linear contrast methods. All methods of analysis stressed the value of comparing the variability of the data for an individual patient, as well as the variability between patients.

Results

At this time, we are reporting only the preliminary findings obtained. As this is a very early stage of a long-term investigation, the information available involves only a fraction of the 30 patients to be included (17 evaluated preoperatively, 14 at six months postoperatively and 4 at one year postoperatively). For the 17 patients included in the preoperative evaluation, the diagnosis at presentation was primarily osteoarthritis ($n = 11$), although a few patients had diagnoses of avascular necrosis of bone ($n = 3$) and rheumatoid arthritis ($n = 3$). This group included 11 males and 6 females. The age of the patients ranged from 34 to 77 years with a mean of 58 years. Most of the patients resided in Maryland ($n = 13$), primarily in Baltimore and its suburbs ($n = 7$); the others lived in New York ($n = 2$), Pennsylvania ($n = 1$), and Illinois ($n = 1$). The majority of the patients were not smokers or excessive alcohol consumers. All but 1 patient had jobs in which occupational exposure is not known to occur; 1 patient was a machinist.

All values for urinary metal ion excretion are reported as the mean plus or minus the standard error. All the statistical analyses weighted the comparison of each individual to himself. As no significant differences were seen in the specific gravities of the samples for each individual patient, no conversion of the results from micrograms per litre was attempted.

Preoperative and one-week-postoperative specimens were collected and analyzed for 17 of the study patients (Fig. 1). For these patients, the average urinary levels of the preoperative samples were 0.95 $\pm$ 0.28 μg/L for cobalt, 1.41 $\pm$ 0.24 μg/L for chromium, and 4.31 $\pm$ 1.14 μg/L for nickel. These averages and variances are similar to those reported by previous investigators [10–13,16,17]. Postoperatively (at one week), values of 1.08 $\pm$ 0.28 μg/L for cobalt, 2.85 $\pm$ 1.34 μg/L for chromium, and 11.8 $\pm$ 6.74 μg/L for nickel were determined.

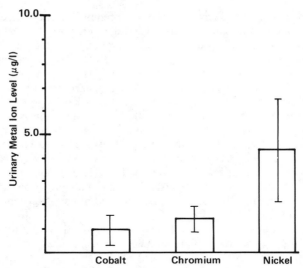

FIG. 1—*Mean control values for urinary metal ion levels for 17 patients scheduled to undergo total hip replacement. The bar lines indicate the 95% confidence limits.*

Although increases were seen for each metal ion, these increases were not statistically significant. The urinary chromium and nickel levels for one patient were significantly elevated one week postoperatively (17.3 and 85.2 μg/L, respectively). As these values are considerably outside the range seen for the other patients, it was suggested that these results may have been due to contamination of the specimen. As is shown in Table 1, this patient had normal levels at the six- and twelve-month time periods. The average values, if this patient is excluded, are 1.53 ± 0.27 μg/L for chromium and 5.08 ± 1.00 μg/L for nickel. Three patients underwent subsequent THR for the contralateral hip and those results are not included in the six- and twelve-month results reported herein.

Figure 2 graphically illustrates the results for these patients evaluated at six months (n = 14). At six months, a slight increase was detected for cobalt (1.57 ± 0.50 μg/L) and chromium (1.52 ± 0.27 μg/L), while a larger average increase was found for nickel (11.43 ± 2.92 μg/L). This elevation in nickel ion level is primarily a reflection of the elevated values found for two patients (29.2 and 41.0 μg/L). One of the patients returned to normal levels (10.6 μg/L) by one year. The one-year follow-up for the other patient is not available at this time. However, using statistical methods that allow comparison of the variance within one patient to that of all patients (that is, ANOVA for related measurements), it was determined that these differences were not statistically significant. No relationships between the elevated values and age, occupation, alcohol consumption, or cigarette smoking were detected.

At this time, results are available for only 4 patients for the one-year follow-up period (Table 1). This group included 2 males and 2 females with an average age of 68 years (the range was 61 to 74). None of the patients was a smoker and their alcohol consumption ranged from abstinence to one drink per day. No change in the level of urinary chromium (1.20 ± 0.41 μg/L) was detected. Cobalt was observed to increase in 3 of the 4 patients, with an average value of 2.90 ± 0.92 μg/L (Fig. 3). With respect to nickel, an average value of 13.58 ± 2.33 μg/L was seen, also reflecting an increase in 3 of the 4 patients. Although these data reflect a threefold increase for cobalt and nickel in comparison with the preoperative values, this increase was not found to be statistically significant.

TABLE 1—*Urinary metal ion results determined for four patients followed up to one year subsequent to total hip replacement.*[a]

Patient No.	Metal Ion	Preoperative	Postoperative		
			1 Week	6 Months	1 Year
1	cobalt	0.52	1.40	0.26	5.13
	chromium	1.9	17.3[b]	1.5	0.56
	nickel	0	85.2[b]	7.4	17.33
2	cobalt	0.49	1.01	1.56	3.69
	chromium	0.8	3.4	4.7	1.23
	nickel	5.2	0	29.2	10.6
3	cobalt	0.15	2.89	0.56	1.57
	chromium	0.7	1.4	1.1	2.35
	nickel	0	12.1	11.1	17.78
4	cobalt	1.54	0.92	2.32	1.20
	chromium	3.7	0.56	1.18	0.65
	nickel	14.29	6.94	12.42	8.6
Mean ± standard error	cobalt	0.68 ± 0.30	1.56 ± 0.46	1.18 ± 0.47	2.90 ± 0.92
	chromium	1.78 ± 0.70	5.67 ± 3.92	2.12 ± 0.86	1.20 ± 0.41
	nickel	4.87 ± 2.44	26.06 ± 19.87	15.03 ± 4.84	13.58 ± 2.33

[a] The values are in micrograms per litre.
[b] Possible contamination is not ruled out. The mean nickel value for one week postoperatively is 6.35 ± 3.04 µg/L when this value is deleted. The mean chromium value is 1.79 ± 0.84 µg/L when excluding this value.

521

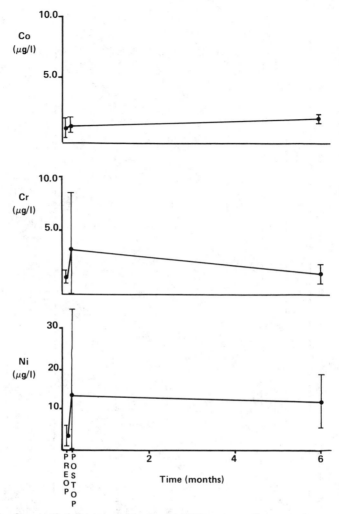

FIG. 2—*Six-month follow-up results for 14 THR patients, demonstrating no increase in urinary cobalt or chromium content, in comparison with preoperative levels, and an elevation in nickel content due, in part, to large increases observed in 2 patients. Excluding these 2 patients, the average urinary nickel level was 7.49 ± 1.16 μg/L at six months. The bar lines encompass the 95% confidence limits.*

Discussion

Although a variety of metals have been inserted into patients since the late 1700s, there remains a paucity of information available on the biological effects of these implants on the human body. Evaluation of retrieved specimens, as well as experimental findings, indicates that metallic implants exhibit corrosive behavior [18–21]. With the advent of metal-on-metal prostheses, accumulation of wear debris surrounding the joint has also been observed [22,23]. The development of ultrahigh-molecular-weight polyethylene (UHMWPE) for plastic-to-metal articulating surfaces has reduced the quantities of metallic products observed in the evaluated joints [24]. However, an unacceptable rate of loosening of the components of the prosthesis in patients with total knee and total hip arthroplasty has been seen with the use of methacrylate cement for implant fixation. It is for this reason that porous-coated prostheses, which allow biologic ingrowth for implant fixation, were developed. Renewed concerns about the toxicity and carcinogenicity of implant material have been primarily related to the increased surface area of the prosthesis, the intimate contact between the bone and the metal, and the technology involved in the sintering of the beads. It is important to relate these concerns to the clinical situation.

The issue of biocompatibility of orthopedic implants involves the study of potential local and systemic effects of the metal ions released from the surfaces of each prosthetic component. In the laboratory, corrosion products of Co-Cr alloy prostheses have demonstrated

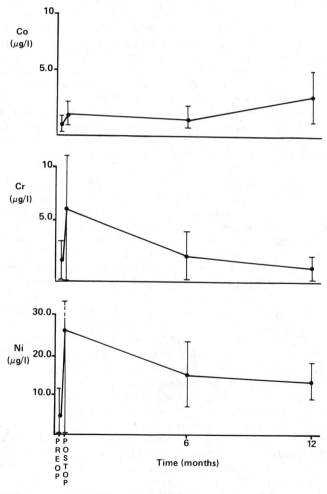

FIG. 3—*One-year results for 4 THR patients, showing increases in urinary cobalt and nickel, in comparison with preoperative levels, for 3 of the 4. No changes were seen in the urinary chromium levels over time. The 95% confidence limits are shown by the bar lines.*

cytotoxic or carcinogenic potential [1–5]. Concerns pertaining to cytotoxicity are usually in reference to the local reaction of the biologic tissue in juxtaposition with the implant material. Not as well studied are the possible toxic effects on the metabolic functions of other tissues in the body. Carcinogenesis may be either a local or a systemic effect. The evaluation of the possible local carcinogenic effect is dependent on the collection of retrieved specimens from deceased patients. Although this project has been initiated at our institute, the results are not available for report. In order to assess, indirectly, the level of corrosion and, directly, the level of systemic response to potential corrosion products of orthopedic implants, serum and urine samples have been monitored by previous investigators [14,25–28]. Both methods of biological monitoring have been validated for use as indicators of exposure to cobalt, chromium, and nickel [7–13]. Collection and analysis of urine samples offer the advantages of ease of collection, minimum discomfort to the patient, and a minimum of matrix complications. For these reasons, we enlisted patient volunteers for the evaluation of 24-h urine specimens subsequent to THR with a PCA total hip system. There is an increased risk of specimen contamination during collection of urine samples; however, care was taken to minimize possible contamination by supplying uncontaminated containers to each patient and by increasing the awareness of each patient of possible sources of contamination.

Animal studies have indicated that there is an increase in urinary metal ion levels subsequent to implantation of a metal substrate. Using Sprague-Dawley rats, Wapner et al. evaluated urine chromium levels in animals implanted with Co-Cr microspheres [ASTM Specification for Cast Cobalt-Chromium-Molybdenum Alloy for Surgical Implant Applications (F 75-82)] of several surface areas [29]. They saw a significant increase in urine chromium only in the animals implanted with spheres having 300 times the basic ratio of surface area to body weight. This increase was most dramatic at 10 days, returning to slightly elevated levels at 100 days. Woodman et al. found a significant elevation of urinary nickel levels in rabbits implanted with porous cast Co-Cr alloy at six months (63.01 ± 1.94 ng/mL) in comparison with solid implants (39.23 ± 1.23) and controls (28.51 ± 1.52) [30]. However, limitations in the experimental design and in animal models make it difficult to assess the clinical relevancy of these experiments. Metallic microspheres implanted loosely in bone are not an adequate representation of the clinical situation. Animals, which have diets, metabolisms, and kinematics that are different from those of humans may not sufficiently model the human situation. As porous-coated prostheses are being implanted in humans, biological monitoring of humans appears to be the best alternative.

Only a few laboratories have attempted to analyze the concentrations of metal ions in urine in patients subsequent to arthroplasty in which metallic components have been used. Coleman et al. evaluated the metal concentrations in the urine, blood, and hair of patients with hip arthroplasties [25]. They found that, preoperatively, the average value for urinary cobalt excretion was 0.5 μg/L, with 0.4 μg/L for urinary chromium excretion. In patients with metal-to-metal THR prostheses, Coleman et al. detected increases in urinary cobalt and chromium levels (24.0 and 6.2 μg/L, respectively), with a very high increase in cobalt (73.0 μg/L) and chromium (26.0 μg/L) in one patient. This is in contrast to their findings for patients with metal-to-plastic THR prostheses (0.7 μg/L of cobalt, 1.2 μg/L of chromium). Jones et al. found an increase in urinary levels ranging from 20 to 55 μg/L in hypersensitive individuals with loosened metal-to-metal THR prostheses [31]. In a study that evaluated the urinary content of cobalt in patients with cementless porous and nonporous Austin-Moore prostheses, Jorgenson et al. found no significant difference between the two groups [26]. However, the level in both groups appeared to be slightly elevated when compared with those values accepted as the normal range [25]. In a retrospective study of patients undergoing total knee replacement, we found no significant difference in urinary cobalt or chromium in patients with cementless as opposed to cemented PCA total knee prostheses in comparison with controls [32]. The results of this study lend support to the possibility of increases in urinary cobalt excretion in some patients by six and twelve months. We were unable to detect any difference in urinary chromium excretion between our preoperative and postoperative specimens, regardless of the time that had elapsed. However, none of our patient results for urinary cobalt or chromium were in the range of the elevations detected by Coleman et al. or Jones et al. for metal-to-metal prostheses [25,31].

Urinary nickel levels for patients after THR have not been documented by previous investigators. However, our six-month and one-year averages for urinary nickel are within the range established by Adams et al. for normal values in a comparison of the results of seven laboratories (10.5 ± 5.1 µg/L) [16]. Despite this, urinary nickel levels at six months for two patients were in the range seen in some individuals exposed occupationally (5.0 to 36.0 µg/L)—a high-risk group for carcinogenesis [16]. However, in one of the two patients this level returned to the preoperative level by one year. The one-year result for the other patient is not available at this time.

The difficulty of documenting a statistically significant increase in urinary cobalt and nickel levels may be due to the relatively small number of patients included. It is also a reflection, however, of the variability of response from individual to individual. The finding of elevations in 3 of the 4 patients studied at one year lends support to the importance of continuing investigation.

The biologic effect of a systemic increase in cobalt and nickel at the levels we have seen is unknown. Cobalt supplementation in patients at levels of 20 to 50 mg/day may result in polycythemia, transient hyperglycemia, and hyperplasia of bone marrow [33]. Increases of nickel in the diet have lead to dermatitis and hypersensitization [34]. Allergy to nickel has been documented by Deutman et al. to occur in 5.8% of patients prior to THR [35]. In addition, these authors also suggest that THR with a nickel-containing prosthesis may trigger hypersensitivity reactions in some patients. There is a lack of information on abnormal accumulation of cobalt, chromium, and nickel in specific tissues throughout the body. Accumulations of metal ions in the soft tissue and bone surrounding metal-on-metal prostheses have been reported [22,23]. No information is available clinically as to the level of exposure of the bone cells in juxtaposition with porous implants. However, Woodman et al. were unable to detect increases in metal ion levels in cortical bone surrounding solid implants of several alloy types in an animal model [36]. It is clear that analysis of various tissues at autopsy is warranted in order to determine the full extent of the consequences of metal ion release.

The PCA total hip prosthesis is made of 63.0% cobalt, 28.0% chromium, and 0.7% nickel. The increase of nickel in some patients despite the low amounts present in the composition of the prosthesis, is not easily explained. The differences in the changes of excretion of these metal ions may be due to differences in (1) corrosion processes, (2) solubility coefficients, (3) binding to proteins, or (4) excretion mechanisms. Additional *in vitro* and *in vivo* research is necessary to elucidate this enigma.

Our results suggest that there is an increase in metal ions released from porous-coated total hip implants. However, the location and mechanism of release is not known. In order to distinguish whether the increase is a consequence of wear of the articulating surfaces or a result of increased corrosion of the porous surfaces, a comparison of our results with those obtained from cemented porous-coated prostheses, where the implant design is the same but the porous surface is not exposed to bone, is necessary.

The response of the body to porous-coated cobalt-chromium prostheses needs to be defined. Because of the limited number of clinical trials, little is known about the variables affecting biologic ingrowth and possible metal ion release from these prostheses. The rate of metal ion release is not known, which thereby clouds the issue of whether the patient will be subjected to a level of metal ion concentration that could lead to potential carcinogenic or toxicologic effects. Two approaches to this problem include the monitoring of patients for tumor formation and the monitoring of urinary metal ion levels. We found no statistically significant increases in urinary metal ion levels in any of the patient groups studied. Although increases in metal ion levels have been detected in some of the patients with cementless porous-coated total hip prostheses, the levels seen are not sufficient to cause immediate alarm. Recent reports of tumor formation juxtaposed to Co-Cr implants suggest that these metals may create a carcinogenic environment in some patients [37,38]. The prostheses of the two patients cited were both of the metal-to-metal articulating type. This leads to the generation of extremely fine particulate debris, itself a chronic irritant, as well as exposure to a surface area of metal many times that of the rigid implant. The reporting of two cases that indirectly point to carcinogenesis out of hundreds of thousands of cases suggests that this potential exists for only a very small minority. Because of the seriousness of this issue, however, it is important that these studies be continued.

Conclusions

We have reported our preliminary findings in a long-term clinical study of the urinary excretion of metal ions subsequent to THR with a cementless porous-coated prosthesis. Based on the study results available to date, there were no statistically significant increases in urinary cobalt, chromium, or nickel levels postoperatively (1, 26, and 52 weeks) in comparison with preoperative levels. However, increases were detected in several patients by six months and in 3 of the 4 patients studied at one year for urinary cobalt and nickel. This suggests that corrosion of the implant may be occurring and that it can be detected in some patients one year postoperatively. We would like to emphasize that this is a preliminary report of our results. Only with longer term follow-up and an increased number of patients recruited will these results be validated and the implications of these findings determined.

Acknowledgments

We wish to express our appreciation to A. Hester and K. Connor for their technical assistance. We also wish to thank J. M. Frazier for his guidance and the use of his laboratory during the study.

References

[1] Heath, J. C., "The Production of Malignant Tumors by Cobalt in the Rat," *British Journal of Cancer,* Vol. 10, 1956, pp. 668–673.

[2] Hueper, W. C., "Experimental Studies in Metal Carcinogenesis: I. Nickel Cancers in Rats," *Texas Report of Biological Medicine,* Vol. 10, 1952, pp. 167–186.

[3] Hueper, W. C., "Experimental Studies in Metal Carcinogenesis: VII. Tissue Reactions to Parenterally Introduced Powdered Metallic Chromium and Chromite Ore," *Journal of the National Cancer Institute,* Vol. 16, 1955, pp. 447–470.

[4] Memoli, V. A., Woodman, J. L., Urban, R. M., and Galante, J. O., "Malignant Neoplasms Associated with Orthopaedic Implant Materials," *Transactions of the Orthopaedic Research Society,* Vol. 7, 1982, p. 164.

[5] Rae, T., *Journal of Bone and Joint Surgery,* Vol. 57B, 1975, pp. 444–450.

[6] Evans, E. M., Freeman, M. A. R., and Vernon-Roberts, V., *Journal of Bone and Joint Surgery,* Vol. 56B, 1974, pp. 626–642.

[7] McNeely, M. D., Nechay, M. W., and Sunderman, F. W., Jr., *Clinical Chemistry,* Vol. 18, 1972, pp. 992–995.

[8] Schroeder, H. A., Nason, A. P., and Tipton, I. H., "Essential Trace Metals in Man: Cobalt," *Journal of Chronic Diseases,* Vol. 20, 1967, pp. 869–890.

[9] Mitman, F. W., Wolf, W. R., Kelsay, J. L., and Prather, E. S., "Urinary Chromium Levels of Nine Young Women Eating Freely Chosen Diets," *Journal of Nutrition,* Vol. 105, 1975, pp. 64–68.
[10] Hambridge, K. M., *American Journal of Clinical Nutrition,* Vol. 27, 1974, pp. 505–514.
[11] Sjogren, B., Hedstrom, L., and Ulfvarson, U., *International Archives of Occupational and Environmental Health,* Vol. 51, 1983, pp. 347–354.
[12] Pedrix, A., Pellet, F., Vincent, M., De Gaudemaris, R., and Mallion, J. M., *Toxicological European Research,* Vol. 5, 1983, pp. 233–240.
[13] Aitio, A., *IARC Science Publications,* Vol. 53, 1984, pp. 497–505.
[14] Black, J., Maitin, E. C., Gelman, H., and Morris, D. M., *Biomaterials,* Vol. 4, 1983, pp. 160–164.
[15] Guthrie, B. E., Wolf, W. R., and Veillon, C., *Analytical Chemistry,* Vol. 50, 1978, pp. 1900–1902.
[16] Adams, D. B., Brown, S. S., Sunderman, F. W., Jr., and Zachariasen, H., *Clinical Chemistry,* Vol. 24, 1978, pp. 862–867.
[17] Underwood, E. J., "Nickel," *Trace Elements in Human and Animal Nutrition,* 4th ed., Academic Press, New York, 1977, Chapter 6, pp. 159–169.
[18] Cahoon, J. R., *Journal of Biomedical Materials Research,* Vol. 7, 1973, pp. 375–383.
[19] Rose, R. M., Schiller, A. L., and Radin, E. L., *Journal of Bone and Joint Surgery,* Vol. 54A, 1972, pp. 854–862.
[20] Cohen, J., *Journal of Bone and Joint Surgery,* Vol. 44A, 1962, pp. 307–316.
[21] Oron, U. and Alter, A., *Clinical Orthopaedics and Related Research,* Vol. 185, 1984, pp. 295–300.
[22] Winter, C. D., "Tissue Reactions to Metallic Wear and Corrosion Products in Human Patients," *Prosthesis and Tissue: The Interface Problem,* Wiley, New York, 1974, pp. 11–26.
[23] Michel, R., Hofmann, J., Loer, F., and Zilkens, J., "Trace Element Burdening of Human Tissues Due to the Corrosion of Hip-Joint Prostheses Made of Cobalt-Chromium Alloys," *Archives of Orthopaedic and Traumatic Surgery,* Vol. 103, 1984, pp. 85–95.
[24] Swanson, S. A. V., Freeman, M. A. R., and Heath, J. C., *Journal of Bone and Joint Surgery,* Vol. 55B, 1973, pp. 759–773.
[25] Coleman, R. F., Herrington, T., and Seales, J. T., "Concentration of Wear Products in Hair, Blood, and Urine After Total Hip Replacement," *British Medical Journal,* Vol. 1, 1973, pp. 527–529.
[26] Jorgensen, T. J., Munno, F., Mitchell, T. G., and Hungerford, D., *Clinical Orthopaedics and Related Research,* Vol. 176, 1983, pp. 124–126.
[27] Pazzaglia, V. E., Minoia, C., Ceciliani, L., and Riccardi, C., *Acta Orthopaedica Scandinavica,* Vol. 54, 1983, pp. 574–579.
[28] Linden, J. V., Hopfer, S. M., Grossling, H. R., and Sunderman, F. W., Jr., *Annals of Clinical and Laboratory Science,* Vol. 15, 1985, pp. 459–464.
[29] Wapner, K. L., Black, J., and Morris, D., "Chromium Release by Cast Co-Cr Alloy in Vivo: Ionic Valence and its Implications for Morbidity," *Transactions of the Orthopaedic Research Society,* Vol. 8, 1983, p. 240.
[30] Woodman, J. L., Urban, R. M., Lim, K., and Galante, J. O., "Cobalt, Chromium and Nickel Release from Porous Coated Cast Cobalt Chromium Alloy," *Transactions of the Orthopaedic Research Society,* Vol. 9, 1984, p. 150.
[31] Jones, D. A., Lucas, H. K., O'Driscoll, M., Price, C. H. G., and Wibberley, B., *Journal of Bone and Joint Surgery,* Vol. 57B, 1975, pp. 289–296.
[32] Jones, L. C., "Biocompatibility of Implant Materials" *Total Hip Arthroplasty: A New Approach,* D. S. Hungerford, A. Hedley, E. Habermann, L. Borden, and R. V. Kenna, Eds., University Park Press, Baltimore, 1984, pp. 108–121.
[33] Carlberger, G., *Kinetics and Distribution of Radioactive Cobalt Administered to the Mammalian Body,* Karolinska Institutet, Stockholm, 1961.
[34] Norseth, T., *IARC Science Publications,* Vol. 53, 1984, pp. 395–401.
[35] Deutman, R., Mulder, T. J., Brian, R., and Nater, J. P., *Journal of Bone and Joint Surgery,* Vol. 59A, 1977, pp. 862–865.
[36] Woodman, J., Shinn, W., Urban, R., and Galante, J., *Journal of Biomedical Materials Research,* Vol. 18, 1984, pp. 463–466.
[37] Penman, H. G. and Ring, P. A., *Journal of Bone and Joint Surgery,* Vol. 66B, 1984, pp. 632–634.
[38] Swann, M., *Journal of Bone and Joint Surgery,* Vol. 66B, 1984, pp. 629–631.

DISCUSSION

Z. Glaser[1] *(written discussion)*—In regard to the high nickel ion content in the urine of patients after implantation of porous prostheses, is there a possibility that the source of the nickel could be another implant or a dental bridge? Is the body possibly excreting nickel from tissue or cells other than the implant site?

L. C. Jones, D. S. Hungerford, R. V. Kenna, G. Braem, and V. Grant (authors' closure)— As *in vivo* corrosion of metallic orthopedic implants occurs at a very low rate, other factors that may contribute to temporal changes in urinary metal ion levels have been sought. In this study, only patients undergoing a primary surgery were included (that is, no revisions). None of the patients had any other orthopedic implants. It is somewhat unlikely that corrosion and wear of dental implants contributed to the increase demonstrated in some individuals. Wear debris from dental implants would primarily be digested. Cobalt, chromium, and nickel are poorly absorbed by the intestines. Therefore, the contribution of this potential source would probably not have a significant impact on the total urinary excretion of these metal ions. No relationship between the epidemiological data gathered (age, sex, occupation, health, and so forth) and the urinary measurements was detected. However, this may be partly due to the low numbers of patients evaluated at this time. It is possible that nickel is released from the cells at the implant site due to trauma. Increases in circulating and excretory nickel levels have been correlated with other types of trauma, including myocardial infarction, acute stroke, and severe burns. The fact that increases in nickel have been observed in the areas surrounding implants, which do not contain nickel, lends support to this hypothesis. However, this hypothesis remains to be tested. A likely source of metal ion release is the debris from implantation of the components. This might vary from patient to patient, depending on the extent of lavage of the surgical site and the effectiveness of the individual's body in removing this debris from the joint cavity and implant interface.

[1] Food and Drug Administration, Rockville, MD.

Stanley A. Brown,[1] Katharine Merritt,[1] Lillian J. Farnsworth,[1] and Timothy D. Crowe[1]

Biological Significance of Metal Ion Release

REFERENCE: Brown, S. A., Merritt, K., Farnsworth, L. J., and Crown, T. D., **"Biological Significance of Metal Ion Release,"** *Quantitative Characterization and Performance of Porous Implants for Hard Tissue Applications, ASTM STP 953,* J. E. Lemons, Ed., American Society for Testing and Materials, Philadelphia, 1987, pp. 163–181.

ABSTRACT: The issue of the biologic significance of metal ion release is addressed by reviewing the answers to five questions: what is released, how much is released, where do the ions go, what can they do, and what have they done? Data are presented to demonstrate that ions are released but not necessarily in proportion to the alloy composition. Nickel and cobalt bind to serum albumin, whereas chromium is released with a valence of $+6$ and binds to red blood cells. Tissue levels of chromium and titanium are often very high, indicating that these elements may form stable complexes that remain in local tissues. The results of chemical analysis of urine from animals exposed to metal ions from injection of salts or *in vivo* corrosion demonstrate a rapid excretion of nickel and cobalt; excretion of chromium is slow and represents a small percentage of the total to which the animal is exposed. The biological reactions of toxicity, metal sensitivity, and oncogenicity are discussed from the point of view of what they have been shown to do in laboratory and animal experiments. The limited number of clinical reports of these reactions are also reviewed. These issues are discussed in light of the need for future research and for the development of standard test methods for characterization of porous implant materials and the biological reactions to these materials.

KEYWORDS: porous implants, metal ions, corrosion, toxicity, metal sensitivity, oncogenicity

There are five major questions or issues that should be addressed in a discussion of the biological significance of metal ion release from dense and porous-coated metallic implants. This paper will include a review of the literature and the results from the authors' research.

1. *What is released?* What metal ion species are released from implants *in vivo,* and in what form are they released? Do corrosion products remain as metal ions or metal oxides, or do they form protein or cell-bound complexes? Do corrosion products behave in the same way as metal ions from metal salts? The discussion will be limited to nickel, cobalt, chromium, titanium, aluminum, and vanadium.

2. *How much is released?* This second major question concerns the differences in corrosion rates due to alloy composition and form—that is, dense versus porous-coated alloys and different thermal heat treatments. This issue also involves different mechanisms of corrosion—for example, crevice and fretting corrosion and *in vivo* versus *in vitro* corrosion rates. The question of whether the release of elements is in proportion to the alloy composition will also be addressed.

3. *Where do they go?* Where do the metal ions from corrosion products go? The answer to the question is critical to the accurate answering of all the other questions. *In vivo* rates

[1] Associate professors of biomedical engineering, student, and research assistant, respectively, Department of Biomedical Engineering, Case Western Reserve University, Cleveland, OH 44106.

are frequently determined by chemical analysis of local tissue, blood, organs, and excretion. Since the forms in which elements are released are different, their distributions will be different. This phenomenon must be considered in determination of *in vivo* metal ion release rates.

4. *What can they do?* From a scientific, laboratory, and animal research point of view, what are the possible biological effects of metal ions and corrosion products? Is the reaction dose response related on an individual implant basis, or is it a cumulative dose response on a chronic basis?

5. *What have they done?* What biological reactions have been associated with metal ion release from implants in humans? Clinical problems caused by metal ion exposure from other sources may or may not imply that similar reactions are due to ion release from implants.

What is Released?

From the point of view of porous-coated orthopedic implants, the authors' major concern is with cobalt-chromium-molybdenum alloy [ASTM Specification for Cast Cobalt-Chromium-Molybdenum Alloy for Surgical Implant Applications (F 75-82)], titanium [ASTM Specification for Unalloyed Titanium for Surgical Implant Applications (F 67-83)], and Ti-6Al-4V alloy [ASTM Specification for Wrought Titanium 6Al-4V ELI Alloy for Surgical Implant Applications (F 136-84)]. This discussion will also include results with Type 316L stainless steel [ASTM Specification for Stainless Steel Bars and Wire for Surgical Implants (Special Quality) (F 138-82)], since its corrosion rate is high enough to facilitate accurate determination of weight loss, which can be correlated with metal ion release measurements. While these results may not relate directly to porous implants, they do give insight into the release and biological activity of nickel and chromium.

The answer to what is released *in vivo* comes by inference from chemical analysis of tissues or solutions such as blood and urine. It is safe to say, based on such analyses, that all elements are released to some degree.

The question of in what form elements are released has to do with the formation of complexes. The studies of Black and his co-workers [1–5], Clark and Williams [6], Williams et al. [7], and Merritt and her co-workers [8–12] have investigated the specific protein binding of metal ions. Salts and corrosion products show albumin binding of nickel and cobalt with a valence of $+2$. Metal salt and corrosion product studies show red cell binding of chromium with a valence of $+6$; salts of Cr^{3+} bind to albumin. Gray and Stirling [13] have shown that Cr^{6+} will cross the red cell membrane, whereas Cr^{3+} will not. Once in the red cell, Cr^{6+} may be reduced to Cr^{3+}. The results with titanium, aluminum, and vanadium are not as well documented.

The form in which the elements are released is of course, critical to all the other questions. If they form protein complexes, they can be transported in the blood and either accumulate in organs or be excreted. Red cell complexes will also be transported, but their excretion may not be as effective. Complexes formed with implant site cells and with the nonreactive oxides will remain in the local tissue area or possibly be transported in the lymphatics, and therefore blood and excretion analyses will not be indicators of release rates. Subsequent studies and ASTM standard test methods to be developed must therefore address this critical issue of the form in which corrosion products are released *in vivo*.

How Much is Released?

The question of how much is released deals with the rates of corrosion, the rates of release, and the proportionality of release of individual elements. The authors [14,15] and others

[16–18] have shown that corrosion rates can be significantly affected by small amounts of relative motion, leading to fretting corrosion. Evidence of abrasion on uncemented hemihip arthroplasties could imply increased release rates due to direct bone-to-metal contact and fretting [19]. This issue could be of concern with uncemented porous-coated devices and therefore should be addressed in standard test methods for these materials.

The question of increased rates due to the surface porosity itself is discussed by others in this volume. There are, however, two studies that should be mentioned. Woodman et al. [20] compared the serum, urine, and fecal levels of nickel released from control rabbits with those for rabbits that had received solid and porous-coated implants and cobalt-chromium-molybdenum powder (ASTM Specification F 75-82). Their results demonstrated increases in concentration indicative of increased release rates due to increased surface areas. Ducheyne and his co-workers have shown substantial increases in the release of titanium due to porosity *in vivo* [21], and Ducheyne has demonstrated this phenomenon *in vitro* [22].

The ratio of nickel to chromium in 12% nickel, 17% chromium stainless steel is 0.7; that in cast cobalt-chromium-molybdenum (F75) alloy is in the range of 0.03 to 0.09, depending on the nickel content. Woodman et al. [4,5,23] and Koegel and Black [2] have reported a disproportionately high release of nickel from microspheres of both these alloys in serum solutions. They have reported data with a nickel/chromium ratio of 3.6 to 3.9 from stainless steel and 0.83 to 1.23 from F75 alloy. In a study of *in vitro* release from stainless steel and MP35N [ASTM Specification for Wrought Cobalt-Nickel-Chromium-Molybdenum Alloy for Surgical Implant Applications (F 562-84)] in Ringer's solution, Treharne and Marek [24] also reported a disproportionately high release of nickel.

The authors have been studying corrosion rates and nickel/chromium release ratios for two types of *in vitro* corrosion. The first study utilized a fretting corrosion simulator [14,25,26] [ASTM Practice for Measuring Fretting Corrosion of Osteosynthesis Plates and Screws (F 897-84)], which generates an oscillatory motion between a two-hole osteosynthesis plate and two screws. The corrosion rates were determined by measurements of weight loss and by determination of the metal ion release with graphite furnace atomic absorption spectroscopy (GFAAS). This device was used to study nickel/chromium ratios of corrosion products from the fretting corrosion of stainless steel plates and screws in saline, albumin, and gamma globulin at different pH values. These experiments were designed to study the effects of these proteins at pH values acidic to, equal to, and alkaline to their isoelectric points.

The second study utilized an accelerated anodic corrosion method, in which 4 by 25-mm cylindrical specimens were held at a potential of 500 mV [by saturated calomel electrode (SCE)] with a potentiostat for 30 min. The corrosion rates were determined by measuring the weight loss of the specimens, by integrating the corrosion current over time to determine the net charge transfer in coulombs, and by GFAAS. The weight loss data were also used to calculate or predict the amount of metal that should be detected by GFAAS if the release was proportional to the alloy composition. These experiments have been done in saline and in 10% serum, with a Luggin probe salt bridge to a calomel electrode and with graphite rods as counter electrodes. Accelerated corrosion experiments were also conducted *in vivo* in hamsters, with rods placed subcutaneously in the back. A saline-soaked sponge on the back served as a salt bridge to a calomel reference electrode and the animals were placed on a metal counter electrode coated with electrode paste.

The results of the fretting corrosion experiments, shown in Table 1, indicate that the variations of pH in the saline did not result in a significant elevation in the nickel/chromium ratio. Significantly, more nickel was released in the presence of both proteins, especially in the physiologic pH range. The amount of nickel and chromium in the accelerated-corrosion *in vitro* test solutions and urine samples from *in vivo* corrosion, and their respective nickel/chromium ratios are shown in Table 2. The results indicate a close agreement between the

TABLE 1—*Nickel/chromium ratios of fretting corrosion products from stainless steel plates and screws after two weeks of fretting in saline, 0.5% albumin, and 0.5% gamma globulin at pH values less than, equal to, and greater than their isoelectric points.[a]*

Corrosion Medium	pH 3	pH 5 to 6	pH 8	pH 11
Saline	0.94 (0.35)	0.78 (0.27)	0.64 (0.13)	0.59 (0.01)
Albumin	1.02 (0.15)	1.68 (0.10)	1.46 (0.54)	...
Gamma globulin	...	1.77 (0.32)	1.59 (0.46)	1.30 (0.33)

[a] Results are expressed as mean values with standard deviations in parentheses.

chemical analysis of the solutions and predictions based on weight loss in saline, but there was a shift toward a higher release of nickel in 10% serum. Similar agreement was observed when the metal ion release was calculated from the total charge transfer using Faraday's law. It is apparent that the nickel/chromium ratio in saline was close to that expected, whereas that in 10% serum was somewhat elevated. The amount of chromium excreted in urine was very low, as will be discussed subsequently, and therefore the ratio was not calculated.

Due to the random nature of pit initiation in response to the applied anodic potential, there was substantial variation in the amount of weight loss and net charge transfer for individual stainless steel test specimens. As a result it was possible to examine the correlation between weight loss and charge transfer. As is shown in Table 3, there was a high degree of correlation between the weight loss and the net charge. These correlations between weight loss, metal ion release, and calculations based on charge transfer validate the use of charge transfer for predicting metal ion release rates when the actual weight loss is below the detection limits of a microbalance.

It is also apparent that the corrosion rates in 10% serum are not significantly different from those *in vivo*. Both are significantly less than those in saline. Other experiments have shown that the relationship between the corrosion rates in saline and those in 10% serum reverses at potentials above 1 V (SCE); at higher potentials, the corrosion rate in serum is significantly greater than that in saline [27].

TABLE 2—*Weight loss and metal ion release from accelerated corrosion of 4 by 25-mm stainless steel rods held at 500 mV (SCE) for 30 min in saline and in 10% serum, and the total amount in 24-h urine samples collected for four days after* in vivo *corrosion.[a]*

Corrosion Medium	Weight Loss, mg	Ion Release				Nickel/Chromium Ratio[c]
		Nickel		Chromium		
		μg	%[b]	μg	%[b]	
Saline	1.85 (0.34)	221 (103)	98	342 (46)	110	0.63 (0.22)
Serum	0.58 (0.57)	57 (37)	112	90 (83)	86	0.86 (0.4)
In vivo	0.62 (0.23)	73 (34)	96	8 (4)	7	...

[a] Results are expressed as mean values with standard deviations in parentheses.
[b] Percentage of the amount predicted from the weight loss of 12% nickel, 17% chromium stainless steel.
[c] Ratios for accelerated corrosion *in vitro*. (The nickel/chromium ratio of a 12:17 stainless steel is 0.71.)

TABLE 3—*Linear regression equations, correlation coefficients, weight loss (wl) data, and net charge transfers for the accelerated corrosion data in Table 2.*[a]

Corrosion Medium	Regression Equation	Correlation Coefficient	Weight Loss, mg	Net Charge Transfer, coulombs
Saline	$C = 3.74wl + 0.44$	0.999	1.85 (0.34)	6.93 (1.1)
Serum	$C = 3.69wl - 0.04$	0.998	0.58 (0.57)	2.10 (2.1)
In vivo	$C = 3.63wl - 0.03$	0.992	0.62 (0.23)	2.19 (1.0)

[a] Results are expressed as mean values with standard deviations in parentheses.

Accelerated corrosion experiments have also been done with dense and porous-coated F75 alloy specimens. The results of these experiments are shown in Table 4. For 1% nickel, 28% chromium alloy, we would expect a nickel/chromium ratio of 0.035. The results show that with the exception of one dense rod *in vivo,* the ratio was much higher than expected. The weight losses of the specimens were close to the limits of the five-place microbalance used and therefore did not provide a reliable data base for release calculations. However, calculations based on Faraday's law and charge transfer indicated that the release of nickel was greater than predicted, while the release of chromium was less than predicted. The release of cobalt *in vitro* was close to that predicted, while the amount excreted was low. Additional experiments are in progress to determine whether the nickel release is high or the chromium release is low, and to determine the release rates in serum *in vitro.*

The final issue to be considered on the subject of how much is released is the question of metal ion release as indicated by the total corrosion current versus the current density. Data presented elsewhere in this volume indicate that the total current of a porous-coated specimen is greater than that of a solid specimen, although the local current density may not be greater. The biological significance of these differences will depend on the level of concern. The response of a cell adjacent to a surface will be a function of current density; the response of the organism on a systemic level will be a function of total current and ion release. The response of the cells within the interstices of a porous coating will be a function of the corrosion current density of all the surfaces that surround them. This most critical issue must be addressed in biological tests and in standards.

Where Do They Go?

Given that metal ions are released, do they stay in the local tissues or are they transported in the blood? If they are transported, do they accumulate in other organs, or are they eliminated? We will address these questions by first examining the local implant site and then the blood cells, serum, organs, and excretions.

Local Tissue Levels

The classic studies on metal ion release and where they go are those by Ferguson et al. [28,29] and Laing et al. [30] in the 1960s. Cylinders of metal were implanted in the muscles of rabbits, and chemical analysis of local tissues and organs was performed. These authors found elevated local levels of nickel and very high levels of chromium around stainless steel. Nickel and chromium, as well as cobalt, were in tissues surrounding F75 alloy implants. High levels of nickel and chromium were reported by Michel and Zilkins [31]. Local elevations of cobalt and chromium around a metal-on-metal Co-Cr total hip at necropsy have also been reported by Dobbs and Minski [32]. They did not find such elevation around a

TABLE 4—*Weight loss, total charge, and metal ion release data and nickel/chromium ratios for accelerated corrosion of 4 by 25-mm F75 alloy rods at 500 mV (SCE) for 30 min.*[a]

Number of Rods	Corrosion Medium	Weight Loss, mg	Total Charge, coulombs	Ion Release, μm			Nickel/Chromium Ratio[b]
				Nickel	Cobalt	Chromium	
Dense Rods							
6	saline	0.03 (0.03)	0.021 (0.005)	0.7 (0.5)	5.1 (2.6)	0.6 (0.2)	1.05 (0.74)
6	10% serum	0.13 (0.04)	0.029 (0.005)	0.2 (0.1)	9.7 (4.9)	0.9 (0.2)	0.20 (0.08)
7	*in vivo*	0.06 (0.05)	0.026 (0.003)	0.7 (0.7)	4.8 (0.9)	0.9 (0.3)	0.80 (0.72)
1	*in vivo*	1.63	5.005	3.2	949	52.6	0.06
Porous Rods							
2	saline[c]	0.11/0.05	0.09/0.09	0.3/0.5	12.4/11.3	2.8/2.7	0.11/0.20
4	*in vivo*	0.06 (0.02)	0.15 (0.11)	1.01 (0.29)	7.37 (3.7)	2.32 (1.0)	0.49 (0.21)

[a] Results are expressed as mean values with standard deviations in parentheses.
[b] The nickel/chromium ratio for a 1% nickel, 28% chromium alloy is 0.035.
[c] The values for the two individual experiments are given.

metal-on-plastic prosthesis. Evans et al. [33] also reported high cobalt and chromium levels around metal-on-metal prostheses. High levels of titanium in local tissues have been reported by Meachim and Williams [34], Ducheyne et al. [21], and Willems et al. [35]. A mechanism of microfibrils has been proposed by Solar et al. [36] as an explanation for these high local levels of titanium.

Some answers for the surprisingly high levels of chromium and titanium in local tissues may be gleaned from examination of the data from Simpson et al. [37]. This work was a combined study of implant and tissue analysis of a series of patients with metal plates and screws for fracture fixation. As shown in Table 5, the plates and screws used were F138 stainless steel (SS), cobalt alloy [ASTM Specification for Wrought Cobalt-Nickel-Chromium-Molybdenum-Tungsten-Iron Alloy for Surgical Implant Applications (F 563-78)], and F67 and F136 titanium. These were also used in the combination titanium plates with stainless steel screws. In combination with the histological analysis of the local tissues, it was concluded that titanium and chromium may have been in high concentration because they remained at the implant site and were not transported away in the blood.

Blood Levels of Metal Ions

A number of studies have examined the question of blood serum and cell concentrations of metal ions from implants. Perhaps the original study was that by Coleman et al. [38] in which they analyzed blood from patients with metal-on-metal and metal-on-plastic F75 total hips for cobalt and chromium. The results showed significant increases in both elements in the blood from nine patients with metal-on-metal prostheses and a progressive increase in concentration in three patients studied from before implantation to over a year after. The blood levels of cobalt and chromium were in proportion to the alloy composition. No elevation was observed in patients with metal-on-plastic prostheses. In contrast, Woodman et al. [4,5] and Koegel and Black [2] have shown a disproportionate increase in serum nickel from microspheres of stainless steel and F75 cobalt-chromium-molybdenum alloy. Pazzaglia et al. [39] have reported a significant increase in blood and serum levels of nickel and chromium from patients with stainless steel Charnley total hips. Woodman et al. [40] have reported a significant increase in serum aluminum but no increase in titanium and vanadium from porous-coated Ti-6Al-4V implants in baboons.

We have measured serum and cell concentrations of metal ions in hamsters injected with metal salts or *in vitro* generated corrosion products [9,10] and in hamsters exposed to accelerated *in vivo* corrosion. Within 1 h of a single injection or 30 min of accelerated corrosion, the blood levels reached a maximum; within three days the levels of nickel and cobalt had returned to normal. Analysis of the proteins has demonstrated, as has also been reported by others [2,4–7], that nickel and cobalt bind to albumin. Chromium with a valence of $+3$ or chromium from corrosion in saline also binds to albumin and is rapidly cleared. Chromium with a valence of $+6$ or from *in vitro* corrosion in serum binds to red cells and is not as rapidly cleared, as will be discussed in more detail subsequently. Blood levels of chromium from *in vivo* accelerated corrosion of stainless steel only account for 14% of the predicted chromium release.

Metals in Organs

Ferguson et al. [28,29] and Laing et al. [30] also conducted detailed studies of elemental concentrations in the organs of the rabbits. With stainless steel rods these authors found high levels of nickel in muscle, liver, spleen, and lung. High levels of nickel and cobalt were found in the kidney and lung in animals with F75 alloy implants. A surprising result was

TABLE 5—Chemical analysis of the fibrous capsule over fracture fixation plates for stainless steel with stainless steel screws (SS/SS), F563 cobalt alloy plates and screws (Co/Co), titanium and Ti-6Al-4V alloy plates and screws (Ti/Ti), and titanium plates with stainless steel screws (Ti/SS) [37].

Plate and Screw Material	Number of Implants Removed	Average Time of Implantation, months	Ion Release, µg/g or ppm							
			Nickel		Cobalt		Chromium		Titanium	
			Average	Maximum	Average	Maximum	Average	Maximum	Average	Maximum
SS/SS	16	18	157	910	...	...	1595	10464	...	...
Co/Co	25	18	85	786	371	2510	806	10219	...	...
Ti/Ti	16	15	...	...	...	...	15	150	1308	9960
Ti/SS	23	16	99	640	...	...	1129	9727	66	255

the high levels of titanium in the spleens and lungs of animals with titanium or titanium-bearing implants [28]. Woodman et al. [40] have also reported high levels of titanium in lung, spleen, and liver tissue, as well as regional lymph nodes. They also reported a substantial and progressive increase in aluminum content in lungs and lymph nodes of baboons with porous titanium alloy implants. We have observed high levels of chromium in the lungs, spleens, and kidneys of hamsters injected with potassium dichromate (+6 valence) but no increase in nickel or cobalt after injections of chlorides of these elements.

Metal Ion Excretion

While chemical analysis of hair has demonstrated increases in metal ions related to metal implants [38,41], the major emphasis has been on analysis of urine. As has been previously mentioned, Woodman et al. [20] have reported significant increases in the urine levels of nickel, cobalt, and chromium from rabbits with solid F75 alloy implants. These levels were even higher with porous-coated implants or metal powder. They have also reported [40] increases in the urine titanium of baboons with porous titanium implants, but no significant increases in the urine vanadium or aluminum.

A number of studies have examined the urine excretion of patients with total joint arthroplasties. Coleman et al. [38] reported no increase in cobalt and chromium after metal-on-plastic total hip arthroplasty, but a significant increase in both after metal-on-metal total hip replacement (THR). After preoperative values of less than 1 μg/L, they observed postoperative values of 24 ± 25 μg/L for cobalt and 6 ± 25 μg/L for chromium. One patient with a unilateral metal-on-metal prosthesis had urine cobalt and chromium levels of 73 and 26 μg/L at one year after implantation. Pazzaglia et al. [39] reported a significant increase in urine nickel after stainless steel Charnley THR but no increase in urine chromium. Bartolozzi and Black [1] reported urine chromium data from patients with cobalt-chrome total hips that reflected serum chromium levels and that showed a trend indicating an increase with time. Jorgenson et al. [42] reported a 69% increase in urine cobalt in patients with porous-coated Austin-Moore prostheses over those with cemented devices, although the differences were not statistically significant.

The difficulty in interpreting urine metal levels is the lack of knowledge of the actual levels of release from implants. To address this issue, we have studied the urinary excretion of metals from hamsters injected with known amounts of metal ions, or exposed to known amounts of metal ion release from *in vivo* accelerated corrosion. Three sets of experiments will be discussed.

In the first set of experiments, two groups of four hamsters received intramuscular injections of nickel chloride, cobalt chloride, and potassium dichromate. The first group, with a "1×" dose, received 0.1 cm³ of the standard 2% solution of the salts used for skin testing, or about 5 to 6 μg of each. The second group, with a "20×" dose, received 20 times this dose, or about 90 to 100 μg. Twenty-four-hour urine samples were collected from each animal and analyzed separately for three days. The results from GFAAS in micrograms per millilitre were multiplied by the total volume excreted in each of the 24-h specimens and were expressed as the total amount in micrograms. As shown in Table 6, virtually all the nickel and cobalt were recovered in the urine, whereas less than half of the chromium was excreted.

The second set of experiments used 40 hamsters to study the blood and urine levels of metal ions after repeated injections of high doses of metal salts. The forty hamsters were divided into two groups. The first group received monthly injections of 20× metal salts for four months, the first of which was in Freund's complete adjuvant; this group is referred to as immunized. The second, nonimmunized group received no injections during the im-

TABLE 6—*Urine excretion levels of nickel, cobalt, and chromium from hamsters after single 1× and 20× injections of nickel chloride, cobalt chloride, and potassium dichromate.*[a]

Metal Ion	Dose Injected	Excretion Level, µg			
		24 h	48 h	72 h	Control
Nickel	1 × 5.18	4.27 (2.27)	0.61 (0.21)	0.28 (0.02)	0.096 (0.030)
	20 × 90.0	96.0 (35)	4.11 (2.90)	3.72 (4.47)	
Cobalt	1 × 5.40	3.57 (1.65)	0.33 (0.16)	0.09 (0.02)	0.058 (0.036)
	20 × 94.0	80.2 (20)	6.29 (2.6)	5.78 (5.5)	
Chromium	1 × 6.91	2.30 (1.20)	0.50 (0.20)	0.20 (0.06)	0.009 (0.003)
	20 × 117.0	41.5 (14)	8.60 (3.0)	6.30 (2.9)	

[a] Results are expressed as mean values with standard deviations in parentheses.

munization period. After an eight-week pause, four animals from each group were selected for collection of blood and 24-h urine control samples. Animals in both groups then received six weekly injections of 20× metal salts. Four hamsters from each group were then selected at weekly intervals for collection of 24-h urine samples and were then sacrificed for blood and organ harvesting. The analyses were done on individual blood and urine samples, and the results were averaged.

The results for the blood cell chromium concentration in micrograms per litre or parts per billion and the urine chromium content in micrograms are presented in Table 7. It can be seen that the blood cell chromium was below the detection level in the preinjection (control) samples taken prior to the six weekly injections, eight weeks after the first series of injections. In the weekly samples taken after the six weekly injections, the blood cell chromium levels were still high, even after four weeks. The preinjection (control) urine chromium levels were still elevated in the immunized group. Neither group returned to the control level by four weeks after the last of the second set on injections. Clearly, the clearance of chromium was slow.

The data in Table 8 are a summary of the results from several of the previously described

TABLE 7—*Blood cell chromium concentrations and 24-h urine chromium contents from immunized and nonimmunized hamsters in the metal overload experiment.*[a,b]

	Control	Period After Second Set of Injections			
		Week 1	Week 2	Week 3	Week 4
Blood cell concentration, µg/L or ppb					
Nonimmunized	<0.001	0.85 (0.36)	0.91 (0.34)	0.84 (0.35)	0.34 (0.24)
Immunized	<0.001	1.13 (0.19)	1.00 (0.57)	0.15 (0.07)	0.29 (0.06)
24-h urine content, µg					
Nonimmunized	0.025 (0.018)	1.53 (0.28)	0.54 (0.10)	0.68 (0.13)	0.54 (0.25)
Immunized	0.41 (0.14)	1.17 (0.06)	0.65 (0.13)	0.64 (0.17)	0.26 (0.12)

[a] The immunized group received four monthly 20× injections. After an eight-week pause, control samples were taken from the immunized and nonimmunized groups. Both groups then received six weekly 20× injections, and then animals were selected at one, two, three, and four weeks after the last injection for 24-h urine collection and blood samples. There were four animals in each sample group.
[b] Results are expressed as mean values with standard deviations in parentheses.

TABLE 8—*Percentages of recovery of metals from accelerated corrosion and metal salt injection studies.*[a,b]

Metal Ion	Recovery, %				
	Accelerated Corrosion[c]			Salt Injection, Urine[d]	
	In Vitro				
	Saline	Serum	*In Vivo*, Urine	1× Dose	20× Dose
Nickel	98 (33)	112 (92)	96 (11)	119 (31)	121 (33)
Cobalt	NA[e]	NA	NA	82 (29)	99 (13)
Chromium	110 (7)	86 (35)	7 (3)	48 (18)	47 (9)

[a] Accelerated corrosion of stainless steel rods was conducted *in vitro* in saline and 10% serum, and *in vivo* in hamsters.

[b] Results are expressed as mean values with standard deviations in parentheses.

[c] Metal concentrations in the solutions and 24-h urine samples collected for four days are expressed as percentages of the amounts predicted from the measured weight loss based on a 12% nickel, 17% chromium alloy.

[d] Percentages of metal ions recovered in 24-h urine samples collected over a four-day period are shown.

[e] NA = not applicable.

accelerated corrosion and metal salt injection experiments. The percentages of nickel and chromium in the saline and serum corrosion solutions are given, based on calculations from the amount of implant weight loss. Similarly, the amounts of nickel and chromium recovered in the urine from *in vivo* corrosion are also expressed as percentages of the calculated release from the weight loss data. These data are compared with the urine recovery of metal after 1× and 20× intramuscular injections of metal salts. Within experimental error, all the nickel and cobalt were recovered in all the experiments. This means that urine levels reflect release rates.

In contrast, less than half the chromium from injections and only 8% of the chromium from *in vivo* corrosion was recovered. When red cell chromium in the *in vivo* corrosion data are included, based on a 50% hematocrit and a blood volume of 10% of body weight, the recovery level is brought up only to 22%. The linear regression data for *in vivo* versus *in vitro* accelerated corrosion of stainless steel shown in Table 2 indicate striking similarities and permit us to conclude that the chromium release rates are in proportion to the alloy composition. It can be concluded that chromium is not rapidly excreted in the urine under these conditions. The question that is currently under investigation is this: does the chromium from *in vivo* corrosion remain at the implant site, or is it stored in the blood or in organs? The data presented for F75 alloy suggest that the release rates may not be in proportion to the alloy composition. Preliminary results of chemical analysis of these tissues indicate the presence of elevated levels of chromium in the implant site as well as in the systemic organs. Urine chromium levels may reflect blood levels of chromium, but they do not reflect the total chromium release from implants.

What Can They Do?

The purpose of this discussion is to address the issues of what the presence of metal ions or corrosion products and wear debris has been shown to do in laboratory studies or in other circumstances in which animals or humans are exposed to metals. This discussion will focus on *in vitro* and *in vivo* cytotoxicity, metal sensitivity, and oncogenicity in animals.

In vitro cell and organ culture studies have been conducted to study both metal ions and wear debris. Pappas and Cohen [43] and Mital and Cohen [44] have reported a cytotoxic response of cells exposed to F138 stainless steel powder and a greater response for cells exposed to F75 alloy powder. Rae [45–47] has shown a cytotoxic response to cobalt and nickel powder and to particulate debris from F75 alloy. Rae [48] has also shown an hemolytic response to nickel and F75 powder, a response that is virtually eliminated by serum protein adsorption. The question of whether these laboratory-generated debris are similar in morphology to those generated *in vivo* has not been resolved.

However, the fact still remains that ions would be released. Daniel et al. [49] showed a toxicity to low concentrations of cobalt ions. It is interesting that the cells that survived exposure to 5 μg of cobalt could tolerate higher levels in subsequent exposure. Bearden and Cooke [50] showed a cell growth inhibition to both nickel and cobalt. Gerber and Perren [51] have shown a dose-response toxicity to vanadium and to a lesser degree to nickel, cobalt, and copper. Lucas et al. [52] have also shown a dose-related *in vitro* toxicity to metal ions mixed to simulate the chemical composition of stainless steel. These debris and metal ion studies do demonstrate the toxicity of the ions; future studies and standards must relate the dose response *in vitro* to actual implant release rates and local concentrations.

Toxicity responses to solid metals have also been observed *in vitro*. Gerber and Perren [51] also demonstrated an inhibition of growth of fetal rat femora in the presence of metal pins of cobalt and to a lesser degree for nickel. Johnsson and Hegyeli [53] have reported an inhibition of cell growth on titanium surfaces *in vitro*.

The question of *in vivo* cytotoxicity was investigated by Lucas et al. [52] in a study in which they injected stainless steel implant sites with stainless steel ions. The tissue showed a definite dose response to these injections. Geret et al. [54] and McNamara and Williams [55,56] have shown local toxic responses to implants of pure cobalt and nickel, but not to surgical alloys. It is possible that the release rates from the pure metals are unrealistically high and that those from the alloys are low enough to avoid toxicity.

However, tissue damage due to mechanisms other than toxicity can occur. It is well known in the contact dermatitis literature [57] that nickel, cobalt, and chromium can cause a cell-mediated delayed hypersensitivity. This can arise from industrial contact with metal salts or corroded metallic surfaces, from jewelry, or from the activities of daily living [57–59]. We [8,60,61] have shown that, in animals, sensitivity to these three elements can be induced by injection of metal salts with Freund's complete adjuvant. Rabbits with metal sensitivity demonstrate a reduction of bone formation and local necrosis in tissues adjacent to stainless steel and cobalt-alloy implants. The healing of tibial fractures was delayed and long-term inhibition of bone formation was also observed in sensitized rabbits with stainless steel intramedullary rods [62]. We have also observed [11] similar findings in rabbits injected with titanium salts and implanted with titanium screws.

The guinea pig has been a good animal model to demonstrate the acquisition of metal sensitivity from the injection of metal salts. A great deal is known about its response to potassium dichromate and chromic chloride [59], and cobalt has been used successfully to sensitize animals [63]. However, the results with nickel have been confusing, and the guinea pig is not a good animal to model nickel contact sensitivity, although they apparently do have immune responses to nickel [58,59,63]. In our studies, guinea pigs receiving intra-muscular injections of dichromate demonstrated delayed hypersensitivity, whereas those receiving intraperitoneal injections developed antibody-mediated sensitivity [8]. Both types of sensitivity were associated with adverse cellular reactions in tissues adjacent to stainless steel plates used on femoral fractures. Similarly, guinea pigs made sensitive to nickel, although no longer displaying contact dermatitis reactions but still having antibodies, showed adverse cellular reactions in tissue adjacent to stainless steel screws.

Biological responses to metal ions acquired from sources other than implants have been noted. Some individuals react to nickel, cobalt, and chromium in food and drink [64], and other metal salts not part of this discussion, such as gold, can cause reactions. The one of great concern in the population at this time is aluminum, which can be acquired from dietary sources, medications, and from the water in dialysis and has been associated with Alzheimer's disease. The elements we are concerned with in this volume are, in general, trace elements (titanium is not) required for normal function of the physiologic system. However, all of these elements have been shown to be toxic in high doses or in the wrong form. We still do not have adequate information to define what a high dose or the wrong form is.

The issue of oncogenicity in animals is even less clear. Heath et al. [65] demonstrated an increased incidence of sarcoma in hooded Strangeways rats exposed to metal powder of cobalt, nickel, and F75 alloy. Similar responses were not seen with other metal debris of the same structure. In contrast, Meachim et al. [66] did not observe an increase in sarcomas in Liverpool rats or guinea pigs. The difference may be due in part to the particle size: those powders used by Heath et al. were mostly in the submicrometre size, whereas those used by Meachim et al. were 0.5 to 5 μm in size. The question of tumors induced from the physical structure of the metals versus those induced from the elements in the composition of the metals remains unanswered in these studies.

It is evident that compounds of nickel or chromium and, to a lesser extent, of cobalt can cause neoplasms in laboratory animals [67]. Most of the forms of these metallic salts are not those that come from wear, degradation, or corrosion products of implant alloys. Nickel subsulfide and nickel carbonyl are known carcinogens in both man and animals but are not known compounds generated *in vivo* from surgical alloys. Chromium is a well-known cause of tumors in individuals exposed to high levels of chromium in the lungs from occupational exposure. Cobalt is known as a carcinogen mostly in powder form. It is evident that tumors can be caused by some forms of these alloys and some forms of the metallic compounds. If standard test methods for assessing the biological response to wear debris and metal ions are to be relevant, they must ensure that the particulate debris and released ions are in the forms in which they would be released *in vivo*.

There have been several reports on tumors in animals associated with the use of metallic implants in experimental and clinical situations. Memoli et al. [68] have reported a significant incidence of neoplasms associated with the implantation of cylinders of surgical alloys. Stevenson et al. [69] have reported a number of osteosarcomas associated with fracture fixation devices in dogs. While the number of cases is small, their locations were significant because of the normally low incidence of that particular type and site of tumor. However, the correlation between the implant and carcinogenicity is difficult to draw since there is a relatively high incidence of infections associated with these implants. The combination of implant and infection leading to carcinogenicity remains to be carefully addressed. There is some evidence that particular devices may be more associated with tumor development than others. However, once again, the presence of underlying infection remains a complicating issue. This was also evident from the report of Sinibaldi et al. [70] which reports a number of tumors associated with Jonas pins. These studies also included implant analysis, which showed a high degree of corrosion of these multicomponent devices. Thus, there may be a correlation between the oncogenicity observed in these animals and the high levels of metal ion release from the corrosion of these implants.

From this review it can be concluded that metal ions can be cytotoxic, can cause metal sensitivity, and can cause an increase in tumor formation. The question remains as to whether or not implants and their corrosion, wear, and degradation products have caused any of these problems in humans.

What Have They Done?

The question of *in vivo* toxicity of surgical implants in humans is difficult to address because of the limited number of cases, other circumstances such as low-grade infection, and the apparent mobility of the ions released from the implant. Thus, toxic responses might not (probably will not) occur at the site of the implant but would be more likely to occur in the end organ of elimination or accumulation, such as the lung or the kidney. The correlation of symptoms in these organs with the presence or past presence of an implant would be difficult to determine. In a study of tissues adjacent to titanium implants, Meachim and Williams [34] reported that of 19 implants with adverse tissue reactions "metal release from the implant was a probable contributory factor in five, and it was an unlikely factor in three." In a study of fracture fixation plates, Simpson et al. [37] presented anecdotal evidence of adverse clinical reactions and noted the presence of increased tissue reactions around cobalt alloy plates (ASTM Specification F 563–78), an alloy that is not used in this country. Tissue reactions consistent with toxicity were noted around some of the other implants also but the extent of reaction was small. French et al. [71] have identified a correlation between corrosion and local tissue response. The interesting thing is that the correlation was positive in the asymptomatic patient group; there was no correlation in the symptomatic group.

Perhaps the strongest case for a toxic response is from the studies of tissue response to wear debris from revised total hips by Willert et al. [72,73]. These authors have made a strong case for increased inflammatory response and bone resorption due to increased wear. However, it is not clear whether this is a reaction to metal ions, to particulate debris, most of which was polymeric, or to a combination of the two.

There are a number of clinical reports on metal sensitivity reactions to fracture fixation plates [74], intramedullary rods [75], and bone screws [76]. The bulk of the clinical data on metal sensitivity comes from studies of patients with metal-on-metal total joint arthroplasties [33,77,78]. The metal ion release from a metal-on-metal prosthesis is much greater than that from metal-on-plastic prostheses, and this has been implicated as the cause of the large number of reactions. Discussion of the problem of sensitivity reactions in orthopedic patients has been dismissed by some with the statement that these devices are no longer used. However, although the actual devices may no longer be used, devices with high corrosion and wear rates and the use of large-surface-area porous-coated prostheses with extensive bone-to-metal contact may pose problems similar to those seen with the metal-on-metal prostheses.

The incidence of metal sensitivity in the normal population is high, with up to 15% of the population sensitive to nickel and perhaps up to 25% sensitive to at least one of the common sensitizers Ni, Co, Cr. The incidence of metal sensitivity reactions requiring premature removal of an orthopedic device is probably small (less than the incidence of infection). Clearly there are factors not yet understood that cause one patient but not another to react. The role of implant wear, corrosion, and degradation has been discussed and clearly is an important issue. When one attempts to define a dose-response curve, problems arise. The sensitivity response behaves more like an on-off switch than a metered response. If the allergen is encountered at a sufficient concentration, a response occurs and doses above this concentration do not usually result in a measurably increased response. Similarly, individual patients will have different threshold doses. It is evident that these doses can be very low [79] and can approach the limits of detection of our analytical instruments. The introduction of devices with increased corrosion, degradation, and wear rates will be associated with an increased percentage of the population responding and not necessarily an increase in the severity of symptoms in an individual patient. On the other hand, toxicity is a dose-related response. As the level of metals released into the tissues increases, the local toxicity will increase.

The question of whether or not implants cause sensitivity is a separate issue from whether or not sensitivity reactions to implants occur. The studies on sensitivity reactions to metal-on-metal total joint arthroplasties were done by using the skin test on both patients requiring revision and those not requiring revision. The results indicated that metal sensitivity could be causing the symptoms in some cases. Studies on patients receiving total joint arthroplasties of metal-on-plastic have also indicated an acquisition of sensitivity in some patients. Whether this acquisition was due to the implant and not the skin test remains unresolved. We have conducted a study of the metal sensitivity status of orthopedic patients at the time of implantation and at the time of implant removal [11,19]. An *in vitro* test [the leucocyte inhibition factor (LIF) test] was used to document the sensitivity without the possible complication of the acquisition of sensitivity being due to repeated skin testing rather than to the implant.

In a clinical study of sensitivity, we [11] have tested 629 patients at the time of implantation and 283 patients at the time of implant removal. While this may represent a biased population, examination of the results does shed some light on the problem of sensitivity associated with orthopedic implants. The incidence of sensitivity in the preimplantation group was 25%, which is consistent with other studies on the normal population. The incidence of sensitivity in the patients having implants removed was 43%, which is a marked difference from the preimplantation data and indicates that the implants do cause sensitivity. Of these 283 patients having implants removed, 187 were documented to have had stainless steel implants, while 55 were documented to have had cobalt-chromium-molybdenum (F75) alloy implants. The test results were reported as no sensitivity; a sensitivity to nickel, cobalt, chromium, or any combination thereof; or a sensitivity reaction to the implant. The lowest incidence of no sensitivity, 26%, and the highest incidence of a sensitivity reaction, 38%, were in the F75 alloy group. The patient population was select in that many of the removals were for relief of symptoms, and not all patients with routine removals were studied; therefore, these patients do not represent the total normal population. Fifteen of the 55 patients with F75 alloy implants had uncemented hemihip arthroplasties—the largest incidence of sensitivity to be associated with noncemented F75 hemihip arthroplasties [19]. This may be a significant observation in discussion of noncemented porous-coated devices.

The issue of morbidity associated with the sensitivity reaction to the implant remains as coincidental evidence. Documentation of metal sensitivity in these patients and the observation that removal of the devices alleviated the symptoms is taken as strong evidence that metal sensitivity reactions to the implant caused the symptoms. It can never be proven that other influences were not also important, and reinsertion of the device to prove its role [80] is not a warranted procedure from the ethical or scientific point of view since producing the same corrosion or wear in the same circumstances is unlikely.

The final biological response to be discussed is that of oncogenicity. There is no question that tumors have arisen in patients with implants [81]. There is also no question that some of the tumor tissues have had elevated levels of nickel, cobalt, and chromium in comparison with normal tissues. However, there is great question about cause and effect. There is evidence that these tumors are implant associated but no evidence that the implants caused the tumors. Some of these case reports involve many variables, including old alloys with low corrosion resistance and 20 to 30 years' duration [82–84] and tumors in elderly patients that arose within 2 to 4 years after implantation [85–87], which is a short time for tumor development. Finally, the number of cases reported is few. There is also no evidence that there is a greater incidence of tumors in patients with existing or previous implants than in the normal population. Careful, thorough documentation of patients receiving an implant and lifetime follow-up of these patients would be necessary to answer the question of incidence. This is an issue of great concern to the medical profession and it needs to be carefully examined.

Similarly, tumor development could, again, be associated with infection [88]. The role of the implant, the role of the infecting agent, and the role of the combination of the two need to be addressed. Only careful examination of the patient and careful examination of the patient's record will reveal the importance of the various factors.

Summary

In summary, metal ions are released; some stay in the local implant site while others are transported in the blood. Blood-borne elements may remain in the blood as cell complexes, accumulate in organs, or be excreted. Metal ions are known to be toxic, sensitizing, and oncogenic. The increase in metal-ion release rates from porous-coated implants can be expected to increase the risks of local cellular reactions and systemic effects. The developments of standard test methods to address these issues must ensure that the methods accurately model the *in vivo* situation. Methods for testing porous coatings for intramedullary bony ingrowth should include testing in serum solutions and should consider the release rates within the pores as well as the total release. The accelerated corrosion method presented permits determination of release rates *in vitro* and *in vivo* and thus is a model for determining the systemic distribution and fate of corrosion products. Tests for bearing surface materials and wear rates should consider using synovial fluid and the generation and testing of wear debris in the form in which they would be produced *in vivo*. However, it is only with careful clinical follow-up that we will find the answer to the question of what the biological significance of metal ion release is.

Acknowledgments

The authors wish to acknowledge financial support from U.S. Public Health Service NIH Grants AM20271, AM29648, AM32138, and AM35590. The donation of implant materials by Zimmer, Synthes, Ltd., and Bio-Vac that were used in the experiments described is gratefully acknowledged.

References

[1] Bartolozzi, A. and Black, J., *Biomaterials,* Vol. 6, 1985, pp. 2–7.
[2] Koegel, A. and Black, J., *Journal of Biomedical Materials Research,* Vol. 18, 1984, pp. 513–522.
[3] Smith, G. K. and Black, J. in *Corrosion and Degradation of Implant Materials: Second Symposium, ASTM STP 859,* American Society for Testing and Materials, Philadelphia, 1985, pp. 223–247.
[4] Woodman, J. L., Black, J., and Jimenez, S. in *Biomaterials 1980,* G. D. Winter, D. F. Gibbons, and H. Plenk, Jr., Eds., Wiley, Chichester, England, 1982, pp. 245–250.
[5] Woodman, J. L., Black, J., and Jiminez, S. A., *Journal of Biomedical Materials Research,* Vol. 18, 1984, pp. 99–114.
[6] Clark, G. C. F. and Williams, D. F., *Journal of Biomedical Materials Research,* Vol. 16, 1982, pp. 125–134.
[7] Williams, D. F., Askill, I. N., and Smith, R., *Journal of Biomedical Materials Research,* Vol. 19, 1985, pp. 313–320.
[8] Merritt, K. and Brown, S. A. in *Biomaterials and Biomechanics 1983,* Elsevier, Amsterdam, 1984, pp. 221–226.
[9] Merritt, K., Brown, S. A., and Sharkey, N. A., *Journal of Biomedical Materials Research,* Vol. 18, 1984, pp. 1005–1015.
[10] Merritt, K., Brown, S. A., and Sharkey, N. A., *Journal of Biomedical Materials Research,* Vol. 18, 1984, pp. 991–1004.

[11] Merritt, K. and Brown, S. A. in *Corrosion and Degradation of Implant Materials: Second Symposium, ASTM STP 859,* American Society for Testing and Materials, Philadelphia, 1985, pp. 195–207.

[12] Merritt, K., Wortman, R. S., Millard, M., and Brown, S. A., *Biomaterials, Medical Devices, and Artificial Organs,* Vol. 11, 1983, pp. 115–124.

[13] Gray S. J. and Stirling, K., *Journal of Clinical Investigation,* Vol. 29, 1950, pp. 1604–1613.

[14] Brown, S. A. and Merritt, K., *Journal of Biomedical Materials Research,* Vol. 15, 1981, pp. 479–488.

[15] Brown, S. A. and Simpson, J., *Journal of Biomedical Materials Research,* Vol. 15, 1981, pp. 867–878.

[16] Cook, S. D., Gianoli, G. J., Clemow, A. J. T., and Haddad, R. J., Jr., *Biomaterials, Medical Devices, and Artificial Organs,* Vol. 11, 1984, pp. 281–292.

[17] Thull, R., *Orthopaede* Vol. 7, 1978, pp. 29–42.

[18] Cohen, J. and Lindenbaum, B., *Clinical Orthopaedics,* Vol. 61, 1968, pp. 167–175.

[19] Brown, S. A. and Merritt, K. in *Implant Retrieval: Material and Biological Analysis,* NBS SP 601, National Bureau of Standards, Gaithersburg, MD, 1981, pp. 299–322.

[20] Woodman, J. L., Urban, R. M., Lim, K., and Galante, J. O., *Transactions of the Orthopaedic Research Society,* Vol. 9, 1984, p. 150.

[21] Ducheyne, P., Willems, G., Martens, M., and Helsen, J., *Journal of Biomedical Materials Research,* Vol. 18, 1984, pp. 293–308.

[22] Ducheyne, P., *Biomaterials,* Vol. 4, 1983, pp. 185–191.

[23] Woodman, J. L., Black, J., and Nunamaker, D. M., *Journal of Biomedical Materials, Research,* Vol. 17, 1983, pp. 655–668.

[24] Treharne, R. W. and Marek, M. in *Biomaterials 1980,* G. D. Winter, D. F. Gibbons, and H. Plenk, Jr., Eds., Wiley, Chichester, England 1982, pp. 251–254.

[25] Brown, S. A. and Merritt, K. in *Corrosion and Degradation of Implant Materials: Second Symposium, ASTM STP 859,* American Society for Testing and Materials, Philadelphia, 1985, pp. 105–116.

[26] Brown, S. A. and Merritt, K. in *Clinical Applications of Biomaterials,* Wiley, Chichester, England, 1982, pp. 195–202.

[27] Brown, S. A. and Merritt, K., *Journal of Biomedical Materials Research,* Vol. 14, 1980, pp. 173–175.

[28] Ferguson, A. B., Jr., Laing, P. G., and Hodge, E. S., *Journal of Bone and Joint Surgery,* Vol. 42A, 1960, pp. 77–90.

[29] Ferguson, A. B., Jr., Akahoshi, Y., Laing, P. G., and Hodge, E. S., *Journal of Bone and Joint Surgery,* Vol. 44A, 1962, pp. 323–336.

[30] Laing, P. G., Ferguson, A. B., Jr., and Hodge, E. S., *Journal of Biomedical Materials Research,* Vol. 1, 1967, pp. 135–149.

[31] Michel, R. and Zilkens, J., *Zeitschrift für Orthopaedie,* Vol. 116, 1978, pp. 666–674.

[32] Dobbs, H. S. and Minski, M. J., *Biomaterials,* Vol. 1, 1980, pp. 193–198.

[33] Evans, E. M., Freeman, M. A. R., Miller, A. J., and Vernon-Roberts, B., *Journal of Bone and Joint Surgery,* Vol. 56B, 1974, pp. 626–642.

[34] Meachim, G. and Williams, D. F., *Journal of Biomedical Materials, Research,* Vol. 7, 1973, pp. 555–572.

[35] Willems, G. L., Palmans, R. A., Colard, J., Ducheyne, P., and Martens, M. in *Biomaterials and Biomechanics 1983,* P. Ducheyne, G. Van der Perre, and A. E. Aubert, Eds. Elsevier, Amsterdam, 1984, pp. 237–242.

[36] Solar, R. J., Pollack, S. R., and Korostoff, E. in *Corrosion and Degradation of Implant Materials, ASTM STP 648,* 1979, pp. 161–172.

[37] Simpson, J. P., Geret, V., Brown, S. A., and Merritt, K. in *Implant Retrieval: Material and Biological Analysis,* NBS SP601, National Bureau of Standards, Gaithersburg, MD, 1981, pp. 395–422.

[38] Coleman, R. F., Herrington, J., and Scales, J. T., *British Medical Journal,* Vol. 1, 1973, pp. 527–529.

[39] Pazzaglia, U. E., Minoia, C., Ceciliani, L., and Riccardi, C., *Acta Orthopaedica Scandinavica,* Vol. 54, 1983, pp. 574–579.

[40] Woodman, J. L., Jacobs, J. J., Galante, J. O., and Urban, R. M., *Journal of Orthopaedic Research,* Vol. 1, 1984, pp. 421–430.

[41] Owen, R., Meachim, G., and Williams, D. F., *Journal of Biomedical Materials Research,* Vol. 10, 1976, pp. 91–99.

[42] Jorgensen, T. J., Munno, F., Mitchell, T. G., and Hungerford, D., *Clinical Orthopaedics,* Vol. 176, 1983, pp. 124–126.
[43] Pappas, A. M. and Cohen, J., *Journal of Bone and Joint Surgery,* Vol. 50A, 1968, pp. 535–547.
[44] Mital, M. and Cohen, J., *Journal of Bone and Joint Surgery,* Vol. 50A, 1968, pp. 547–556.
[45] Rae, T., *Journal of Bone and Joint Surgery,* Vol. 57B, 1975, pp. 444–450.
[46] Rae, T., *Archives of Orthopaedic and Traumatic Surgery,* Vol. 95, 1979, pp. 71–91.
[47] Rae, T., *Journal of Bone and Joint Surgery,* Vol. 63B, 1981, pp. 435–440.
[48] Rae, T., *Journal of Pathology,* Vol. 125, 1978, pp. 81–89.
[49] Daniel, M., Dingle, J. T., Webb, M., and Heath, J. C., *British Journal of Experimental Pathology,* Vol. 44, 1963, pp. 163–176.
[50] Bearden, L. J. and Cooke F. W., *Journal of Biomedical Materials Research,* Vol. 14, 1980, pp. 289–309.
[51] Gerber, H. and Perren, S. M. in *Evaluation of Biomaterials,* Wiley, Chichester, England, 1980, pp. 307–314.
[52] Lucas, L. C., Bearden, L. J., and Lemons, J. E. in *Corrosion and Degradation of Implant Materials: Second Symposium, ASTM STP 859,* American Society for Testing and Materials, Philadelphia, 1985, pp. 208–222.
[53] Johnsson, R. I. and Hegyeli, A. F., *Annals of the New York Academy of Sciences,* Vol. 146, 1968, pp. 66–76.
[54] Geret, V., Rahn, B. A., Mathys, R., Straumann, F., and Perren, S. M. in *Evaluation of Biomaterials,* Wiley, Chichester, England, 1980, pp. 351–359.
[55] McNamara, A. and Williams, D. F., *Biomaterials,* Vol. 3, 1982, pp. 165–176.
[56] McNamara, A. and Williams, D. F., *Journal of Biomedical Materials Research,* Vol. 18, 1984, pp. 185–206.
[57] Fisher, A. A., *Contact Dermatitis,* Lea and Febiger, Philadelphia, 1973, p. 87.
[58] Merritt, K. and Brown, S. A. in *Systemic Aspects of Biocompatibility,* Vol. II, CRC Press, Boca Raton, FL, 1981, pp. 33–48.
[59] Polak, L., Turk, J. L., and Frey, J. R., *Progress in Allergy,* Vol. 17, 1973, p. 145.
[60] Merritt, K. and Brown, S. A., *Acta Orthopaedica Scandinavica,* Vol. 51, 1980, pp. 403–411.
[61] Merritt, K. and Brown, S. A. in *Clinical Applications of Biomaterials,* Wiley, Chichester, England, 1982, pp. 85–93.
[62] Brown, S. A., Devine, S. D., and Merritt, K., *Biomaterials, Medical Devices, and Artificial Organs,* Vol. 11, 1983, pp. 73–81.
[63] Wahlberg, J. C. and Boman, A., *Contact Dermatitis,* Vol. 4, 1978, pp. 128–132.
[64] Fisher, A. A., *Cutis,* Vol. 17, 1976, pp. 229–233.
[65] Heath, J. C., Freeman, M. A. R., and Swanson, S. A. V., *Lancet,* Vol. 1, 1971, pp. 564–566.
[66] Meachim, G., Pedley, R. B., and Williams, D. F., *Journal of Biomedical Materials Research,* Vol. 16, 1982, pp. 407–416.
[67] Sunderman, F. W., Jr., *Federation Proceedings,* Vol. 37, 1978, pp. 40–46.
[68] Memoli, V. A., Woodman, J. L., Urban, R. M., and Galante, J. O., *Transactions of the Orthopaedic Research Society,* Vol. 7, 1982, p. 164.
[69] Stevenson, S., Hohn, R. B., Pohler, O. E. M., Fetter, A. W., Olmstead, M. L., and Wind, A. P., *Journal of the American Veterinary Medical Association,* Vol. 180, 1982, pp. 1189–1196.
[70] Sinibaldi, K. R., Pugh, J., Rosen, H., and Lie, S.-K., *Journal of the American Veterinary Medical Association,* Vol. 181, 1982, pp. 885–890.
[71] French, H. G., Cook, S. D., and Haddad, R. J., Jr., *Journal of Biomedical Materials Research,* Vol. 18, 1984, pp. 817–828.
[72] Willert, H. G., Buchhorn, G., Buchhorn, U., and Semlitsch, M. in *Implant Retrieval: Material and Biological Analysis,* NBS SP-601, National Bureau of Standards, Gaithersburg, MD, 1980, pp. 339–367.
[73] Willert, H. G., Semlitsch, M., Buchhorn, G., and Kriete, U., *Orthopaede,* Vol. 7, 1978, pp. 62–83.
[74] Cramers, M. and Lucht, L., *Acta Orthopaedica Scandinavica,* Vol. 48, 1977, pp. 245–249.
[75] McKenzie, A. W., Aitken, C. V. E., and Risdell-Smith, R. I., *British Medical Journal,* Vol. 4, 1967, p. 36.
[76] Barranco, V. P. and Soloman, H., *Journal of the American Medical Association,* Vol. 220, 1972, p. 1244.
[77] Jones, P. A., Lucas, H. K., O'Driscoll, M., Price, C. G. H., and Wibberley, B., *Journal of Bone and Joint Surgery,* Vol. 57B, 1975, pp. 289–296.
[78] Elves, M. W., Wilson, J. N., Scales, J. T., and Kemp, H. B. S., *British Medical Journal,* Vol. 4, 1975, pp. 376–378.

[79] Smeenk, G. and Teunissen, P. C., *Nederlandsch Tijdschrift Voor Geneeskunde,* Vol. 121, 1977, p. 4.
[80] Fisher, A. A., *Journal of the American Medical Association,* Vol. 221, 1972, p. 1279.
[81] Hamblen, D. L. and Carter, R. L., *Journal of Bone and Joint Surgery,* Vol. 66B, 1984, pp. 625–626.
[82] McDougall, A., *Journal of Bone and Joint Surgery,* Vol. 38B, 1956, pp. 709–713.
[83] Delgado, E. R., *Clinical Orthopaedics,* Vol. 12, 1958, pp. 315–318.
[84] Tayton, K. J. J., *Cancer,* Vol. 45, 1980, pp. 413–415.
[85] Bago-Granell, J., Aquirre-Canyadell, M., Nardi, J., and Tallada, N., *Journal of Bone and Joint Surgery,* Vol. 66B, 1984, pp. 38–40.
[86] Swann, M., *Journal of Bone and Joint Surgery,* Vol. 66B, 1984, pp. 629–631.
[87] Penman, H. G. and Ring, P. A., *Journal of Bone and Joint Surgery,* Vol. 66B, 1984, pp. 632–634.
[88] McDonald, I., *Cancer,* Vol. 48, 1981, pp. 1009–1011.

DISCUSSION

W. D. Galloway[1] (*written discussion*)—You said that metal ions are bound to albumin in serum. Do you know if those ions in tissue are bound to protein in organs and, if so, which? This would be important in predicting toxicity.

S. A. Brown, K. Merritt, L. J. Farnsworth, and T. D. Crowe (*authors' closure*)—We are currently trying to determine the answer to this very important question.

Z. Glaser[2] (*written discussion*)—Comment on the possible use of radioisotopes to track the source and repository of ions from implants.

S. A. Brown, K. Merritt, L. J. Farnsworth, and T. D. Crowe (*authors' closure*)—The use of an alloy containing a stable radioisotope of a component, evenly distributed throughout the alloy, would provide a sensitive method for tracking the ions. Such alloys are not available.

[1] Food and Drug Administration, Rockville, MD.
[2] Food and Drug Administration, Rockville, MD.

Performance in Humans and Laboratory Animals

J. Dennis Bobyn,[1] *Charles A. Engh,*[2] *and Robert M. Pilliar*[3]

Histological Comparison of Biological Fixation and Bone Modeling with Canine and Human Porous-Coated Hip Prostheses

REFERENCE: Bobyn, J. D., Engh, C. A., and Pilliar, R. M., **"Histological Comparison of Biological Fixation and Bone Modeling with Canine and Human Porous-Coated Hip Prostheses,"** *Quantitative Characterization and Performance of Porous Implants for Hard Tissue Applications, ASTM STP 953,* J. E. Lemons, Ed., American Society for Testing and Materials, Philadelphia, 1987, pp. 185–206.

ABSTRACT: Undecalcified hard-section histology was performed on twelve canine and eleven human porous-coated femoral hip prostheses implanted without bone cement. The period of implantation varied from four weeks to seven years. All of the canine and nine of the eleven human implants showed regions of substantial bone ingrowth along the porous-coated stem region. Areas of the two human implants ingrown with fibrous tissue were characterized radiographically by a radiopaque line surrounding, but separated from, the implant surface. Resorptive bone modeling was observed with both implant types, particularly proximally. New bone formation and hypertrophy typically occur at regions of proximity to or contact with the endosteal cortex and, in the case of proximally porous coated implants, at the junction of the porous and smooth implant surfaces. The smooth-surfaced distal region of proximally coated implants is typically surrounded by a thin shell of bone that is separated from the implant surface. Overall, resorptive and formative bony change appears more exaggerated in the canine, but the radiographic and histologic pictures of canine and human implants bear resemblances.

KEY WORDS: porous implants, bone ingrowth, histology

A wide variety of porous implant systems designed for biological fixation by bone ingrowth are undergoing clinical trials [1,2]. In North America, the powder-made sintered porous surface has a long and wide history of use both experimentally in animals and clinically in humans [3]. Histological studies of retrieved canine and human porous-coated femoral hip prostheses have enabled comparisons to be made of the bone/implant interface with regard to bone ingrowth and modeling. This type of information could be used for the refinement and improvement of future implant designs.

Materials and Methods

Implants

Both the canine and human femoral prostheses were manufactured from cast cobalt-chromium alloy, developed according to the ASTM Specification for Cast Cobalt-Chromium-

[1] Research director, Orthopaedic Research Laboratory, Montreal General Hospital Research Institute, McGill University, Montreal, Quebec, Canada H3G 1A4.

[2] Medical director, Anderson Orthopaedic Research Institute, Arlington, VA 22206.

[3] Professor of biomaterials, Faculty of Dentistry, University of Toronto, Toronto, Ontario, Canada M5G 1G6.

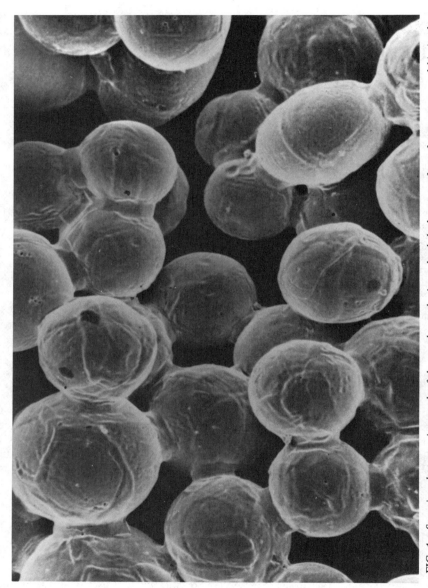

FIG. 1—Scanning electron micrograph of the powder-made sintered cobalt-chrome surface of a porous-coated hip implant (magnification, ×50). From Engh, C. A. and Bobyn, J.D., Biological Fixation in Total Hip Arthroplasty, Slack Inc., Thorofare, NJ, 1985; reprinted with permission.

Molybdenum Alloy for Surgical Implant Applications (F 75-82) by Depuy (Warsaw, Indiana). The implant porous surface was fabricated by metallurgical and sintering techniques (Fig. 1) [4,5]. Computerized image analysis of the porous surface, using methods described elsewhere, was utilized to characterize the porosity. The volume porosity was between 35 and 40% [3]. All of the canine implants and all but two of the human implants possessed an average pore size of 200 to 250 μm (a range is given for average pore size because it varies with the image analysis technique), with about 80% of the pores falling within the range of 45 to 410 μm [3]. Two of the human implants were of an older design possessing a smaller average pore size of 80 to 100 μm [3]. The canine prostheses had a tapered, collarless design, while the human prostheses were of a modified Moore design with a collar and a straight, cylindrical distal stem. Six of the canine prostheses were fully porous coated and the other six were coated on the proximal 40% of the stem. Eight of the human prostheses were fully porous coated and three were coated to within 5 cm of the stem tip. The fully functioning canine hip prostheses were retrieved at 9, 16, and 36 months (four implants per time period) after implantation into adult mongrel dogs [6]. Eleven human hip prostheses were retrieved at various intervals from four weeks to seven years after surgery from individuals ranging from 58 to 87 years of age [7]. The indications for surgery were osteoarthritis in five cases, neck fracture in four cases, rheumatoid arthritis in one case, and avascular necrosis in one case. In ten of these cases, the implant was retrieved intact within the femur. In the remaining case, the implant was electively removed at revision surgery seven years after implantation for a failed cemented acetabular component (and replaced with a larger stem that more closely fitted the intramedullary canal).

Histological Technique

Details of the histological technique have been previously described [3,8]. The specimen is fixed in buffered formalin, dehydrated in ascending solutions of ethanol, degreased and defatted in a 1:1 solution of ether and acetone, infiltrated with polymethyl methacrylate monomer under vacuum, and embedded in polymethyl methacrylate. Sectioning is accomplished with a low-speed, diamond blade cut-off machine. Transverse sections through the entire bone/implant specimen are initially cut about 2 mm thick and then glued to a standard glass histological slide for hand thinning on petrographic grinding wheels. The section is thinned to 50 ± 10 μm, stained with paragon, and viewed under transmitted light.

Results

Bone Ingrowth

With all twelve canine prostheses the histological sections showed the porous region of the implants to be primarily interfaced with cancellous bone. In general, the porous region was ingrown with bone around the majority of the implant circumference (Figs. 2, 3a, and 4c). The degree of the internal porosity of the implant surface that was filled with osseous tissue was not quantified but was estimated by visual inspection to average generally between about 30 and 50%. Isolated portions of the bone/implant interface demonstrated nearly complete filling of the porosity with bone. Eight of the human prostheses with the 200 to 250-μm pore size showed bone ingrowth in regions along the entire porous-coated portion of the implant (Fig. 5). One human implant possessing the 80 to 100-μm pore size was retrieved at revision surgery seven years after implantation and showed an isolated region of bone ingrowth proximally (Fig. 6). In general, in the human implants the bone ingrowth primarily occurred in regions where the implant porous surface was in proximity to or apposition with endosteal cortical bone. Cancellous bone ingrowth was sparser than in the

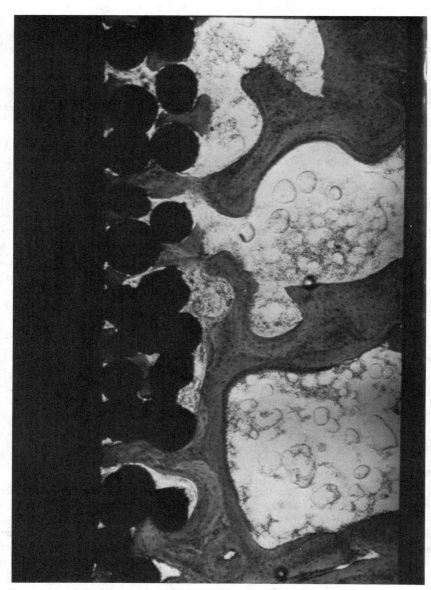

FIG. 2—Histological photomicrograph illustrating cancellous bone growth into a three-year canine implant.

canine implants, but the density of the cancellous bone in the human specimens was markedly lower, as would be expected with older individuals. In contrast to the canine implants, the porous surface was generally ingrown with bone around the minority of the implant circumference (Fig. 7). That is, there were often relatively large regions of no ingrowth at all. Like the canine implants, in regions of ingrowth, the degree of filling of the internal porosity by bone was estimated to average generally between about 30 and 50%, with some regions showing nearly complete filling (Fig. 5). One implant with the larger pore size (2.5-year implant period) showed fibrous tissue ingrowth only, possibly because there was a very poor fit of the stem within the intramedullary canal. Radiographically, this implant was entirely surrounded by a thin radiopaque line that was separated from the porous surface by a space of 0.5 to 1 mm in thickness [9]. The remaining human prosthesis possessed the smaller pore size and showed fibrous tissue ingrowth in the distal region, where a radiopaque line was visible, and bony apposition more proximally, where the bone/implant interface showed no distinguishing radiographic features (Figs. 8a through 8c). All the canine and human specimens but one were sectioned with the implant *in situ,* thus providing a clear indication of the spatial relationship between the cortex, the cancellous bone, the region of ingrowth, and the implant. In all sections, regions of the ingrown bone were continuous with the endosteal cortex, thus providing a clear impression of biological implant fixation.

The distal smooth surface of the six proximally coated canine implants was either partially or completely enveloped by a thin continuous shell of bone (Fig. 3b). An obvious layer of fibrous tissue was interposed between the implant and the shell of bone in four of these six implants. A similar feature was observed in two of the three proximally coated human prostheses (Fig. 9). All of these implants, both canine and human, showed bone ingrowth (and hence probable fixation) in the proximal porous-coated region.

In both the canine and the human there is a common tendency for bone to develop heterotopically around the implant and within the implant porosity, that is, in regions where there is absence of contact between the porous surface and the endosteal cortex or where there normally is no bone (Fig. 7). With implants that are proximally porous coated, there is also a common tendency for bone to develop and hypertrophy at the junction of the porous and smooth surfaces (Fig. 3a). In the human specimens, the consistent regions of bone ingrowth are the proximal-medial and distal-lateral, where the implant originally came in contact or nearly came in contact with the endosteum of the cortex. In the canine prostheses, the regions of bone ingrowth are more uniform around the circumference and along the length of the implant.

Bone Modeling

In the canine studies, the contralateral unoperated femurs were also available for histology, thus permitting a detailed assessment of adaptive bone modeling around the implant stems. Cortical thinning, attributed to stress shielding, was observed with both proximally and fully porous-coated implant types, although inconsistently (Figs. 4a through 4c). As detailed in a separate publication, the extent of cortical thinning varied with the postoperative time period and implant type and ranged from being essentially absent to as much as about 40% [6]. Generalized intracortical porosity in the presence or absence of cortical thinning was not observed in the canine. Hypertrophy of cancellous bone into a thin continuous shell surrounding the implant was common in the canine, as was exaggerated hypertrophy at the junction of the porous and smooth implant surfaces of proximally coated stems (Figs. 3a and 4c).

In the human specimens, similar comparative histology with a normal contralateral femur was only possible in three cases, in which a portion of the contralateral femur was obtained.

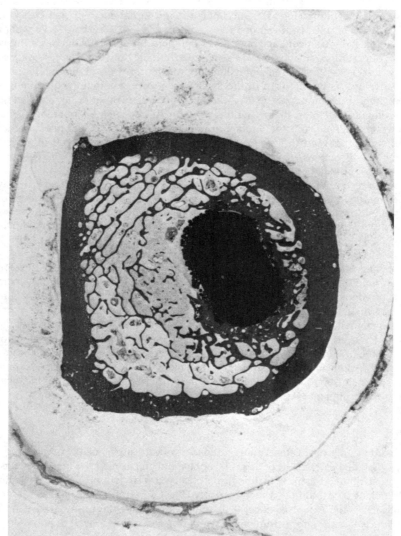

FIG. 3a—*Photograph of a transverse section taken at the junction of the porous and smooth implant surfaces of a proximally porous coated canine implant three years after implantation. There is bone ingrowth in most of the porous region. Note the hypertrophy of bone forming an inner "cortex" around the implant. From Bobyn, J. D., et al., "The Effect of Proximally and Fully Porous Coated Canine Hip Stem Design on Bone Modeling," Journal of Orthopaedic Research, 1987; reprinted with permission.*

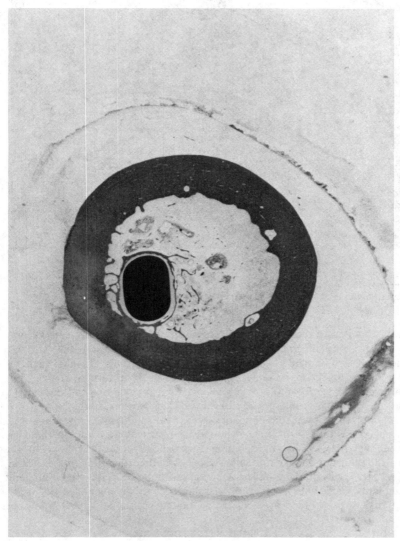

FIG. 3b—Photograph of a transverse section taken near the stem tip, where the implant is smooth surfaced, for a proximally porous coated canine implant three years after implantation. Note the shell of bone around the implant, with an interposed space. From Bobyn, J. D., et al., "The Effect of Proximally and Fully Porous Coated Canine Hip Stem Design on Bone Modeling," Journal of Orthopaedic Research, 1987; reprinted with permission.

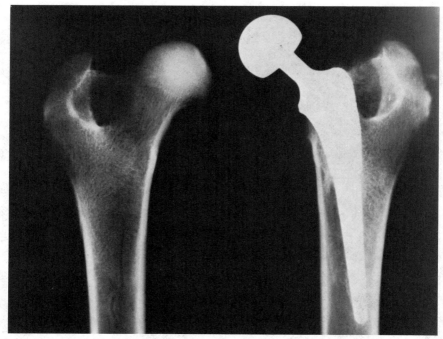

FIG. 4a—*Postmortem radiograph of the femurs of a dog 16 months after surgery. Thinning of the proximal medial cortical wall is noticeable in the femur with the proximally porous coated implant.* From Engh, C. A. and Bobyn, J. D., Biological Fixation in Total Hip Arthroplasty, *Slack Inc., Thorofare, NJ, 1985; reprinted with permission.*

In one of these three cases there was a noticeable difference in the intracortical porosity between the implant-containing and implant-free sections (Figs. 10c and 10d). Of the eight cases in which only the implant or implant-containing femur was obtained, generalized intracortical porosity (especially proximal) was observed in six to varying degrees, but the relative amount due to age (pathologic osteopenia) and that due to the implant (stress shielding causing osteopenia) could not be ascertained. Marked cortical thinning (20 to 40% loss of cortical bone), such as was sometimes observed in the canine studies, was not observed in the human specimens retrieved with a portion of the contralateral femur. Although some cortical thinning may have occurred in the human specimens retrieved without the contralateral femur, they did not appear to demonstrate marked cortical thinning either in histological study (Fig. 7) or in radiographic comparison with the contralateral femurs that did not contain implants. Cancellous hypertrophy was minimal, although increased cortical bone density and mass were observed at the junction of the porous and smooth surfaces of proximally coated stems.

Discussion

Because of differences in the implant design, implant position, implant loading, and type of bone surrounding the implant, direct comparisons between the canine and human studies are difficult. General, qualitative comparisons can be made, however. Histology has con-

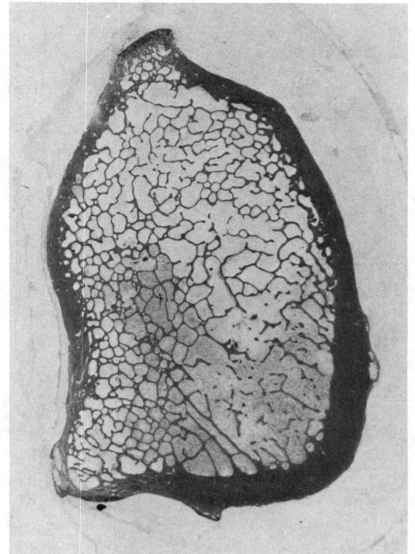

FIG. 4b—*Photograph of a proximal transverse section through the normal femur of the dog in Fig. 4a. From Engh, C. A. and Bobyn, J. D., Biological Fixation in Total Hip Arthroplasty, Slack Inc., Thorofare, NJ, 1985; reprinted with permission.*

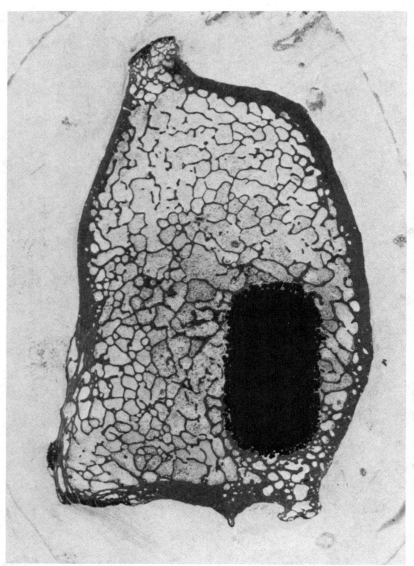

FIG. 4c—*Photograph of a proximal transverse section, at the same level as in Fig. 4a, in the implant-containing femur of the dog in Fig. 4a. Note the cancellous bone ingrowth and also the cortical thinning both medially (left field) and anteriorly (lower field). From Engh, C. A. and Bobyn, J. D., Biological Fixation in Total Hip Arthroplasty, Slack Inc., Thorofare, NJ, 1985; reprinted with permission.*

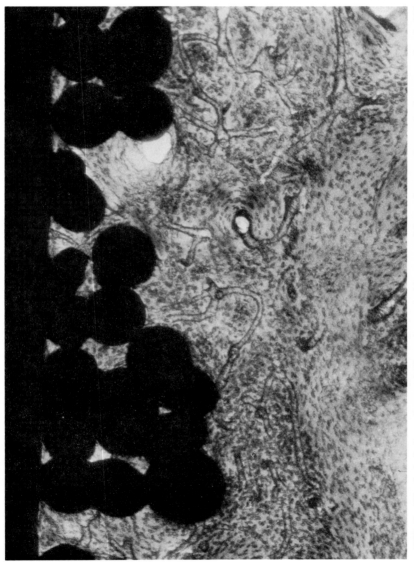

FIG. 5—*Histological photomicrograph illustrating cortical bone growth into a three-year human implant possessing the 200 to 250-μm average pore size. From Engh, C. A. and Bobyn, J. D., Biological Fixation in Total Hip Arthroplasty, Slack Inc., Thorofare, NJ, 1985; reprinted with permission.*

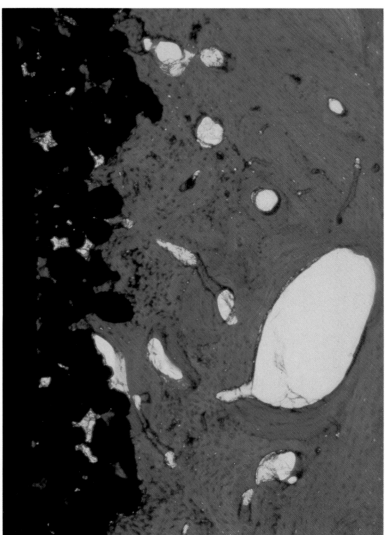

FIG. 6—*Histological photomicrograph illustrating bone growth into a seven-year human implant possessing the 80 to 100-μm average pore size. In this region of the implant, bone had penetrated down to the implant substrate. The majority of the bone/implant interface showed more superficial ingrowth or bone apposition to the outer contour of the porosity. From Engh, C. A., Bobyn, J. D., and Glassman, A. H., "Porous Coated Hip Replacement: A Study of Factors Governing Bone Ingrowth, Stress Shielding, and Clinical Results," Journal of Bone and Joint Surgery (London), Vol. 69-B, January 1987; reprinted with permission.*

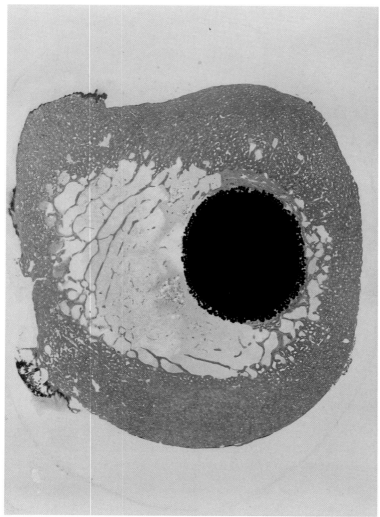

FIG. 7—*Photograph of a transverse section taken near the lesser trochanter of a human femur retrieved three years after surgery. The posterior aspect (upper field) shows no bone ingrowth. Note the new bone that has filled the space between the endosteum and the implant both medially (right field) and anteriorly (lower field). Although the opposite femur in this case was not available for comparison, the thickness and density of the cortex in this section suggest that marked cortical thinning had not occurred. From Engh, C. A. and Bobyn, J. D., Biological Fixation in Total Hip Arthroplasty, Slack Inc., Thorofare, NJ, 1985; reprinted with permission.*

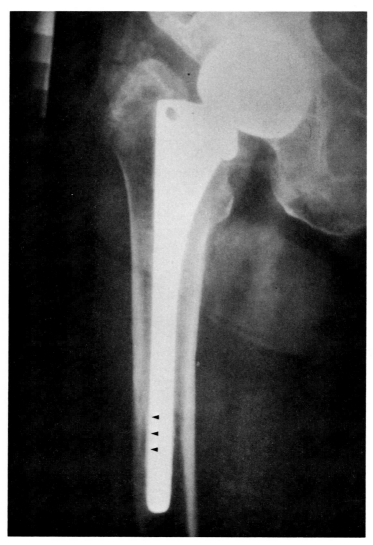

FIG. 8a—*Radiograph of a six-year fully coated human implant with the 80 to 100-μm average pore size. A faint radiopaque line is visible around the distal 2 in. of the stem* (see arrows). *The remainder of the bone/implant interface is unremarkable.*

firmed the suitability of the powder-made, cobalt-alloy, sintered porous surface with an average pore size of 200 to 250 μm for the fixation of load-bearing hip prostheses by bone ingrowth. It was also interesting to observe bone ingrowth in the seven-year retrieved human implant, which possessed an average pore size of 80 to 100 μm, since the design has been modified out of concern that the pores were too small for bone ingrowth in the human (Fig. 6). In regions of ingrowth near cortical bone, the degree of filling of the porosity is similar

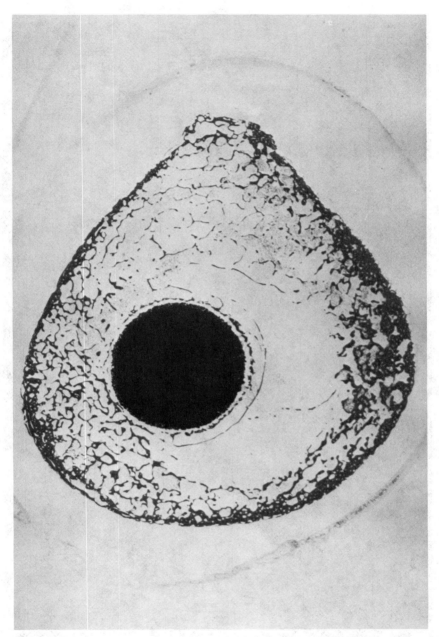

FIG. 8b—*Photograph of a histological section of the implant in Fig. 8a, prepared through the region of the radiopaque line. A thin shell of bone surrounds the implant. The marked osteopenia in the cortex was also present in the opposite femur, a portion of which was also retrieved at autopsy.*

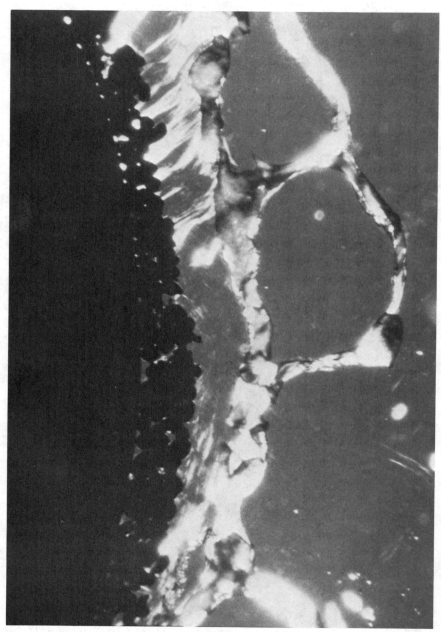

FIG. 8c—*A higher-power view of the interface in Fig. 8b under polarized light, showing the shell of bone separated from the implant with some interposed fibrous tissue.*

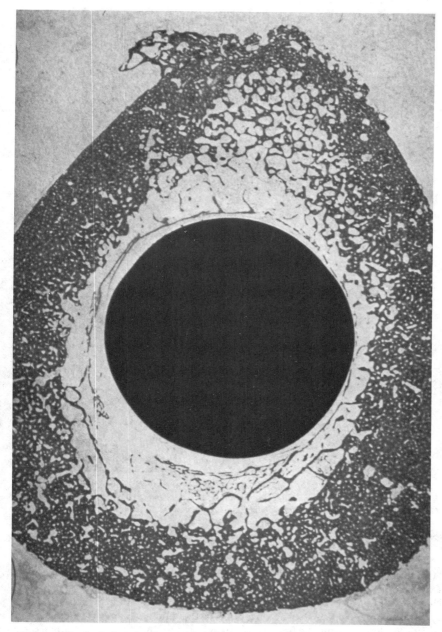

FIG. 9—*Histological photomicrograph of the bone/implant interface of the smooth portion of a proximally coated human prosthesis retrieved six months after surgery. A shell of bone surrounds the prosthesis with an interposed fibrous tissue layer. From Bobyn, J. D. and Engh, C. A., "Human Histology of the Bone–Porous Metal Implant Interface," Orthopaedics, Vol. 7, 1984, p. 1410; reprinted with permission.*

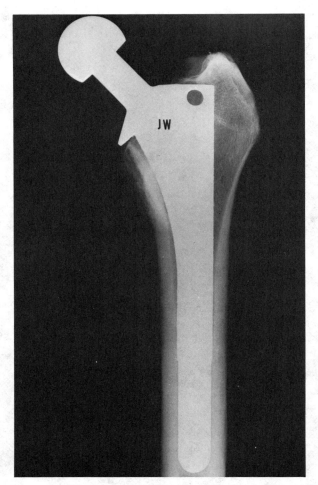

FIG. 10a—*Postmortem anteroposterior (AP) radiograph of an implant-containing human femur retrieved seven months after implantation. From Engh, C. A., Bobyn, J. D., and Glassman, A. H., "Porous Coated Hip Replacement: A Study of Factors Governing Bone Ingrowth, Stress Shielding, and Clinical Results,"* Journal of Bone and Joint Surgery (*London*), *Vol. 69-B, January 1987; reprinted with permission.*

in both the human and the canine but, overall, the degree of bone ingrowth is generally less extensive in the human. This may possibly be a function of the age of the individuals studied rather than a reflection of the absolute potential for human bone to repair and calcify within implant porosity. The dogs used in the animal study were between two and four years old, that is, within the first third of their life span. In contrast, the human specimens were retrieved from individuals in the last third of their life span. A greater osteogenic response can probably be expected in younger individuals, especially with regard to cancellous bone.

The radiographic appearance of the bone/implant interface, as illustrated in Fig. 10a, is characterized by bone ingrowth at multiple points along the stem when examined histolog-

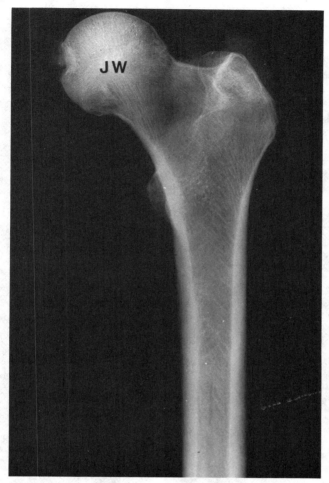

FIG. 10b—*Postmortem AP radiograph of the femur contralateral to that in Fig. 10a. No difference in cortical density can be perceived radiographically. From Engh, C. A., Bobyn, J. D., and Glassman, A. H., "Porous Coated Hip Replacement: A Study of Factors Governing Bone Ingrowth, Stress Shielding, and Clinical Results," Journal of Bone and Joint Surgery (London), Vol. 69-B, January 1987; reprinted with permission.*

ically. In general, it has been observed that radiographic indicators of bone ingrowth include evidence of new endosteal bone formation in contact with the stem, cortical or cancellous bone densification at the level where the porous coating ends, no evidence of subsidence, no radiopaque lines around the porous-coated portion of the stem, and some degree of proximal resorptive bone modeling [7,10]. If there are radiopaque lines around the porous-coated region of the stem, our experience has been that the porous surface is infiltrated with fibrous, not osseous, tissue [9].

In both the canine and the human, it has been observed that proximally coated stems can be well fixed in the porous region by bone ingrowth and be surrounded by a radiopaque line with an interposed fibrous tissue layer in the smooth stem region [6,8]. The radiopaque

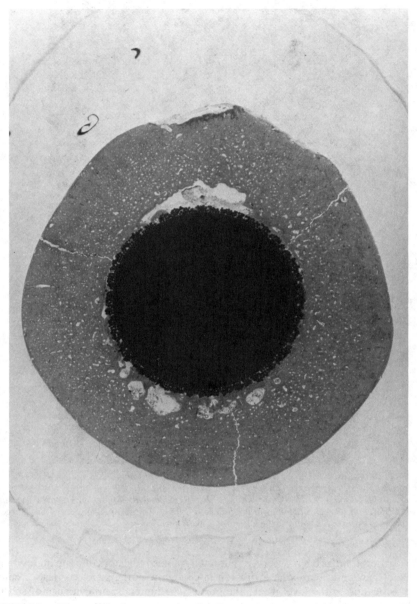

FIG. 10c—*Photograph of a section taken distally near the junction of the porous and smooth implant surfaces in the femur in Fig. 10a. Endosteal bone has grown into the implant porosity medially, anteriorly* (lower field), *and laterally. The cortex in the implant-containing section has a noticeably higher intracortical porosity than in Fig. 10d. From Engh, C. A., Bobyn, J. D., and Glassman, A. H., "Porous Coated Hip Replacement: A Study of Factors Governing Bone Ingrowth, Stress Shielding, and Clinical Results,"* Journal of Bone and Joint Surgery (London), Vol. 69-B, January 1987; reprinted with permission.

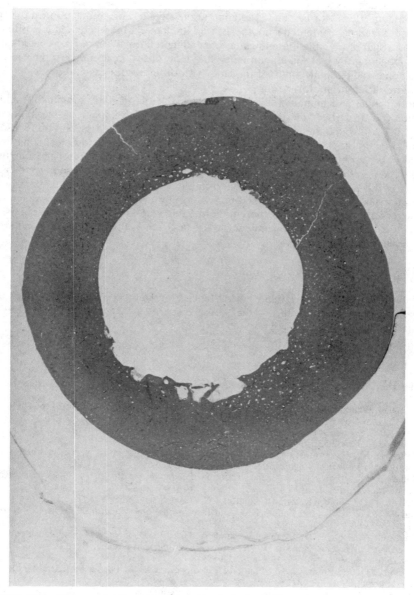

FIG. 10d—*Photograph of a section taken distally in the femur in Fig. 10b, at the same level as in Fig. 10c. From Engh, C. A., Bobyn, J. D., and Glassman, A. H., "Porous Coated Hip Replacement: A Study of Factors Governing Bone Ingrowth, Stress Shielding, and Clinical Results,"* Journal of Bone and Joint Surgery (*London*), *Vol. 69-B, January 1987; reprinted with permission.*

line distally is suggestive of relative motion between the femur and the stem, and the fact that the stem appears well fixed by bone ingrowth proximally suggests that the motion results from the femur flexing or bending in relation to the stem during ambulation. In the context of cemented and noncemented smooth-surfaced implants, a radiopaque line is often interpreted as a sign of possible loosening and impending failure. With porous-coated implants fixed by bone ingrowth at some point along the stem, this is apparently not the case, which indicates that radiographic study of implants with different designs and modes of fixation can require different interpretations.

In both canine and human studies, bone resorption (particularly proximally) and bone formation (particularly at the junction of porous and smooth implant surfaces) commonly occur [3,6,7,10]. Radiographs are generally poor indicators of the degree of bone resorption or formation. In other words, radiography is not sensitive enough to detect some features that are observed histologically. From a clinical perspective, this is most meaningful with regard to bone resorption, since one of the primary concerns with the use of porous-coated implants is the potential for serious bone atrophy due to stress shielding. The optimum method for quantifying implant-related bone resorption is through histological comparison with the contralateral femur. Further studies of this type are needed and are particularly important in view of the current tendency to use large stems that fill the canal and are more likely to cause stress shielding.

Acknowledgments

The authors are grateful for financial assistance from Depuy (Warsaw, IN), the Anderson Orthopaedic Research Institute, and the Natural Sciences and Engineering Research Council of Canada.

References

[1] Morscher, E., Ed., *The Cementless Fixation of Hip Endoprostheses*, Springer Verlag, New York, 1983.
[2] Bobyn, J. D. and Engh, C. A., "Biologic Fixation of Hip Prostheses: Review of the Clinical Status and Current Concepts," *Advances in Orthopaedic Surgery*, Vol. 7, 1983, p. 137.
[3] Engh, C. A. and Bobyn, J. D., "Biological Fixation in Total Hip Arthroplasty," Slack Inc., Thorofare, NJ, 1985, pp. 9, 191.
[4] Pilliar, R. M., Cameron, H. U., and Macnab, I., "Porous-Surfaced Layered Prosthetic Devices," *Journal of Biomedical Engineering*, Vol. 10, 1975, p. 126.
[5] Pilliar, R. M., "Powder Metal-Made Orthopaedic Implants with Porous Surface for Fixation by Tissue Ingrowth," *Clinical Orthopaedics*, Vol. 176, 1983, p. 42.
[6] Bobyn, J. D., Pilliar, R. M., Binnington, A. G., and Szivek, J. A., "The Effect of Proximally and Fully Porous Coated Canine Hip Stem Design on Bone Modeling," *Journal of Orthopaedic Research*, Vol. 5, 1987, p. 393.
[7] Engh, C. A., Bobyn, J. D., and Glassman, A. H., "Porous Coated Hip Replacement: A Study of Factors Governing Bone Ingrowth, Stress Shielding, and Clinical Results," *Journal of Bone and Joint Surgery* (London), Vol. 69-B, 1987, p. 45.
[8] Bobyn, J. D. and Engh, C. A., "Human Histology of the Bone–Porous Metal Implant Interface," *Orthopaedics*, Vol. 7, 1984, p. 1410.
[9] Pilliar, R. M., Cameron, H. U., Welsh, R. P., and Binnington, A. G., "Radiographic and Morphologic Studies of Load-Bearing Porous-Surfaced Structured Implants," *Clinical Orthopaedics*, Vol. 156, 1981, p. 249.
[10] Engh, C. A. and Bobyn, J. D., "Biological Fixation of a Modified Moore Prosthesis—II. Evaluation of Adaptive Femoral Bone Modeling," *Proceedings of the Hip Society*, 1984, p. 110.

Thomas A. Gruen[1]

Radiographic Criteria for the Clinical Performance of Uncemented Total Joint Replacements

REFERENCE: Gruen, T. A., "**Radiographic Criteria for the Clinical Performance of Uncemented Total Joint Replacements,**" *Quantitative Characterization and Performance of Porous Implants for Hard Tissue Applications, ASTM STP 953,* J. E. Lemons, Ed., American Society for Testing and Materials, Philadelphia, 1987, pp. 207–218.

ABSTRACT: Enthusiasm, concern, and controversy exist around the clinical application of porous materials for skeletal fixation of uncemented total joint replacements. The clinical and radiographic evaluations of such prosthetic components involve numerous measurements, which are necessary to communicate quantitative information on the patient's response, as well as the implant's performance. Just as communication requires a commonly accepted language for efficiency and effectiveness, clinical and radiographic analyses require commonly accepted criteria and terminology. The importance of radiography, the radiological factors influencing its interpretation, its evaluation methodology, and proposed criteria for sequential evaluation of the implant/bone interface and bone remodeling responses by using a zonal approach have been described to establish objectivity in the performance assessment of uncemented total joint replacements.

KEY WORDS: Porous implants, porous implant materials, total joint replacement, radiography, implant loosening, bone remodeling

Cementless skeletal fixation in total joint replacement is a topic of great interest to orthopedic surgeons and research scientists in biomaterials and biomechanics. The clinical experience in the United States has been relatively short, but there is enthusiasm, as well as concern, about the use of cementless joint prostheses. The enthusiasm comes from the extensive clinical experience in Europe and from the numerous experimental fixation studies in animals [1–3]. Nevertheless, concern and controversy persist regarding the clinical application of cementless fixation because of complex specific details of prosthesis design, multiple design changes, apparent bone stress-shielding remodeling phenomena associated with porous-coated implants, and potential local and systemic effects from ion release from the increased surface area of the porous matrix.

Numerous clinical investigations to assess the *in vivo* performance of uncemented total joint arthroplasty components are in progress. These investigations are supplemented with radiography, analysis of retrieved implants obtained from revision procedures, and postmortem studies involving ground hard plastic histology [3,4].

The continuing clinical assessment of this evolutionary process often includes radiographic examination, and it is therefore particularly important to understand what these studies can and cannot do; to identify radiographic manifestations and their interpretative significance; to emphasize awareness of the numerous radiological factors influencing the radiographic

---

[1] Director, Clinical Research, Institute for Bone and Joint Disorders, Phoenix, AZ 85012.

features; and, ultimately, to understand how these factors relate to the clinical performance. These data are critical to the application of uncemented prosthetic components since, at this date, there are no acceptable standardized criteria for radiographic assessment of component loosening and bone remodeling associated with cemented prostheses, which have been in clinical use since 1958 [5,6]. Recent communications have urged a new set of uniform and objective parameters in radiographic evaluation of patients with hip replacement prostheses [7,8].

The objective of this paper is to address the need for a common language for radiographic assessment of uncemented total joint replacements and to understand how such radiographic measurements can be useful or misleading.

Radiography

The purpose of radiography is to provide a record of maximum information from which performance criteria can be identified and established. To attain that purpose, every step in the radiological technique must be thoroughly understood and carefully carried out [10–12]. Many radiological factors are involved with specific exposure techniques, which vary not only from one radiology department to another, but from one unit to another within the same department, and also from one exposure to another using the same X-ray unit. A brief list of inherent radiological factors influencing the exposure quality within one X-ray unit is enumerated in Table 1.

TABLE 1—*Factors influencing the exposure quality within one X-ray unit.*

X-ray machine
  Electrical output (mA, kV, time)
  X-ray tube
  Radiation characteristics
  Cone/diaphragm sizes
  Filtration

X-ray film
  Film types
  Screen types
  Potter-Bucky diaphragms

Exposure technique
  Beam centering versus obliquity
  Tube-to-film distance
  Object-to-film distance
  Exposure setting (kV, mA, time)

X-ray film processing
  Fixer/developer
  Processing time and temperature
  Original versus duplicate

Patient-related factors
  Motion
  Positioning
    Supine versus standing
    Internal versus external rotation
    Flexion versus extension tilt
  Tissue absorption/penetration
    Thin, emaciated, demineralized bone
    "Normal"
    Large, muscular, obese, edema

With this multitude of radiological technique factors for a single exposure, a retrospective or prospective study using sequential X-ray film evaluation can result in controversial interpretation of radiographic findings. Nevertheless, the use of radiography is still indispensable in the *in vivo* assessment of all types (porous-coated, press-fit, and cemented) of prosthetic components with regard to loosening and bone remodeling.

For the skeletal fixation of total joint prostheses with radiopaque acrylic bone cement, the assessment of component performance was facilitated by using three types of primary radiographic evidence associated with loosening—cement-bone radiolucency, implant-cement radiolucency, and fractured cement. These features and various proposed mechanisms of failure for cemented prostheses have been recognized [9] and analyzed from retrospective rather than prospective studies since the initial concern about the use of acrylic cement was related to the material's setting characteristics, in particular, the local and systemic cardiovascular responses to the monomer methyl methacrylate and its subsequent polymerization.

The absence of the radiopaque grouting material in uncemented applications obviously changes the radiographic picture. The relatively homogeneously radiopaque acrylic bone cement is replaced by surrounding bone, consisting of cancellous and cortical bone of varying quality and quantity. The proximal metaphyseal part of the femur consists primarily of nonhomogeneous anisotropic cancellous bone surrounded by a very thin cortical shell, except for thick cortices at the femoral neck. The medial cortex, as observed on routine anteroposterior radiographs has often been referred to as the "calcar"; however, despite the term's appealing and apparently misleading use, this area should not be confused with the primary compression buttress of the femoral neck [13–15].

The true calcar femorale has been identified by anatomists as a spur of densified cancellous bone positioned deep in relation to the lesser trochanter but posterior to the neutral axis of the femoral neck. The best radiographic appearance of the calcar femorale is found when the X-ray film is taken at a right angle to the neutral axis of the femoral neck [13].

The calcar femorale appears to be a cortical extension from the femoral neck into the diaphyseal shaft region. Nearby there is a transition from the heterogeneous cancellous metaphysis, which decreases distally as the cortex of the femoral diaphyseal shaft predominates. This transition in bone stock varies considerably among femurs, and each bone type has unique mechanical and biological variations; for example, cortical bone is approximately ten times as dense as cancellous bone, whereas the trabeculae of cancellous bone have a higher metabolic turnover rate (approximately eight times that of cortical bone). Cancellous bone is highly responsive to metabolic and mechanical stimuli.

Qualitative assessment of bone quality from standard radiographs is primarily subjective, since this method is greatly affected by body habitus, radiographic exposure factors, the presence and type of overlying soft tissues, and patient positioning. Relatively large amounts of bone mineral must be lost (about 30%) before differences can be assessed from visual examination of radiographs [16]. Therefore, during the evaluation of sequential films, it was found to be very helpful to read the different films at the same time, in order to compare and appreciate the relative differences in bone densities between specific locations on one film and between the same location on subsequent films in order to distinguish real time-related osseous changes from density differences attributed to radiological technique. Density differences are often the result of varying radiological techniques.

Radiographic Criteria

The proposed terminology for radiographic criteria for assessing uncemented prostheses is derived from an ongoing prospective clinical and radiographic study of the porous-coated anatomical (PCA) total hip replacement. The radiographic study is being performed inde-

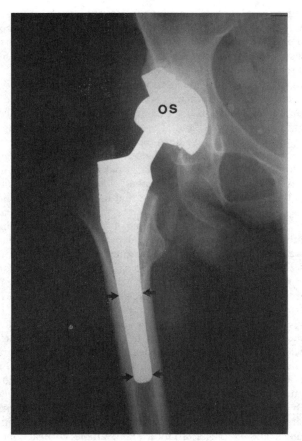

FIG. 1—*Radiograph of a porous-coated anatomical total hip joint replacement consisting of an acetabular socket component with a hemispherical porous surface in articulation with a stem-type femoral component with a proximal porous coating. The arrows delineate levels at which comparative structural dimensions are assessed.*

pendent of the clinical study—that is, it is a single-blind test—in order to avoid or reduce bias in the outcome of measurement.

First, a standard protocol of radiographic follow-up evaluation should include a preoperative X-ray film, an immediately postoperative film (prior to discharge from the hospital), and subsequent serial postoperative follow-up X-rays at three, six, and twelve months, Thereafter, annual follow-up evaluations are recommended. The standard anteroposterior pelvis-centered and hip-centered views should be supplemented with a reliably consistent lateral view. Precision positioning of patients for such serial evaluations remains a technical dilemma [17].

The prosthesis/bone interface on the anteroposterior film is divided into seven zones: two proximal zones for the porous surfaces on the medial and lateral sides, two equidistant zones each on the medial and lateral nonporous surfaces, and one zone at the stem tip (Fig. 1). Application of the zonal analysis scheme has helped to provide a means of achieving objective

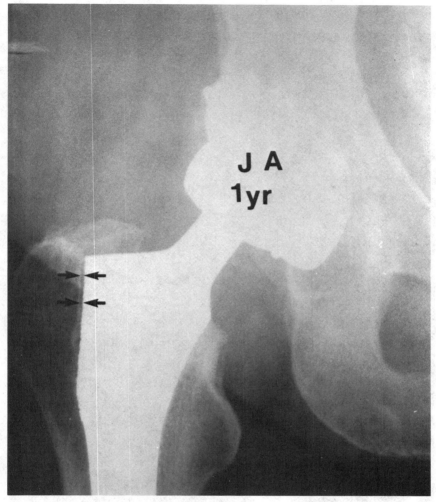

FIG. 2—*Radiograph of a porous-coated anatomical total hip prosthesis with evidence of a thin radiolucent line and a new sclerotic bone reaction (between arrows) at the proximal-lateral implant/bone interface.*

assessment of the prosthesis/bone interface as well as bone remodeling in critical regions [9].

For other designs of cementless and cemented femoral components, different zone designations may be useful, but the author suggests that the femoral component be divided into at least three pairs of zones between the proximal and distal regions and one at the distal tip region. In addition, the nature of the implant surface (whether porous or smooth) should be identified for each zone.

The minimum essential information from the earliest postoperative radiograph pertaining to the implantation technique should include the level of osteotomy from the top of the

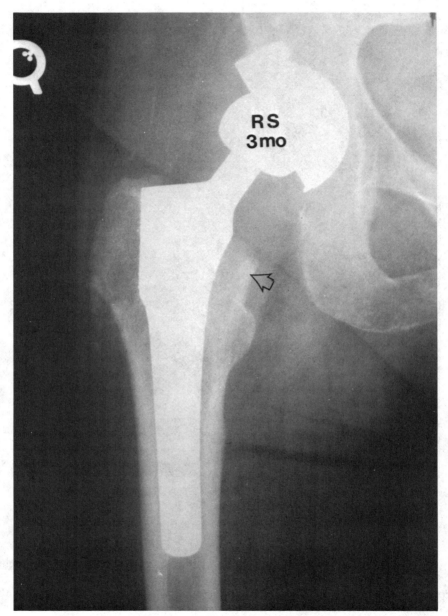

FIG. 3—*Three-month and one-year radiographs of a porous-coated femoral implant illustrating apparent corticocancellization of the proximal-medial femoral neck (open arrow) and apparent cancellous hypertrophic densification (solid arrow).*

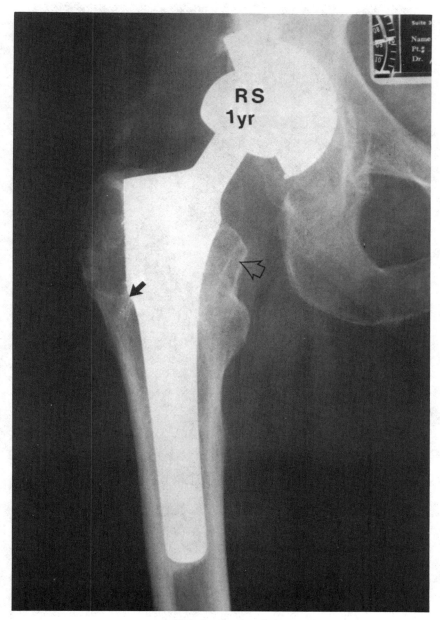

FIG. 3—*Continued*.

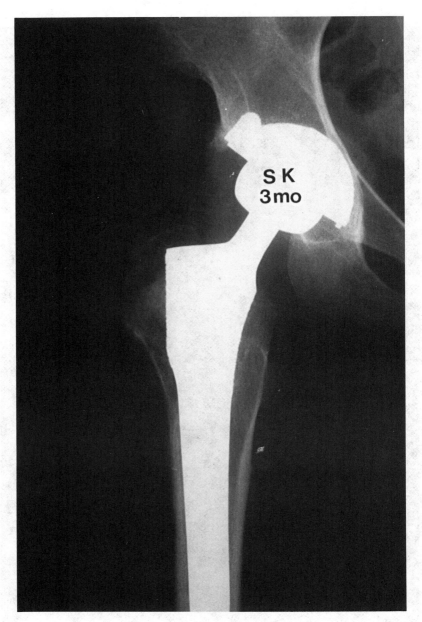

FIG. 4—*Three-month and one-year radiographs of a porous-coated femoral implant illustrating distal periosteal hypertrophy, shown in the femoral cortex adjacent to the stem tip of the component.*

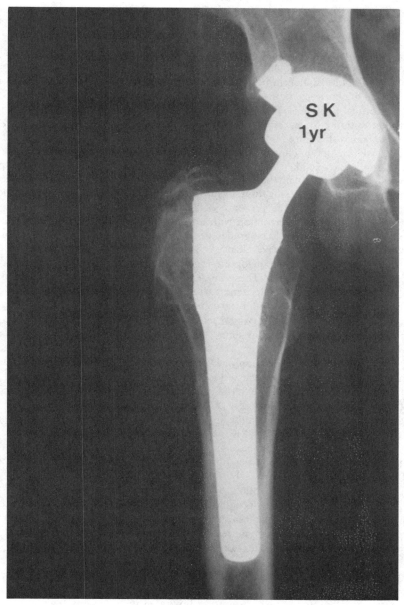

FIG. 4—*Continued.*

lesser trochanter, the implant's orientation, position, and relationship to the osteotomy level and mechanical axis (for example, the varus-valgus angle), and the comparative structural dimensions of the femoral component, intramedullary canal, and external cortical width and their corresponding ratios (that is, the percentage of the intramedullary canal filled by the implant and the relative cortical thickness) at several cross-sectional levels (Fig. 1). The percentage filled, or the stem/canal ratio, is an index for assessing the extent of stress shielding, which may be related to the implant's structural stiffness. The radiographic assessment of the cortical bone quality is based on the ratio of the outside femoral dimension to the outside femoral dimension minus the intramedullary canal dimension; that is, the ratio of cortical thickness to the total femoral width [18]. As the ratio increases, the greater the relative cortical bone mass becomes; femurs with low ratios may have innate structural weakness due to thin cortices.

Since bone ingrowth and remodeling is a two-phase dynamic process (consisting of initial healing and subsequent functional adaptive remodeling) the importance of consistent sequential radiographic examinations with efforts to maximize the patient's follow-up compliance is obvious. The implant/bone interface is evaluated within each zone and assessed according to the following classification: (1) normal state, (2) new bone directly in apposition to the implant's surface, (3) the presence of a radiolucent gap, and (4) new radiopaque lines, that is, a sclerotic bone reaction with an interposing radiolucent gap (Fig. 2). The widths of the radiolucency should be graded as: less than 1 mm, 1 mm, 2 mm, or more than 2 mm. On the basis of progressive changes in the width, extent, and location of the radiolucent gaps, femoral component loosening is then classified as subsidence (axial distal migration) or a varus-valgus shift. These modes of loosening are graded as incipient, progressive, or stabilized. Progressive loosening would be manifested by a component migration of 2 mm or more in the absence of any structural stabilization.

The structural integrity of the proximal femur, as seen on the radiograph, is examined and classified as showing (1) normal status, (2) medial neck round-off, (3) minimal (≤3-mm) medial neck osteolysis, (4) severe (>3-mm) medial neck osteolysis, (5) corticocancellization, or (6) cancellous densification or hypertrophy (Fig. 3). The radiographic status of the distal region at the stem tip is assessed for evidence of the following: (1) normal status, (2) periosteal hypertrophy, that is, external thickening of the cortex (Fig. 4), (3) endosteal consolidation, and (4) endosteal atrophy.

Medial neck round-off occurs when the angle of the surface of osteotomy cut in the medial femoral cortex begins to round off at its sharp margin. This is believed to represent maturation of the osteotomy with evidence of an "amputation phenomenon" [19]. This rounding off of the cut end of a long bone is a common finding, and it is believed that the rounding off is of no clinical significance.

Medial neck osteolysis, formerly called "calcar" resorption, is manifested by loss of medial neck cortex initiating at the cut surface and this often progresses distally toward the lesser trochanter. This phenomenon was often thought to represent stress shielding, but it is now being recognized as being an osteolytic response to component loosening, possibly related to the presence of acrylic bone cement or polyethylene wear debris [20].

Corticocancellization, a phenomenon not previously reported, occurs when the medial cortex of the femoral neck loses radiological density while the cancellous bone occupying the interval between the cortex and the medial surface of the prosthesis becomes denser. The appearance is therefore one of a uniform bone density extending from the periosteal surface to the surface of the porous coating of the prosthesis. This unique radiographic evidence is suggestive of functional adaptive remodeling.

Endosteal consolidation occurs when new endosteal bone forms within the intramedullary canal adjacent to the distal end of the prosthesis. In the case of noncemented prostheses there is no cement mantle to prevent this bone from forming adjacent to the prosthesis.

There is a wide variation in how this bone formed—in some cases it initiated on the medial cortex; in other cases it initiated on the lateral side. In some cases the bone was confined to one or the other side of the intramedullary canal, whereas in still other cases a solid bony bridge formed from the medial to the lateral cortex, fully bridging the medullary canal and appearing to be in close contact with the distal tip of the stem.

This proposed schema recommended for radiographic evaluation of uncemented total joint prosthetic components would provide realistic parameters, which can be assessed in a reliable and reproducible manner. The described terminology would not only represent a common language but also permit diagnosis of actual and potential complications, as well as assessment of the risk factors associated with the fixation mechanism, the implant, the host bone, and the patient. The *in vivo* interface and bone remodeling responses should be correlated with theoretical, experimental, and animal study data to provide effective explanation of the biological and mechanical phenomena associated with porous implant materials. This will then simplify the interpretation of radiographic data by independent observers, facilitate comparisons, and supplement clinical research, which in turn will serve the best interests of the orthopedic community and the patient.

People see only what they are prepared to see.

Ralph Waldo Emerson, 1863

References

[1] Galante, J., "Total Joint Arthroplasty Without Cement," *Clinical Orthopaedics and Related Research*, No. 176, New York, 1983, pp. 2–114.
[2] Morscher, E., *The Cementless Fixation of Hip Endoprostheses*, Springer-Verlag, New York, 1984.
[3] Gruen, T. A. and Sarmiento, A., *Journal of Biomedical Materials Research*, Vol. 18, No. 5, 1984, pp. 577–599.
[4] Engh, C. A. and Bobyn, J. D., *Biological Fixation in Total Hip Arthroplasty*, Slack, Thorofare, NJ, 1985.
[5] Stauffer, R. N., *Journal of Bone and Joint Surgery*, Vol. 64A, No. 7, September 1982, pp. 983–990.
[6] Harris, W. H., McCarthy, J. C., and O'Neill, D. A., *Journal of Bone and Joint Surgery*, Vol. 64A, No. 7, September 1982, pp. 1063–1067.
[7] Galante, J. O., Editorial, *Journal of Bone and Joint Surgery*, Vol. 67A, No. 4, April 1985, pp. 511–512.
[8] Hamilton, H. W., *Journal of Bone and Joint Surgery*, Vol. 68A, No. 4, April 1986, p. 635.
[9] Gruen, T. A., McNeice, G. M., and Amstutz, H. C., *Clinical Orthopaedics and Related Research*, No. 141, 1979, pp. 17–27.
[10] Thompson, T. T., *Cahoon's Formulating X-Ray Techniques*, 9th ed., Duke University Press, Durham, NC, 1979.
[11] Merrill, V., *Atlas of Roentgenographic Positions*, 3rd ed., C. V. Mosby, St. Louis, 1967, p. 13.
[12] Pennock, P. W. in *Bone in Clinical Orthopaedics*, G. Sumner-Smith, Ed., Saunders, Philadelphia, 1982, pp. 253–260.
[13] Griffin, J. B., *Clinical Orthopaedics and Related Research*, No. 164, 1982, pp. 211–214.
[14] Harty, M., *Journal of Bone and Joint Surgery*, Vol. 39A, 1957, pp. 625–630.
[15] Koch, J. C., *American Journal of Anatomy*, Vol. 21, 1917, p. 177.
[16] Meema, H. E., and Meema, S., *Journal of the American Geriatric Society*, Vol. 11, 1963, p. 1170.
[17] Kirkpatrick, J. S., Clarke, I. C., Amstutz, H. C., and Jinnah, R. H., *Clinical Orthopaedics and Related Research*, No. 174, 1983, pp. 158–163.
[18] Barnett, B. and Nordin, B. E. C., *Clinical Radiology*, Vol. 11, 1960, pp. 166–174.
[19] Bocco, F., Lanyan, P., and Charnley, J., *Clinical Orthopaedics and Related Research*, No. 128, 1977, pp. 287–296.
[20] Johnston, R. C. and Crowninshield, R. D., *Clinical Orthopaedics and Related Research*, No. 181, 1983, pp. 92–98.

DISCUSSION

I. Clarke[1] (*written discussion*)—Looking at your clinical X-rays, it occurred to me that one good thing about cement is that it is *easily discernible* in most X-rays. Stems with noncemented interfaces rely much more on X-ray techniques for consistent visualization. Based on your current work, what are the problems you've seen and what are your recommendations for others on the following problems:

    (*a*) stem rotation modifying/eliminating interface detail,
    (*b*) variations in X-ray density at different follow-ups, obfuscating trends, and
    (*c*) standardizing the anteroposterior X-rays?

T. A. Gruen (*author's closure*)—The absence of acrylic bone cement does indeed render interpretation of noncemented interfaces more difficult. There is no doubt that radiographic visualization of noncemented interfaces for detailed assessment could be improved with consistently precise radiological techniques. However, this requires cooperation between the radiologist and the technicians and the use of relatively simple patient positioning devices with appropriate compensatory adjustments in radiological exposures and processing techniques for each and every radiographic follow-up.

There is definitely a need for effective standardized radiological procedures and effective patient positioning devices to provide radiographs reliably for accurate measurements of interface radiolucencies or component migration or radiological density changes. There is often major variation within a large number of cases or in a multiple-center study. A major problem when assessing total hip replacements relates to the centering of the X-ray beam being offset by rotation of the femur or the pelvis. During radiographic review, the rotational effects can be easily assessed from reliable roentgenographic landmarks on the prosthetic component or the skeletal morphology, the latter being highly variable among hips being treated with arthroplasty procedures.

The best skeletal indicators for assessment of femoral rotation or flexion are the profiles of the greater and lesser trochanters. The obturator foramen profiles are similarly used for assessing pelvic rotational asymmetry.

Reproducible and reliable radiographic evaluations of sequential follow-up exposures are then subjected to another variable not previously mentioned, which is intraobserver and interobserver variability. Only recently, as far as I am aware, has the first published study appeared that indicated substantial interobserver variability in the interpretation of radiographic lucencies about *cemented* total hip replacements from a limited number of X-ray films[2].

Unlike clinical assessments, radiographs can be reviewed years later, and a consensus of basic terminology can be formed and more objective and accurate methods of assessment can be applied.

[1] Bioene Research Institute, Los Angeles, CA 90025.
[2] Brand, R. A., Yoder, S. A., and Pedersen, D. R., *Clinical Orthopaedics and Related Research*, Vol. 192, 1985, pp. 237–239.

John C. Keller[1] and Franklin A. Young, Jr.[1]

Histomorphometric Analysis of Bone Ingrowth into Porous-Coated Dental Implants

REFERENCE: Keller, J. C. and Young, F. A., Jr., **"Histomorphometric Analysis of Bone Ingrowth into Porous-Coated Dental Implants,"** *Quantitative Characterization and Performance of Porous Implants for Hard Tissue Applications, ASTM STP 953,* J. E. Lemons, Ed., American Society for Testing and Materials, Philadelphia, 1987, pp. 219–232.

ABSTRACT: Histomorphometric analyses were used to determine quantitatively the patterns of bony ingrowth that resulted from the placement of porous-surfaced dental implants. Twenty-one implants were placed in the mandibles of rhesus monkeys for up to 74 months by utilizing a two-stage approach. Quantitative histopathological evaluations were made using ground section microscopy. Bone remodeling and ingrowth from adjacent cortical bone and medullary trabeculae resulted in stable implant roots. Quantitative histomorphometric analyses revealed that approximately 60% of the available internal pores of the implants were occupied by bone. No fibrous connective tissue ingrowth or encapsulation was observed in the implant crypts.

KEY WORDS: porous implants, bone remodeling, bone ingrowth, quantitative histology, porous dental implants

Previous work has established that bone ingrowth into porous-surfaced implants is a reliable method of fixation for biomedical implants [1–6]. Important parameters that may alter the effectiveness of porous-surfaced implants have been identified [1–8], including (1) close apposition of the implant to bone within the implant site, which helps prevent mobility by allowing bone ingrowth to occur [9–11]; (2) the employment of a nonfunctional healing period to enhance the probability of bone ingrowth before the implant receives the functional stress of service life [9–12]; and (3) particular implant design parameters, such as the type and extent of porosity [2,13–17], the size of the pores and pore interconnections [2,5,8,13,16,18,19], and the alloys best suited for use as implants [4,5,20–23].

Descriptions in the literature of bone remodeling that occurs in response to the presence of porous dental implants have been largely qualitative in nature, involving histopathological descriptions of tissue changes as a function of time after implantation [5,9–11,24]. Recently, however, quantitative histological techniques have been employed to provide an improved description of the tissue response to porous dental and orthopedic implant materials. In earlier work, Klawitter et al. [25] reported that there was no direct correlation between the tissue structure at the bone/implant interface and the resultant mechanical properties of five porous Co-Cr implants after two years of load-bearing function. Spector et al. [23] reported significant bony ingrowth into porous polysulfone-coated Co-Cr implants by five months after implantation (30% of the total pore volume). Following bone remodeling at twelve months after implantation, the levels of bone ingrowth had reached 60% of the total

[1] Associate professor and professor, respectively, Department of Materials Science, Medical University of South Carolina, Charleston, SC 29425.

pore volume. Cook et al. [26] observed thickened cortical plates and greater area fractions of cancellous bone within the vents of low-temperature isotropic (LTI) carbon blade implants, in comparison with carbon-coated and uncoated alumina (Al_2O_3) blade implants. These results were apparently due to the effects of stress shielding, which accompanied the higher modulus Al_2O_3 materials. Similarly, in a study employing porous acetabular implants in dogs, Harris et al. [4] observed bone ingrowth at a level of 53 ± 19% at eight months after implantation.

In an attempt to understand better the characteristics of bone remodeling that occur during the service life of a porous dental implant, the objective of this project was to study the effects of bone remodeling, especially bone ingrowth, in two-stage porous-rooted titanium alloy implants using quantitative histomorphometric analyses.

Materials and Methods

Two-stage dental implants consisting of a porous-surfaced endosseous root and a smooth pergingival abutment were fabricated from Ti-6Al-4V alloy. The outer surface was made porous by sintering spherical beads of the alloy onto the central core. The implants were 4 mm in diameter and 7 mm in length. This process produced an interconnecting surface porosity with an average diameter of 125 μm and an overall pore volume of 34% [5,8,27].

Mandibular premolar and molar teeth were extracted in twelve adult rhesus monkeys to serve as sites for implantation. The extraction sites were allowed to heal for six to eight weeks. The porous tooth roots were then implanted in a two-stage approach. In the first stage, a total of 21 porous roots were implanted into the extraction sites. Each implant was placed so that the neck of the implant was level with the superior aspect of the cortical plate. A postimplantation healing period of eight weeks was then employed to allow initial bone ingrowth and gingival healing. A threaded cap was placed into the prethreaded root core to prevent the ingrowth of tissue in this region. In the second stage, a smooth-surfaced abutment was cemented into the prethreaded central core of the root. At this time the implant was in partial function, but not occlusion. Crowns or bridges were later cemented into place on a majority of the abutments to complete occlusion and to allow full functioning of the implants. Prosthetic treatment was not performed on some of the shorter term implants. Pertinent data regarding the number of implant sites, the length of implantation, and whether crowns or bridges were placed are presented in Table 1. The implants were in service for up to 74 months, at which time the animals were sacrificed and the mandibles were retrieved for histopathological and quantitative analyses of bone remodeling.

Quantitative histological analysis was performed utilizing ground section histological techniques. The ground sections were prepared using techniques modified from Anderson [28]. The previously fixed tissues were embedded in low-viscosity methyl methacrylate embedding media with the implant remaining *in situ*. Multiple sections, approximately 150 μm in thickness were cut in cross section through the mandible using a low-speed saw and a diamond wafering blade.[2] These thick sections were hand ground and polished to approximately 40 μm utilizing standard metallographic techniques. These sections were stained with either hematoxylin and eosin or alizarin red S and toluidine blue for the study of calcified tissues adjacent to and within the pores of the implant. Microradiographs of the ground sections were prepared to provide further qualitative evidence of bone ingrowth.

Histomorphometric analyses were performed on a ground section from the central portion of each cylindrical root in order to describe various parameters of bone ingrowth into the porous root quantitatively. The analyses were made using a photomicroscope equipped with

---

[2] Buehler low-speed saw, Buehler, Ltd., Lake Bluff, IL.

TABLE 1—*Percentage of tissue ingrowth into the surface and internal pores of porous-coated dental implants.*

Implant No.	(Animal No.)	Type of Capping[a]	Length of Implantation, years	Surface Pores, %			Internal Pores, %		
				Bone	Fibrous Connective Tissue	Bone-Marrow-Like Tissue	Bone	Fibrous Connective Tissue	Bone-Marrow-Like Tissue
1	(1)	...	0.25	89.6	5.9	4.9	50.2	28.4	19.8
2	(4)	...	0.8	60.8	0	39.1	71.6	0	28.3
3	(1)	...	1.0	58.8	19.7	18.2	54.2	18.6	22.4
4	(8)	...	1.0	53.3	7.9	38.7	67.5	19.6	12.8
5	(2)	...	1.2	58.7	13.6	23.3	57.5	11.9	30.4
6	(3)	...	1.5	68.4	22.4	6.8	37.9	47.2	14.4
7	(3)	...	1.5	65.1	0	34.8	44.6	0.4	56.9
8	(10)	...	1.8	82.5	11.6	5.8	76.4	12.2	11.3
9	(5)	C	2.0	63.5	0	36.8	48.1	0	51.8
10	(5)	C	2.0	79.2	0	20.7	70.1	0	29.8
11	(12)	C	2.0	74.2	8.5	17.2	77.9	9.5	10.3
12	(7)	B	3.6	43.1	7.7	49.1	44.2	12.1	43.6
13	(12)	B	3.7	65.9	6.7	31.6	71.6	2.8	25.1
14	(12)	B	3.8	83.9	6.5	9.4	81.2	4.0	14.7
15	(6)	C	3.8	70.9	2.7	26.3	73.3	1.6	24.9
16	(9)	C	4.9	83.1	2.2	14.2	79.8	6.5	13.6
17	(9)	C	4.9	84.6	2.8	12.5	76.2	2.8	20.8
18	(11)	B	5.0	81.0	9.8	9.0	75.8	8.2	15.9
19	(11)	B	5.0	82.8	3.6	13.5	83.9	1.6	15.4
20	(6)	C	5.5	62.1	17.2	20.6	57.9	17.9	17.6
21	(7)	B	6.2	67.8	13.8	18.2	66.7	18.5	14.7
Average ± standard deviation				70.4 ± 12.2	7.2 ± 6.6	21.4 ± 12.6	65.0 ± 13.9	10.6 ± 11.6	23.5 ± 13.0

[a] C = crown; B = bridge.

a drawing tube assembly and a graphics tablet and stylus linked to an Apple II + computer. In this operation, the investigator moves the stylus on the digitizer board while viewing the focused section through the microscope, in order to trace the histological features of interest (for example, bone, fibrous connective tissue, and so forth). Measurements are taken only on features of interest that are in the plane of focus. These traced areas are superimposed on the histological sections when the drawing tube is focused on the TV monitor. When the tracing procedures are completed, the computer calculates the area in appropriate units, allowing for magnification. The data are then stored on a disk for future use. The computer software necessary to perform the image analyses, including calculations of the measured areas and linear distances, was developed in our laboratory. The analyses proved to be reproducible within 5% error.

In order to determine whether the tissue response to the implant was related to the nature of the surrounding bone, in each analysis the implant site was divided into three specific regions: the base or apical portion of the implant and the two sides that initially had the maximum and minimum contact with the cortical bone (Fig. 1). This was done because it was not possible to implant the specimens uniformly in the same buccolingual orientation because of the variability in the oral anatomy of the animals. As a result, some implants had maximum contact with the lingual endosteal surface of the cortical plate, while others had maximum contact with the buccal endosteal surface of the cortical plate. Parameters that were investigated included the percentage of bone and fibrous connective tissue in the surface pores, as well as in the internal pores of the implants (Fig. 2). Employing these methods of evaluation, it was not possible to grade the inflammatory response to the implants adequately or to study the implant/tissue interface thoroughly over the entire implant surface.

Results

It was evident from the histological analyses of the ground sections and the corresponding microradiographs that extensive bone growth adjacent to and within the pores of the implant roots had occurred in all the implant sites. The microradiograph shown in Fig. 3 illustrates the bone ingrowth observed for an implant 6.2 months after implantation (Implant No. 21).

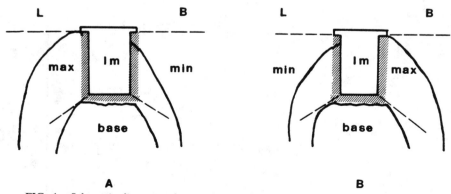

FIG. 1—*Schematic diagrams of a monkey mandible with a porous dental implant cut in cross section to illustrate the three areas used for quantitative analysis: the base and the sides with maximal (max) and minimal (min) contact with cortical bone. The side with maximal cortical bone was in either (a) the lingual direction (L) or (b) the buccal direction (B), depending on the original position of the implant and the natural bone architecture. IM = implant.*

FIG. 2—*Schematic diagram of a cross section of a porous dental implant illustrating the neck area. The diagram illustrates the available surface porosity (in dots at the right-hand side) and the available internal porosity (in diagonal line shading) used in the quantitative analysis of bone ingrowth.*

As can be seen in Table 1, substantial bone remodeling, including ingrowth into both the surface and internal pores was achieved using the implant design and placement protocols employed in this work. In most cases, bone ingrowth appeared to come in contact with the central core of the implant (Fig. 4). The surface pores that were not filled with bone were occupied by connective tissue of a different morphology. A somewhat organized dense fibrous tissue with a moderate inflammatory cell infiltrate was located primarily at the neck of the implant, superior to the cortical plate, as is shown in Fig. 5. This tissue probably originated from tissue underlying the adjacent oral epithelium, as well as from connective tissue regenerating after the surgical procedures. Less organized, looser connective tissue resembling bone marrow was located only within pores that were exposed to the central mandibular canal and devoid of bone ingrowth.

In order to determine whether the original buccolingual orientation affected the parameters of bone ingrowth, further quantitative analyses of three defined areas of the implant (the base and the sides with minimal and maximal cortical contact) were performed. The percentages of bone ingrowth into available internal pores in these areas are shown in Table 2 for all implant sites. Analysis of variance procedures ($\alpha = 0.95$) failed to reveal significant differences among the means between the regions of the implant.

Discussion

Qualitative histological evaluations and the quantitative analyses presented in Table 1 indicate that substantial bone remodeling, including bone ingrowth into the available surface

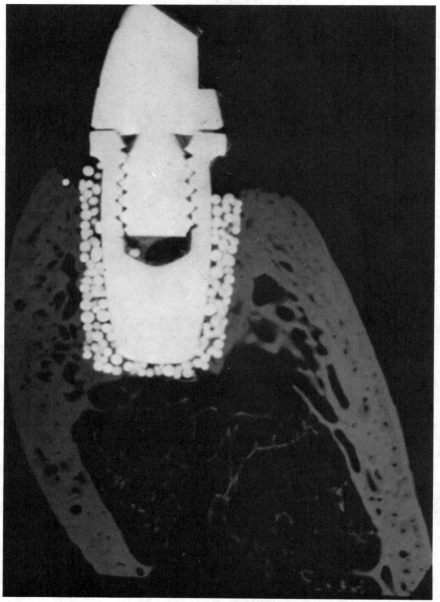

FIG. 3—*Microradiograph of a ground section demonstrating bone ingrowth into the pores of an implant from the cortical plates and adjacent trabeculae. The length of implantation was 6.2 years.*

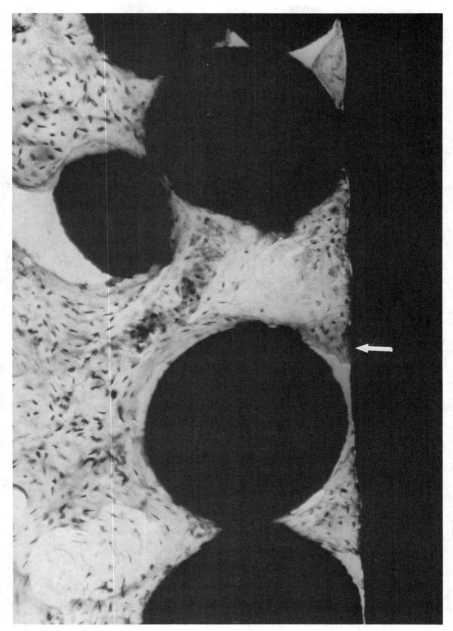

FIG. 4—Ground section demonstrating bone ingrowth into the central core of the implant (arrow).

FIG. 5—*Ground section demonstrating a somewhat organized fibrous connective tissue with moderate inflammatory cell infiltrate superior to the bone/implant interface.*

TABLE 2—*Percentage of bone ingrowth into internal pores of the implant as a function of location on the implant.*

Implant No.	Side with Maximum Cortical Contact, %	Base, %	Side with Minimum Cortical Contact, %
1	39.1	62.8	41.1
2	58.4	71.0	82.2
3	51.6	63.6	50.8
4	33.5	86.0	81.9
5	49.3	38.3	80.8
6	43.8	27.4	41.9
7	62.1	30.3	42.3
8	87.3	71.2	67.2
9	53.7	30.9	58.0
10	98.5	80.9	59.6
11	95.8	82.9	56.6
12	48.4	54.1	32.9
13	82.0	65.8	53.5
14	83.2	85.5	71.4
15	73.4	69.9	75.9
16	86.4	85.9	69.1
17	79.9	66.7	80.4
18	77.5	82.0	68.5
19	89.2	69.7	86.5
20	50.3	59.1	63.8
21	62.2	65.9	72.8
Average ± standard deviation	66.9 ± 19.7	64.2 ± 18.6	63.6 ± 15.6

and internal pores, occurred for all the implants utilized in this project. From Figs. 6*a* and 6*b* it is readily apparent that after placement of the implant root, bone ingrowth from the adjacent buccal and lingual cortical plates, as well as from the cancellous trabeculae within the mandibular canal, encircled the implant. The bone pattern surrounding the implants was closed along the length of most implants. Bone ingrowth into the pores apparently occurred either after or in concert with this remodeling. These qualitative and quantitative histological findings indicate that the phenomenon of stress shielding, which occurs in areas of hypophysiological stress was not present.

The slight loss of bone at the superior aspect of the implants (Fig. 7) occurred during the initial healing periods following tooth extraction and placement of the implant root level with the superior aspect of the cortical bone. Similar observations of bone loss during the healing period following tooth extraction have been reported in a series of articles by Simpson [29,30]. The bone loss observed in our work did not progress with time but was probably responsible, in part, for the presence of fibrous tissue and, in several cases, the exposure of implant beads in the superior aspects of several implants.

It was initially thought that the original buccolingual orientation may have an effect on the overall pattern of bony ingrowth. However, after the quantitative analysis of the three defined areas of interest, it was observed that the percentages of bone ingrowth into these areas were not significantly different on a statistical basis. This finding indicates that the original buccolingual orientation during implant placement is probably not a critical parameter for bone ingrowth into these implants.

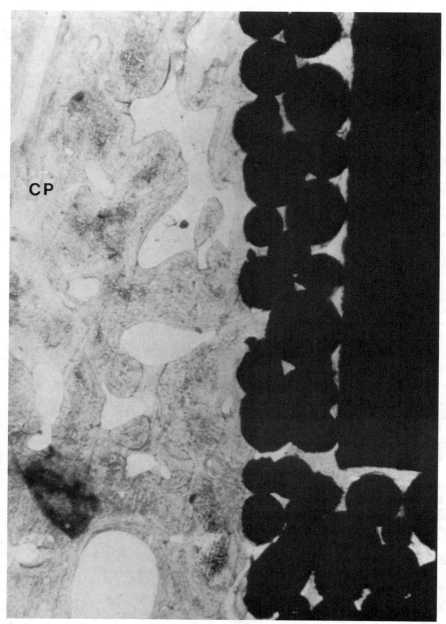

FIG. 6a—*Ground section demonstrating bone remodeling and ingrowth from the adjacent cortical plate (CP).*

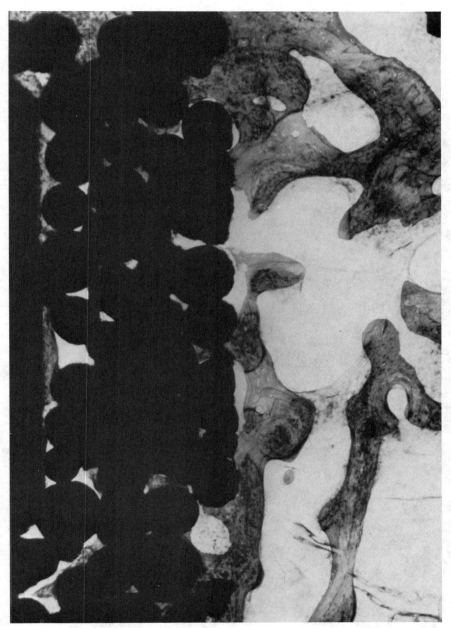

FIG. 6b—Ground section demonstrating bone remodeling and ingrowth from the underlying medullary trabeculae.

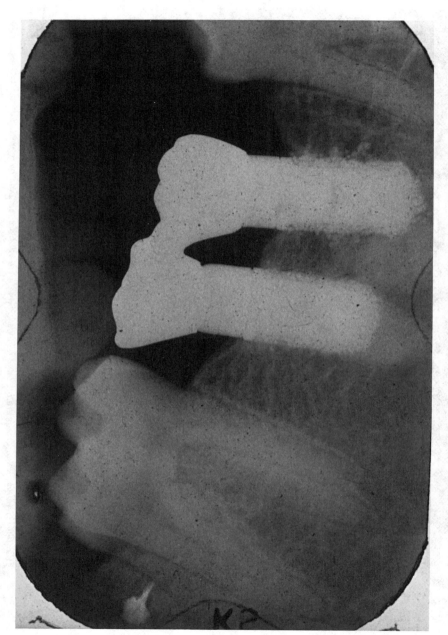

FIG. 7—Clinical X-ray of an implant site showing slight bone loss at the superior aspect of the implant. The bone loss failed to progress with time.

Conclusions

Based on the quantitative histomorphometric analyses of bone ingrowth presented here, it can be concluded that for periods up to 74 months, beneficial bone remodeling occurred in porous-coated implants. An average of 65% of the available internal pores was filled with bone, which provided stable mechanical retention. The absence of significant fibrous connective tissue at the implant/tissue interface within the implant crypt indicated that favorable stress conditions were present for advantageous bone remodeling.

Acknowledgments

The authors are grateful for the financial support of the National Institutes of Health (Grant No. DE03497) and for the technical contributions of N. Reese, C. Kerr, and B. Davis in support of their work.

References

[1] Weiss, M. and Rostoker, W., "Multiple Uses of a New Metallic Endosseous Implant," *Journal of Dental Research,* Vol. 59, Special Issue B, 1980, p. 918.
[2] Pilliar, R. M., "Powder-Metal-Made Orthopedic Implants with Porous Surface for Fixation by Tissue Ingrowth," *Clinical Orthopaedics and Related Research,* Vol. 176, 1983, pp. 42–51.
[3] Ronningen, H., Leveim, P., Galante, J., Rostoker, W., Turner, T., and Urban, R., "Total Surface Hip Arthroplasty in Dogs Using a Fiber Metal Composite as a Fixation Method," *Journal of Biomedical Materials Research,* Vol. 17, 1983, pp. 643–653.
[4] Harris, W. H., White, R. E., McCarthy, J. C., Walker, P. S., and Weinberg, E. H., "Bony Ingrowth Fixation of the Acetabular Component in Canine Hip Joint Arthroplasty," *Clinical Orthopaedics and Related Research,* Vol. 176, 1983, pp. 7–11.
[5] Young, F. A., Spector, M., and Kresch, C. H., "Porous Titanium Endosseous Dental Implants in Rhesus Monkeys: Microradiography and Histological Evaluation," *Journal of Biomedical Materials Research,* Vol. 13, 1979, pp. 843–856.
[6] Weiss, M. B., Gutman, D., and Rostoker, W., "A New Fiber Metal Dental Implant," *Compendium of Continuing Education,* Vol. 5, 1984, pp. 383–394.
[7] Bobyn, J. D., Pilliar, R. M., Cameron, H. U., and Weatherly, G. C., "Osteogenic Phenomena Across Endosteal Bone-Implant Spaces with Porous-Surfaced Intramedullary Implants," *Acta Orthopaedica Scandinavica,* Vol. 52, 1981, pp. 145–153.
[8] Young, F. A., Kresch, C. H., and Spector, M., "Porous Titanium Tooth Roots: Clinical Evaluation," *Journal of Prosthetic Dentistry,* Vol. 41, 1979, pp. 561–565.
[9] Albrektsson, T., Branemark, P. I., Hansson, H. A., and Lindstrom, J., "Osteointegrated Titanium Implants," *Acta Orthopaedica Scandinavica,* Vol. 52, 1981, pp. 155–170.
[10] Branemark, P. I., Adell, R., Albrektsson, T., Lekholm, U., Lundkuist, S., and Rockler, B., "Osteointegrated Titanium Fixtures in the Treatment of Edentulousness," *Biomaterials,* Vol. 4, 1983, pp. 25–28.
[11] Albrektsson, T., Branemark, P. I., Hansson, H. A., Kasemo, B., Larsson, K., Lundstrom, I., McQueen, D. H., and Skalah, R., "The Interface Zone of Inorganic Implants in Vivo: Titanium Implants in Bone," *Annals of Bioengineering,* Vol. 11, 1983, pp. 1–27.
[12] Heck, D. A., Nakajima, I., Chao, E. Y., and Kelly, P. J., "The Effects of Immobilization on Biological Ingrowth into Porous Titanium Fiber Metal Prostheses," *Transactions of the Orthopaedic Research Society,* Vol. 9, 1984, p. 178.
[13] Ducheyne, P., DeMeester, P., Aernoudt, E., Martens, M., and Mulier, J. C., "Influence of a Functional Dynamic Loading on Bone Ingrowth into Surface Pores of Orthopedic Implants," *Journal of Biomedical Materials Research,* Vol. 11, 1977, 811–838.
[14] Clemow, A. J. T., Weinstein, A. M., Klawitter, J. J., Koeneman, J., and Anderson, J., "Interface Mechanics of Porous Titanium Implants," *Journal of Biomedical Materials Research,* Vol. 15, 1981, pp. 73–82.
[15] Walsh, K. A., Cook, S. D., and Haddad, R. J., "Evaluation of Parameters Affecting Porous Alloy Implant Retention," *Transactions of the Orthopaedic Research Society,* Vol. 9, 1984, p. 151.

[16] Hulbert, S. F., Cooke, F. W., Klawitter, J. J., Leonard, R. B., Sauer, B. W., and Moyle, D. D., "Attachment of Prostheses to the Musculoskeletal System by Tissue Ingrowth and Mechanical Interlocking," *Journal of Biomedical Materials Research Symposium*, Vol. 4, 1983, pp. 1–23.

[17] Klawitter, J. J. and Hulbert, S. F., "Application of Porous Ceramics for the Attachment of Load Bearing Internal Orthopedic Applications," *Journal of Biomedical Materials Research Symposia*, Vol. 2, 1971, pp. 161–229.

[18] Walsh, K. A., Cook, S. D., Skinner, H. B., and Weinstein, A. W., "Biomechanical and Histological Evaluation of Bone–Porous CoCrMo Alloy Interfaces," *Transactions of the Society for Biomaterials*, Vol. 6, 1983, p. 76.

[19] Bobyn, J. D., Pilliar, R. M., Cameron, H. U., and Weatherly, G. C., "The Optimum Pore Size for the Fixation of Porous Surfaced Metal Implants by the Ingrowth of Bone," *Clinical Orthopaedics and Related Research*, Vol. 150, 1980, pp. 263–270.

[20] Bobyn, J. D., Pilliar, R. M., Cameron, H. U., Weatherly, G. C., and Kent, G. M., "The Effect of Porous Surface Configuration on the Tensile Strength of Fixation of Implants by Bone Ingrowth," *Clinical Orthopaedics and Related Research*, Vol. 149, June 1980, pp. 291–298.

[21] Cameron, H. U., Pilliar, R. M., and Macnab, I., "Porous Vitallium in Implant Surgery," *Journal of Biomedical Materials Research*, Vol. 8, 1974, pp. 283–289.

[22] Spector, M., Michno, M. J., Smarook, W. H., and Kwiatowski, G. T., "A High-Modulus Polymer for Porous Orthopedic Implants: Biomechanical Compatibility of Porous Implants," *Journal of Biomedical Materials Research*, Vol. 12, 1988, pp. 665–677.

[23] Spector, M., Davis, R. J., Lunceford, E. M., and Harmon, S. L., "Porous Polysulfone Coatings for Fixation of Femoral Stems by Bony Ingrowth," *Clinical Orthopaedics and Related Research*, Vol. 176, 1983, pp. 34–41.

[24] Karagianes, M., Westerman, R., Hamilton, A., Adams, H., and Wilks, R. C., "Investigation of Long Term Performance of Porous Metal Dental Implants in Non-Human Primates," *Journal of Oral Implantology*, Vol. 10, 1982, pp. 189–207.

[25] Klawitter, J. J., Sander, T., Weinstein, A. M., and Peterson, L., "Dental Implant Retention Mechanics," *Transactions of the Society for Biomaterials*, Vol. 2, 1978, p. 48.

[26] Cook, S. D., Weinstein, A. W., Klawitter, J. J., and Kent, J. N., "Quantitative Histological Evaluation of LTI Carbon, Carbon Coated Aluminum Oxide, and Uncoated Aluminum Oxide Dental Implant," *Journal of Biomedical Materials Research*, Vol. 17, 1983, pp. 319–538.

[27] Spector, M., Young, F. A., Marcinak, C. F., and Kresch, C. H., "Porous Coatings for Artificial Tooth Roots," *Proceedings*, International Congress of Implantology and Biomaterials in Stomatology, Kyoto, Japan, 1980, pp. 180–187.

[28] Anderson, C., *Manual for the Examination of Bone*, CRC Press, Boca Raton, FL, 1982.

[29] Simpson, H. E., "Experimental Investigation into the Healing of Extraction Wounds in Macacus Rhesus Monkeys," *Journal of Oral Surgery*, Vol. 18, 1960, pp. 391–399.

[30] Simpson, H. E., "Effects of Suturing Extraction Wounds in Macacus Rhesus Monkeys," *Journal of Oral Surgery*, Vol. 18, 1960, pp. 461–464.

DISCUSSION

U. Gross[1] (*written discussion*)—Did you detect particles of titanium in macrophages of the connective tissue?

J. C. Keller and F. A. Young (*authors' closure*)—No; however, there were several implant sites in which beads comprising the coating had loosened from the implant surface and become surrounded by osseous tissue. There was no evidence of impaired bone response in any of these implant sites.

---

[1] Institute of Pathology, Klinikum Steglitz, Free University of Berlin, Berlin, West Germany.

Franklin A. Young[1] and G. Marcos Montes[2]

Clinical Indicators of Dental Implant Performance

REFERENCE: Young, F. A. and Montes, G. M., **"Clinical Indicators of Dental Implant Performance,"** *Quantitative Characterization and Performance of Porous Implants for Hard Tissue Applications, ASTM STP 953*, J. E. Lemons, Ed., American Society for Testing and Materials, Philadelphia, 1987, pp. 233–238.

ABSTRACT: The traditional clinical indicators of implant performance are mobility, pocket depth, keratinized tissue width, amount of bleeding, amount of plaque accumulation, and gingival appearance.

This report examines measurements of these variables at intervals of 1 to 2 months for 36 months for 27 implants in 11 rhesus monkeys. The implants were cylindrical porous-rooted titanium alloy and were placed in premolar and molar mandibular sites using a two-stage technique. The implants were used to support both single-unit and multiple-unit restorations. Examination of traditional plots of indicator values versus time revealed no readily observable trends either for individual implants or for pooled implant data. The assumption was made that each observation may be expressed in terms of the additive relationship between the mean response of the total group of responses for an individual, the effect of time, and the error associated with the measurements. The data were analyzed using a multivariate analysis of variance with a repeated measures design (MANOVA, SPSS Inc., Chicago, Illinois). The analysis provided a control on differences between individuals because the effect of time was measured relative to the average response of the individual.

The conclusions of this study are (1) that the technique applied can be useful in determining trends in the measured indicators and (2) that a difference can be observed between implants supporting single-unit restorations and those supporting multiple-unit restorations.

KEY WORDS: dental implants, titanium implants, porous implants, clinical implant performance indicators

Many of the new implant designs and materials are tested in laboratory animals before they are marketed for humans. It is generally accepted that all designs should be extensively tested in animals before they are employed in humans [1].

The clinical data presented in this paper were collected in the course of a research project designed to test whether bone ingrowth into porous-surfaced artificial tooth roots could be employed to stabilize them. The implants were placed in the mandibles of rhesus monkeys. Details of the design of the implants and the experimental procedure are presented elsewhere [2]. The implants were employed to support either single-unit or multiple-unit restorations.

The usual variables measured to evaluate implant performance are those that have evolved into the best predictors of the health of natural teeth. It has been assumed by investigators that because these measurements will indicate latent problems in teeth, they will also predict

[1] Professor and chairman, Department of Materials Science, Medical University of South Carolina, Charleston, SC 29425.

[2] Research associate, College of Dental Medicine, Medical University of South Carolina, Charleston, SC 29425.

declining behavior of implants. These variables are mobility, periodontal pocket depth, keratinized tissue width, plaque index, bleeding index, and gingival condition.

Mobility

The mobility of the implants was evaluated using the method of Miller [3], in which movement of the crown under finger and instrument handle pressure is evaluated in millimetres.

Periodontal Pocket Depth

Depth measurements in millimetres were made at six positions around each implant using a standard periodontal probe. The averages of three measurements each for both the buccal and lingual sides of the implant were taken as separate responses in the data analysis, which will be discussed further on in this paper.

Keratinized Tissue Width

Using a periodontal probe, the amount of keratinized tissue was measured to the nearest millimetre from the periphery of the implant abutment in both the buccal and lingual directions.

Plaque Index

The amount of plaque present on the implant abutment was evaluated using the scale of Silness and Loe [4], according to the following scheme:

0 = No plaque.
1 = A film of plaque adhering to the free gingival margin and adjacent area of the tooth. The plaque may be observed *in situ* only after application of a disclosing solution or by using a probe on the tooth surface.
2 = Moderate accumulation of soft deposits within the gingival pocket, or on the tooth and gingival margin, that can be seen with the naked eye.
3 = An abundance of soft matter within the gingival pocket, on the tooth and gingival margin, or in all these areas.

Bleeding Index

The amount of bleeding that occurred from periodontal probing was evaluated using a modification of the method of Loe [5]. The following scale was employed:

0 = No discernible bleeding upon gentle probing.
1 = Light bleeding upon probing.
2 = Heavy bleeding upon probing.
3 = Spontaneous bleeding.

Gingival Condition

The overall subjective clinical impression of the health of the gingiva surrounding the implant was estimated by the evaluator using a three-unit scale. The categories used were good, fair, and poor.

Multivariate Analysis

The raw data in this study were the measured indicators evaluated monthly over a 3-year period by a dentist with specialty training in periodontics. When the data were plotted against time no readily discernible trends were observed. A statistical analysis was employed to determine the effects of time and of implant loading on the variable values for both individual and grouped values. The raw data were coded using the time variable according to the following scheme: values from 0 to 6.5 months were assigned to the 6-months period; values taken after 6.5 months and up to 12.5 months were assigned to 12 months; values after 12.5 and up to 18.5 months were assigned to 18 months, and so forth. The response of an implant over a 6-month period was taken from the average (two to six measurements) of the data collected over this period. A total of 26 implants were included, of which 10 were supporting single-unit restorations (crowns) and 16 were supporting multiple-unit restorations (bridges) in a group of 11 rhesus monkeys.

In this study most of the animals contained from 2 to 4 implants with the implicit assumption that the principal changes in implant performance were dependent on the type of restoration supported by the implant and on the length of time of function. All of the implants investigated were considered clinically successful and the animals were all adult monkeys of equivalent health and activity. This assumption was also justified by the uncorrelated responses with respect to time of the various units contained by the same animal. The indicators were analyzed in two sets of variables with commensurable units, one set containing the buccal and lingual pocket depths and keratinized tissue widths and the other containing the gingival, bleeding, and plaque indices.

Assuming that the observed responses have normal distributions, the design of a multivariate analysis of variance with data from a repeated-measurements experiment[3] allows the testing of the null hypotheses for the coefficients of additive functions such as polynomials against the alternative hypotheses of nonzero coefficients [6]. The linear model for the mean of response X at time t of the jth group (type) is given by

$$X_{tj} = a_{1j} + a_{2j} (t - t_0) + e_{tj}$$

where a_{1j} is the mean response at time t_0, a_{2j} is the estimate of the linear regression parameter, and e_{tj} is a random disturbance term. The parameters of the similar quadratic expression were not estimated; only the acceptance or rejection of this model was determined. Differences between implants supporting crowns and those supporting bridges were also determined within the levels of time.

Results

No implants exhibited measurable mobility by the method employed; this variable was not considered in the statistical analysis. The evaluated means of the indicators at 6-month intervals are given in Tables 1 and 2, showing the limits of the 95% confidence interval as deviations from the mean. A significant difference in implant performance between crowns and bridges was found in the pocket depth measured in the lingual side (Hotelling's multivariate test, 0.05 significance level) evaluated after 6 months. A difference between types in the keratinized tissue width measured in the lingual side (0.05 significance level) was only observed during the initial 12 months of observations. The pocket depth and the keratinized tissue width measured in the buccal side showed no significant effect, and the measured indices were also unaffected by the type of restoration supported by the implant.

[3] MANOVA, SPSS Inc., Chicago, IL.

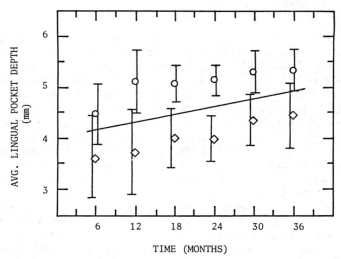

FIG. 1—*Average value of the periodontal pocket depth on the lingual side of implants supporting crowns (◇) or bridges (○) as a function of time. The vertical bars show the 95% confidence intervals.*

The results shown in Fig. 1 for the linear effect of time on the pocket depth on the lingual side (0.05 significance level) illustrate the rejection of no-interaction effects for the type of restoration with time (parallel trends), which had been indicated by the application of the general model. The slope of the estimated continuous line shown was 2.59×10^{-2} mm/month, with a mean value of 4.20 mm at 6 months. A univariate evaluation of the interaction effect of time on the keratinized tissue width at the lingual side (Fig. 2) gave a significant effect (0.05 level) for the linear effect of time on the different implant sites, in which the

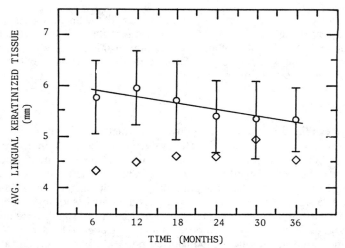

FIG. 2—*Average value of the keratinized tissue width in the lingual direction for implants supporting crowns (◇) or bridges (○) as a function of time. The vertical bars show the 95% confidence intervals.*

TABLE 1—*Means of measured indicators with 95% confidence intervals for single unit (crown) implants, expressed as deviations given by the multivariate analysis.*

Time, Months	Pocket Depth		Keratinized Tissue Width	
	Buccal, mm	Lingual, mm	Buccal, mm	Lingual, mm
6	4.02 ± 0.80 (1.12)[a]	3.68 ± .81 (1.12)	3.54 ± 0.62 (0.87)	4.31 ± 0.85 (1.19)
12	4.09 ± 0.87 (1.21)	3.74 ± .86 (1.20)	3.41 ± 0.67 (0.94)	4.51 ± 1.14 (1.60)
18	4.38 ± 0.45 (0.63)	4.03 ± .58 (0.81)	3.26 ± 0.68 (0.95)	4.67 ± 1.19 (1.66)
24	4.48 ± 0.45 (0.63)	4.03 ± .46 (0.65)	3.36 ± 0.96 (1.34)	4.66 ± 1.17 (1.65)
30	4.34 ± 0.67 (0.94)	4.39 ± .50 (0.70)	3.37 ± 0.84 (1.18)	4.96 ± 1.24 (1.74)
36	3.95 ± 0.87 (1.22)	4.50 ± .64 (0.90)	3.13 ± 0.82 (1.14)	4.58 ± 1.12 (1.56)

[a] The parentheses indicate standard deviations.

TABLE 2—*Means of measured indicators with 95% confidence intervals for multiple unit (bridges) implants, expressed as deviations given by the multivariate analysis.*

Time, Months	Pocket Depth		Keratinized Tissue Width	
	Buccal, mm	Lingual, mm	Buccal, mm	Lingual, mm
6	4.18 ± 0.37 (0.70)[a]	4.52 ± 0.58 (1.10)	3.45 ± 0.45 (0.83)	5.78 ± 0.71 (1.34)
12	4.13 ± 0.45 (0.84)	5.15 ± 0.62 (1.16)	3.35 ± 0.52 (0.98)	5.97 ± 0.72 (1.36)
18	4.16 ± 0.38 (0.71)	5.11 ± 0.36 (0.67)	3.21 ± 0.55 (1.04)	5.72 ± 0.77 (1.45)
24	4.21 ± 0.46 (0.86)	5.17 ± 0.30 (0.57)	3.16 ± 0.52 (0.98)	5.41 ± 0.70 (1.31)
30	4.29 ± 0.59 (1.11)	5.33 ± 1.19 (0.83)	3.35 ± 0.63 (1.19)	5.36 ± 0.76 (1.44)
36	4.43 ± 0.51 (0.95)	5.38 ± 0.41 (0.76)	3.48 ± 0.63 (1.18)	5.35 ± 0.63 (1.18)

[a] The parentheses indicate standard deviations.

bridge type of implant had a slope of -2.03×10^{-2} mm/month with a mean of 5.60 mm at 6 months, and the crown type was constant with an average of 4.61 mm. Examination of the effect of time on the bridge type of unit showed a significant quadratic response on the keratinized tissue width on the buccal side (0.05 significance level), with an initial decrease of about 0.5 mm during the initial 18 months, followed by an increase of equal size at 36 months. The linear effects of time were also significant on the bleeding and plaque indices determined on the entire sample, showing increases from 1.0 to 1.5 in bleeding and from 1.5 to 2.5 in plaque accumulation over the period of 36 months. The gingival condition was not affected by time or by the type of implant and was evaluated on the average as showing fair condition.

Discussion

Implant mobility has been considered to be the primary indicator of potential clinical problems [7,8]. With implants stabilized by bone ingrowth or other immobilization systems, this indicator is of little value because a mobile implant is one which has not achieved initial ingrowth or stabilization. Other clinical indicators of implant health have traditionally been observed as a function of time and as obvious trends used to highlight problems. For the implants cited in this study, these trends were not sufficient to provide precise lifetime predictions without further analysis. Using the analysis given in this paper, trends were observed which, as critical values for the indicators are developed, can be utilized to predict lifetimes and thus to lower the number of implants required and shorten the experimental periods necessary to test new designs or techniques.

Conclusions

A mathematical technique has been applied which allows determination of clinical indicator trends to be determined with acceptable confidence limits for immobile implants. For porous-rooted implants stabilized by bone ingrowth, the analysis revealed trends that were not otherwise observable. Given the observed trends, it will be possible to calculate implant lifetimes when critical values of these indicators are determined.

References

[1] National Institute for Dental Research, "Challenges for the Eighties," NIDR NIH Publication 85-860, National Institutes of Health, Washington, DC, December 1983, p. 177.
[2] Young, F. A., Kresch, C. H., and Spector, M. "Porous Titanium Tooth Roots: Clinical Evaluation," *Journal of Prosthetic Dentistry*, Vol. 41, No. 5, May 1979, pp. 561–565.
[3] Miller, S. C., *Textbook of Periodontia*, 3rd ed., Blakiston, Philadelphia, 1950.
[4] Silness, J. and Loe, H., "Periodontal Disease in Pregnancy," *Acta Odontologica Scandinavica*, Vol. 22, 1964, p. 123.
[5] Loe, H., *Journal of Periodontology*, Vol. 38, 1967, p. 610.
[6] Morrison, D. F., "Multivariate Statistical Methods," 2nd ed., McGraw-Hill, New York, 1976.
[7] McCoy, E. D., "Risk of Vitreous Carbon Implants," *Dental Implants: Benefit and Risk*, NIH Publication 81-1531, National Institutes of Health, Washington, DC, December 1980, p. 211.
[8] Kapur, K. K., "The Literature on Blade Implants," *Dental Implants: Benefit and Risk*, NIH Publication 81-1531, National Institutes of Health, Washington, DC, December 1980, p. 250.

Modeling and Implant Fixation

James B. Koeneman[1]

Fundamental Aspects of Load Transfer and Load Sharing

REFERENCE: Koeneman, J. B., **"Fundamental Aspects of Load Transfer and Load Sharing,"** *Quantitative Characterization and Performance of Porous Implants for Hard Tissue Applications, ASTM STP 953*, J. E. Lemons, Ed., American Society for Testing and Materials, Philadelphia, 1987, pp. 241–248.

ABSTRACT: The objective of this review was to examine some of the critical design considerations for porous-surfaced ingrowth prostheses by looking at the fundamental aspects of load transfer and load sharing. It is shown that prostheses of lower stiffness provide higher average bone stress but can have higher stress concentrations in load transfer regions. There are a number of mechanisms for load transfer. Most load transfer, and thus bone hypertrophy, occurs only in localized regions.

KEY WORDS: porous implants, biomechanics, total joint prostheses

Over the past 15 years many investigators have calculated the change in stress distribution in long bones when the intramedullary stem of an artificial joint prosthesis is inserted. The most thoroughly studied region has been the proximal femur. To simplify the complicated loading in this region and yet provide some measurement of the stress state in the femur, a model of a person standing on one leg has been widely used. The reason for the great interest in analysis of stress in the proximal femur is that bone remodels in response to stress, and the stress distribution has to be changed with the insertion of an intramedullary stem. Clinical experience has shown that there is a reduction of cortical wall thickness in the proximal-medial femoral neck after insertion of a prosthesis. Also, bone remodeling due to stress redistribution has been implicated as one of the contributing factors to loosening of prostheses. With the advent of prostheses held in place with bone growth into porous surfaces instead of with bone cement, some new questions arise: How much porous coating is needed and where should it be placed? What is the difference in stress state between a smooth-surfaced, press-fit prosthesis and a porous-surfaced, ingrowth prosthesis? Should the proximal stem have a large cross-sectional area to completely fill up the intramedullary cavity in the attempt to ensure good initial fixation, or should it be kept small to increase the bone stress? What effect does an adhesive bond from a bioactive surface such as Bioglass or hydroxyapatite have? Should the femoral stem have a collar or not? The objective of this study was to examine these questions by looking at the fundamental aspects of load transfer and load sharing.

Results

Load Sharing

In this general review, methods for determining the strength of materials—for example, beam theory—are used to demonstrate the load sharing between prosthesis and bone. The

[1] Head, Bioengineering Division, Harrington Arthritis Research Center, Phoenix AZ 85006.

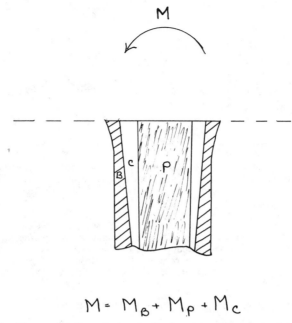

$$M = M_B + M_P + M_c$$

FIG. 1—*Moment loading of a longitudinal section of the proximal femur.*

assumptions of these methods are described in any book on the mechanics of materials, such as Ref *1*. The principal assumption is that plane sections remain plane. This requires that a zero-strain neutral axis exist in bending and that the strain at any point be proportional to its distance from the neutral axis. In tension or compression, strains perpendicular to the load are uniform. Stress concentrations due to irregular geometries cannot be calculated. To determine the magnitude of these stress concentrations, solutions of similar simple geometries in the literature were used.

A moment M applied to a section of a beam consisting of several materials is divided among the constituents according to the product of their elastic moduli and their area moments of inertia. Figure 1 represents a longitudinal section of a femur with an overall moment applied to a plane just below the level of the trochanters. The neutral axes of the central stem, the cancellous bone, and the cortical bone all coincide. The cross-section shapes are all circles. The amount of the total moment carried by the cortical bone is

$$M_b = \frac{M(EI)_b}{(EI)_b + (EI)_c + (EI)_p}$$

where

$$E = \text{elastic modulus of } p, b, \text{ or } c \text{ (see Fig. 1 for an explanation of } p, b, \text{ and } c),$$
$$I = \text{area moment of inertia of } p, b, \text{ or } c,$$
$$M = \text{total moment, and}$$
$$M_b = \text{component of } M \text{ carried by } b.$$

If the neutral axes of the individual components do not coincide, the location of the overall composite neutral axis X can be found using the following formula (assuming bonding between the different materials)

$$X = \frac{(EA)_p X_p + (EA)_b X_b + (EA)_c X_c}{E_b(A_p + A_b + A_c)}$$

where

X = distance from the medial surface to the neutral axis,
$X_{p,b,c}$ = distance from the medial surface to the neutral axis of p, b, or c, and
A = area of p, b, or c.

Similarly, for axial loading P, as shown in Fig. 2, the total load is proportioned among the materials by the ratio of the product of the elastic moduli and the area. The load carried by the bone in Fig. 2 is given by the equation

$$P_b = \frac{P(EA)_b}{(EA)_b + (EA)_c + (EA)_p}$$

A diagram of the pelvis and proximal femur is shown in Fig. 3. The effect of the trochanter load and the head load can be represented by the body weight. This can be shown by doing a free body analysis of the pelvis. Treating the femur as a curved beam, the effective loads on any section can be represented as a shear load, a normal load, and a bending moment. For the moment and axial components, the relative moment and the axial load carried by the bone at a section just below the trochanters are shown in Fig. 4. The bars represent the relative amount of the total moment M_i and total axial load P_i carried by the cortical bone with no prosthesis and with a Co-Cr, titanium, and low-modulus (elastic modulus = 3.5

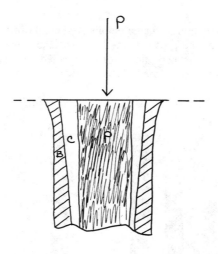

$$P = P_B + P_P + P_C$$

FIG. 2—*Axial loading of a longitudinal section of the proximal femur.*

J Head Load

W Body Weight

M Muscle Load

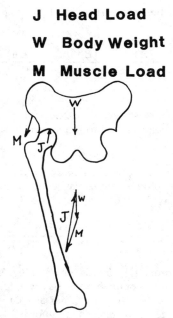

FIG. 3—*Model of the pelvis and proximal femur.*

GPa) prosthesis in the intramedullary cavity. The load-sharing equations just described and a concentric circle model of the bone and prosthesis were used.

Load Transfer

These load-sharing calculations indicate equilibrium conditions. However, the head load is initially carried totally by the prosthesis and the trochanter load is totally carried by the bone initially. For equilibrium conditions to be satisfied, the applied loads have to be transferred from one material to another. There are two main ways load can be transferred—

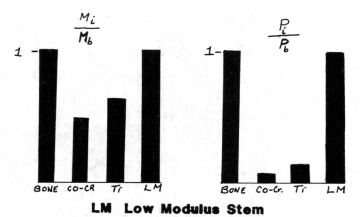

LM Low Modulus Stem

FIG. 4—*Relative axial load and moment carried by the femur.*

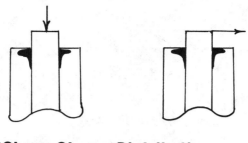

▰Shear Stress Distribution

FIG. 5—*Representations of shear stress in axial and bending load transfer.*

through shear stresses at the interface and through bearing. There are a number of bearing mechanisms.

Many calculations of the shear stress distribution at the interface have been performed in the literature [2–5]. These analyses have been instigated by the interest in load transfer between the fibers and matrix in composite materials. Pushing or pulling one cylinder bonded to another or bending two laminated beams causes a high shear stress where one of the pieces begins or ends. Figure 5 shows schematically the shape of the stress distributions calculated by the models. There is no uniform shear stress along the interface, as is often assumed. The magnitude of the stress concentration factor (SCF) depends on the amount of load to be transferred. Thus, it depends on the relative EA or EI products, as discussed in the previous section. Also, if the model is a three-member cylinder model—which can be used as a simple model of the cortical bone, cancellous bone, and prosthesis—the SCF increases as the inner ring thickness decreases [4].

The intramedullary cavity of the proximal femur tapers. For this reason many prostheses are tapered in some regions. Tapering is a very effective way of transferring axial loads. Figure 6 shows the interface load components and also a graph showing the very high radial loads that can be generated at low-taper angles. This is one reason a femur can be split by a press-fit prosthesis during insertion when wedged within the cortical shell. Just as in shear stress transfer, the highest interface stresses occur at the beginning and end of the interface region for wedge-type loading. Figure 7 shows the interface radial stresses calculated by the finite element method for a simple taper model. The core and outer ring elastic moduli are 110 and 13.8 GPa, respectively. The taper was 6° (half angle). A uniform stress was applied to the top of the central plug, and the outer ring was supported on the bottom.

Another bearing mechanism is a "shoulder," such as the collar on a femoral stem prosthesis. All the load that is to be transferred will be transferred immediately under the collar. Other forms of the shoulder are threads and macrointerlocks. These last two are really collars in series. Thus, if the cylinders are parallel, all the load that is to be transfered will be transferred in the first thread or layer of macrointerlocks.

A lever mechanism of load transfer also exists in intramedullary stem systems. The Austin Moore prosthesis is a good example of this. The stem is supported by the proximal medial cortex of the femoral neck and the lateral cortical wall of the femoral shaft. Reference 6 has an example showing the bone hypertrophy in these regions.

Discussion

Although the loading in the proximal femur is three-dimensional, two-dimensional models can be applied separately to the mediolateral, anteroposterior, and torsion loads. The same

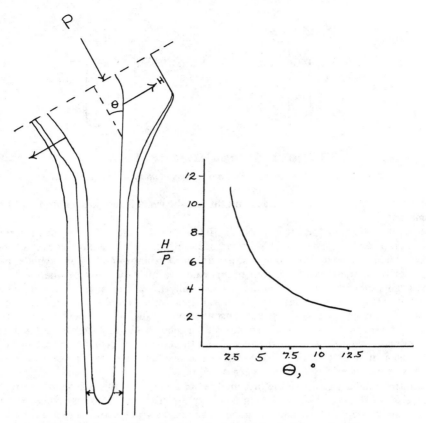

FIG. 6—*Implant/bone load transfer by wedging in the proximal femur.*

mechanisms do work in all types of loads. Stems with low stiffness cause the least reduction of average bone stress. Reference 7 shows an example of bone remodeling, probably due to both a shift in the neutral axis and a lowering of the load carried by the bone, caused by the presence of a prosthesis. However, the lower the stiffness, the greater the load that has to be transferred initially, and the potential exists for higher interface stress concentration factors. Load transfer occurs where there is a change in the relative EI or EA products, or both, or where there is a change in the relative position of the neutral axes of the prosthesis and bone. The transfer only occurs if a mechanism exists. The largest change in relative stiffness usually occurs in locations where the prosthesis enters the bone and where the last contact occurs. Load transfer by microinterlock or macrointerlock occurs over a very short distance. Therefore, in these systems significant bone ingrowth would be expected only in specific regions and not generally throughout the coating area. An example of bone hypertrophy at the proximal and distal load transfer regions is provided in Ref 8.

Examination of retrieved human ingrowth prostheses has shown spotty ingrowth [9]. Some of the ingrowth is at the upper and lower edges of the ingrowth region and some is in more central regions. This could be caused by proximity to the cortical walls or the stiffer cancellous regions. If several load transfer mechanisms exist in parallel, the stiffest path will carry proportionately more load. Also, bone growth for reasons unrelated to the stress state could occur.

If the netural axis of the prosthesis and that of the bone coincide, no shear stresses due to bending exist at the interface. All the load is transferred by bearing at the interface. If the neutral axes do not coincide, shear stresses must exist at the interface or relative motion will occur. In the case of smooth-surfaced prostheses, a large mismatch in neutral axes could possibly cause resorption of bone due to motion.

Load transfer at a collar with no distal high-stiffness mechanism present could lead to wobbling, because all of the load is transferred proximally and there is little stress distally to keep the surrounding bone strong enough. Also, the reverse can occur. A high-stiffness wedging in the distal cortical bone could cause resorption under the collar.

What types of load can the bone handle? From a structure-property argument it has been proposed that the dense cancellous bone of the head carries axial loads but that, in and distal to the neck, the cancellous bone is not stiff enough to carry bending or axial loads [10]. The porous bone in these regions acts as a foam core in a sandwich beam. That is, it carries shear loads and provides local support for the outer shell. The fact that the structure aligns approximately with the principal stress directions is interpreted to mean that the cancellous bone carries shear loads in tension and compression. Gibson [11] has shown the large variation in cancellous bone properties. The open structure that is characteristic of the femoral neck can have a compressive strength as high as 5 MN/m². Using the assumption that this bone will carry shear stress in tension and compression, the shear strength of the bone will be approximately 2.5 MN/m². The shear stress concentration at the neck using the calculations described earlier is about 2.2 MN/m². These calculations can be done for any prosthesis load transfer mechanism. What should be considered in a porous, ingrowth

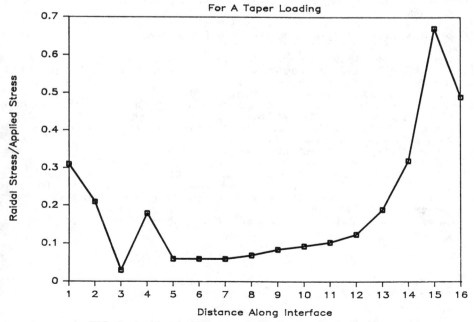

FIG. 7—*Interfacial radial stress distribution in a wedge loading.*

surface is whether the coating has enough porosity in it to accommodate bone of sufficient strength.

Conclusions

Load is shared between the prosthesis and the bone, depending on the relative stiffnesses of the bone and prosthesis. The stiffer the implant, the less the average bone stress will be. Several mechanisms of transferring this load exist. The two major classifications of transfer are by shear stress and by bearing. Most of these mechanisms transfer load in localized regions only. Bone hypertrophy is to be expected in these areas. Load transfer is caused by changes in the relative stiffnesses of the bone and prosthesis—for example, where the prosthesis enters the bone, where the prosthesis ends, and where a trochanter begins or ends—and by changes in the relationship between their neutral axes. If several load transfer mechanisms exist in parallel, the stiffer paths will carry proportionately more load. Prostheses of lower stiffness will make the average bone stress closer to the original bone stress; however, consideration must be given to how the load is to be transferred. The relationship between the neutral axis of the prosthesis and the axis of the bone should be considered. Too large a mismatch will cause relative motion, which could prevent ingrowth in porous systems or cause resorption in smooth systems. The time history should be considered. If a certain type of load transfer is being designed, does that support exist initially or does it have to grow in? If it has to grow in, are the conditions—for example, motion and stress—sufficient to allow this to happen?

References

[1] Timoshenko, S. and Young, D. H., *Elements of Strength of Materials*, 5th ed., Van Nostrand, New York, 1968.
[2] Piggott, M. R., *Load-Bearing Fibre Composites*, Pergamon Press, Toronto, 1980.
[3] Cox, M. A., "The Elasticity and Strength of Paper and Other Materials," *British Journal of Applied Physics*, Vol. 3, 1952, pp. 72–79.
[4] Dow, N. F., "Study of Stress Near a Discontinuity in Fiber-Reinforced Composite Metal," Report No. R63SD61, General Electric Co., Missiles and Space Division, Philadelphia, PA, August 1963.
[5] Levy, A., "Fiber-Matrix Interaction in a Composite Material," Departmental Memorandum RM-467, Grumman Research, Bethpage, NY, February 1970.
[6] Lunceford, E. M., Jr., Weinstein, A. M., and Koeneman, J. B., *Clinical Biomechanics*, J. Black and J. H. Dumbleton, Eds., Churchill Livingstone, New York, 1981, Chapter 7, pp. 140–155.
[7] Bobyn, J. D., Pilliar, R. M., Binnington, A. G., and Szivek, J. A., *Transactions of the Orthopaedic Research Society*, Vol. 10, January 1985, p. 170.
[8] Spector, M., Wigger, W. B., and Buse, M. G., *Clinical Orthopaedics and Related Research*, Vol. 160, October 1981, pp. 242–249.
[9] Koeneman, J., *Transactions of The Society for Biomaterials*, Vol. 8, April 1985, p. 53.
[10] Lee, T. K., Magee, F. P., Weinstein, A. M., and Johnson, R., "Retrieval Analysis of a Plasma Spray Porous-Coated Femoral Hip," *Proceedings*, Tenth European Congress on Biomaterials, Bologna, Italy, 14–17 Sept. 1986.
[11] Gibson, L. J., *Journal of Biomechanics*, Vol. 18, No. 5, 1985, pp. 317–328.

Hank C. K. Wuh,[1] *Lynne C. Jones,*[2] *and David S. Hungerford*[3]

Strain Analysis of the Proximal Femur After Total Hip Replacement

REFERENCE: Wuh, H. C. K., Jones, L. C., and Hungerford, D. S., **"Strain Analysis of the Proximal Femur After Total Hip Replacement,"** *Quantitative Characterization and Performance of Porous Implants for Hard Tissue Applications, ASTM STP 953,* J. E. Lemons, Ed., American Society for Testing and Materials, Philadelphia, 1987, pp. 249–263.

ABSTRACT: Strain analysis of human cadaver femora after cemented total hip arthroplasty (THA) has demonstrated a reduction in stress transfer along the proximal femur. The principal objective of this study was to determine the effect of the cementless application of a press-fit, porous-coated prosthesis on the strain experienced by the proximal femur. Using the photoelastic coating technique (PECT), five human cadaver specimens were subjected to strain analysis before and after cementless arthroplasty with a PCA total hip prosthesis. After total hip replacement, the strain magnitudes were reduced for all points along the medial border when the femur was subjected to loading conditions. A reduction of the level of strain experienced by the calcar ranged from 34.7 to 43.7% under loads ranging from 750 to 2000 N— a considerably smaller reduction than that reported by previous investigators. The region of the greater trochanter was the only area of the lateral surface to demonstrate an increase in strain magnitude after THA; the other, more distal points laterally experienced a reduced level of strain. Increases in strain magnitude, although not statistically significant, were detected along the anterior aspect of the femora. Significant decreases in strain were observed at the two more distal points posteriorly, with no significant change proximally. As this investigation is an evaluation only of the immediate effect of the design of the prosthesis in achieving a press fit, and the specimens are without the benefit of bony ingrowth, additional studies are necessary to determine the effect of biologic ingrowth on the distribution of strain within the proximal femur.

KEY WORDS: porous implants, stress transfer, total hip replacement, porous-coated prostheses, strain analysis, photoelastic coating technique, stress shielding

One of the most important limitations to the long-term success of total hip replacement (THR) is aseptic loosening of the prosthesis. For this reason, a great deal of interest has developed in the application of porous-surfaced implants for cementless THR. The early reported clinical results for cementless fixation have been favorable, both with the macro-porous designs widely used in Europe [1–5] and with the microporous designs extensively studied by a number of investigators in the United States [6–10].

A major concern in the clinical usage of porous-surfaced implants is proximal stress transfer from the femoral prosthesis to the surrounding cortical bone. Sir John Charnley [11] in 1965 and McKee [12] in 1970 have postulated that the cement between the bone and prosthesis

---

[1] Medical student, Johns Hopkins University School of Medicine, Baltimore, MD 21239.
[2] Research associate, Orthopaedic Surgery, Johns Hopkins University School of Medicine, Baltimore, MD 21239.
[3] Associate professor, Orthopaedic Surgery, Johns Hopkins University School of Medicine, and chief, Division of Arthritis Surgery, Good Samaritan Hospital, Baltimore, MD 21239.

distributes the load over a wider area and thus greatly reduces the stress translated to the cortical bone. The design of porous-surfaced implants for biological fixation in the absence of cement may therefore significantly affect bone stresses and subsequent bone remodeling.

The potential for the development of secondary osteoporosis due to significant stress shielding and the resulting loosening of the bone/porous implant interface is a well-recognized phenomenon [13–19]. Engh and Bobyn [20] have evaluated adaptive femoral bone modeling following the biologic fixation of porous-surfaced (modified Moore) prostheses through radiographic analysis. However, direct in vitro stress analysis of the proximal femur following arthroplasty with a porous-coated prosthesis has not been reported. Such analysis may yield important information regarding the initial stress transfer from the porous prosthesis to the surrounding cortical bone and serve to predict subsequent adaptive bone remodeling.

As the clinical experience with cementless THR is limited, there is a lack of specimens retrieved from deceased patients. Therefore, stress analyses of femora implanted with porous prostheses stabilized with biologic ingrowth have not been reported. In vitro studies using fresh cadaver femora should simulate the response to loading of the operated femur after initial implantation. Although the findings would be limited to this application, the information generated should reflect the mechanical response of bone to implantation of a press-fit prosthesis.

The purpose of this study was to evaluate the stress transfer in the proximal femur after the cementless implantation of a porous-surfaced femoral component. Traditionally, four methods have been used to evaluate stress in the loaded proximal femur: (1) theoretical mathematical analysis (including finite element analysis) [21–27], (2) the stress coat (brittle coating) technique [28,29], (3) photoelastic models [30,31], and (4) strain gages [32–37]. However, each of these methods have limitations in their applications to the study of bone, including directional and positional constraints, assumptions of homogeneity, and sensitivity. One technique that overcomes these limitations is the photoelastic coating technique (PECT). Its use toward the study of freshly retrieved bony tissues has only recently been reported by our laboratory [38,39] and will be detailed in the section on Materials and Methods. To carry out our study, physiologic loading conditions simulating a single-limb stance were applied to five freshly retrieved human cadaver femurs. Strain measurements were made using the PECT, first with the femoral head intact and then after cementless implantation of a PCA femoral component.

Materials and Methods

Five freshly retrieved, intact, human cadaver femurs were frozen immediately upon removal at −20°C and thawed to room temperature just prior to testing. It has been established that freezing has no harmful effect on the physical properties of femoral cortical bone [40]. All five femurs were roentgenographically normal. The soft tissues and the periosteal layer were thoroughly debrided. The surface was gently smoothed with fine finishing sandpaper and then wiped with gauze soaked in 30% ethyl alcohol to degrease the surface and to remove any loose particles.

Photoelastic Coating Technique

The photoelastic coating technique (PECT) involves basically four steps: (1) the casting and contouring of photoelastic plastic sheets to the testing area—in this case, the proximal femur; (2) sealing the bone surface with a precoating of bonding cement; (3) bonding of

TABLE 1—*The photoelastic coating technique.*

Step No.	Procedure	Experimental Day
1	casting and contouring	1
2	sealing	2
3	bonding	2
4	measurement	3

the contoured photoelastic plastic to the proximal femur; and (4) measurement of the strain under load, using a reflection polariscope and an attached digital compensator (see Table 1).

Casting and Contouring

A room-temperature-curing, two-component resin/hardener system, PL-1 (Measurements Group, Raleigh, NC), was used for making contourable photoelastic plastic sheets. It has a Poisson's ratio of 0.36, a modulus of elasticity of 0.42×10^6, and a K factor (strain optical constant) of 0.10. The PL-1 was prepared according to instructions [41]. The photoelastic epoxy resin was applied to the proximal femur using two PL-1 photoelastic sheets with a gap of approximately 1 to 2 mm between the two sheets along the anterior and posterior surfaces of the femur. [The seam did not cross the points of measurement.] Care was taken not to stretch the plastic and to ensure that it was in complete contact with the surface of the proximal femur. Once contoured to the shape of the proximal femur, the PL-1 was allowed 18 h to complete its polymerization cycle. At the end of the cycle, the PL-1 was solid and was of the same size and detailed contour as the proximal femur. It was then carefully removed from the proximal femur, and measurements of the coating thickness were made along approximately 1.5 to 2.0 cm intervals using a Starrett micrometer. The thickness of the coating used during this study averaged 1.87 mm.

Sealing

A two-component resin/hardener curing adhesive, PC-1 (Measurements Group, Raleigh, NC), was carefully applied to the proximal femur in a very thin layer to plug the pores of the cortical wall. After the PC-1 had dried for 2 h, the surface of the proximal femur was prepared with fine finishing sandpaper to remove the excess PC-1, thus allowing it to remain only in the pores. This step prevented moisture from escaping during the subsequent bonding procedures.

Bonding

The surface of the sealed proximal femur was cleaned with gauze soaked in 30% ethyl alcohol. The PC-1 was prepared according to instructions and was brushed over the surface of the proximal femur in a thin uniform layer [41]. The contoured PL-1 photoelastic plastic was then carefully placed over the adhesive. Care was taken to ensure total contact between the plastic and the adhesive without underlying air pockets being present. The adhesive was allowed to cure for 12 h, after which the femur was ready for strain measurement.

Measurement

Although the technique of photoelastic coating is capable of measuring the stress distribution over the entire surface of the proximal femur, particular reference points along the femur were noted for comparison with results of other stress analysis studies previously reported. The points were marked with a wax pencil over the photoelastic coating at the calcar, the subtrochanteric region, and the area approximating the distal portion of the stem along the medial, lateral, anterior, and posterior borders. The reference points were at similar locations for each femur tested, for they were determined by the specific ratio of the distance from the femoral head to the reference point over the length of the femur. The ratio was established by the first femur tested.

When load is applied to the femoral head and the femur is viewed under polarized light, the bonded photoelastic coating experiences strain corresponding to that experienced by the underlying bone, and a characteristic colored fringe pattern can be detected. Lines of similar strain, isoclinic lines, can then be used to determine the direction of the principal strains. A digital compensator attached to a reflection polariscope, the source of polarized light, can then be used to measure the magnitude of the strain on the basis of the colored patterns, called isochromatics or fringes. The value obtained is the fringe order N and is directly proportional to the difference between the principal strains of the proximal femur established by the formula

$$\epsilon_x - \epsilon_y = N \frac{\lambda}{2tK}$$

where

$\epsilon_x - \epsilon_y$ = difference in principal strains,
N = fringe order = δ/λ,
δ = light retardation, in.,
λ = the wavelength of light (22.77×10^{-6} in.),
t = plastic coating thickness, and
K = sensitivity of the plastic coating, the K factor (0.10).

The thickness of the PL-1 photoelastic coating t has been measured for all specimens tested and found to be a relatively constant value. K and λ are constants. Therefore, the fringe order N is directly proportional to the difference in principal strains experienced by the proximal femur. We have chosen to report our results as changes in fringe order. However, to convert from N units to the difference in principal strains in microstrain units for this study, the fringe order value can be multiplied by 1.54×10^{-3}. As the purpose of this study was to seek comparisons rather than absolute values, the data have also been presented as percentage changes (Fig. 1). In a previous study, our group found a highly significant correlation between changes in the difference in principal strains ($\epsilon_1 - \epsilon_2$) and the changes in the magnitude of axial strain ($r = 0.996$, $P < 0.001$) [39]. We have, therefore, chosen to use the photoelastic technique as an indirect method of detecting changes in the principal axial strain experienced by the femur. The implications of this are outlined in the section titled Discussion.

Testing Conditions

Each intact femur was fixed distally in a specially designed jig for stabilization. Using a goniometer, the femoral shaft was positioned at a 9° angle from the vertical to simulate the proper anatomical axis. While viewing the medial aspect of the femur, the femoral head was placed directly above the medial epicondyle. Each femur was mounted in the testing rig of a MMED servohydraulic, computer-linked materials tester (Matco, La Canada, CA). Load was applied to the vertical axis in 100 to 200-N increments with strain measurements

Sig.	Ratio	Load			Load	Ratio	Sig.
NS	1.765+0.714	750	A		750	0.653+0.036	p<.001
NS	1.602+0.482	1500		D	1500	0.615+0.042	p<.001
NS	2.422+1.276	2000			2000	0.563+0.035	p<.001
NS	0.938+0.494	750	B		750	0.917+0.203	p<.01
p<.001	0.525+0.161	1500		E	1500	0.818+0.082	p<.001
p<.001	0.484+0.078	2000			2000	0.792+0.065	p<.001
p<.05	0.499+0.064	750	C		750	0.820+0.074	p<.05
p<.01	0.323+0.150	1500		F	1500	0.865+0.052	p<.01
p<.001	0.368+0.276	2000			2000	0.865+0.055	p<.01

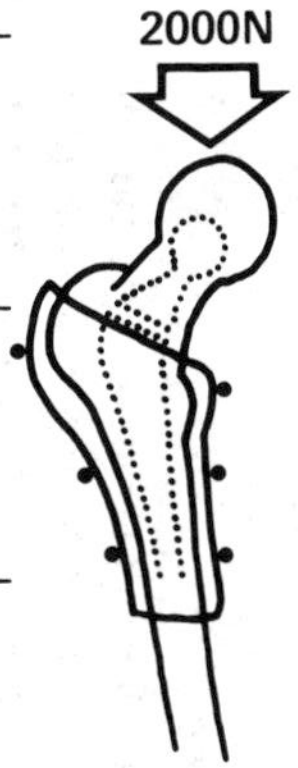

FIG. 1—Summary of averaged data illustrating the change in the magnitude of strain in the proximal femur. The ratios are those of the strain values in the femur after cementless PCA femoral arthroplasty to the strain values in the intact femur. Data are presented for reference points along the medial and lateral borders under loading conditions of 750, 1500, and 2000 N. These cumulative data are based on the percentage changes for each individual femur, which served as its own control. The statistical significance of changes between the intact and implanted femora were determined using two-way analysis of variance for related measurements, with linear contrast methods used for the comparison of groups. NS indicates no significance.

taken at 750, 1500, and 2000 N. The loading procedures and measurements were repeated for each femur. When strain measurements in the intact femurs were completed, each femur was prepared for the cementless implantation of a PCA femoral component. Photographs were taken during each point of measurement.

PCA Implantation

Roentgenograms of each femur were examined for proper sizing of the porous femoral implant. Using the PCA total hip instrumentation system, cementless implantation of a PCA femoral component of the appropriate size, in the neutral position, was performed on each femur. Roentgenograms were then taken to confirm proper positioning. The photoelastic plastic coating around the proximal femur was entirely undisturbed. The femur was then tested under conditions identical to those used for the intact femur.

Statistical Analysis

Each femur was tested twice under each loading condition. As the results for each test were highly reproducible, these values were averaged. Changes in strain magnitude subsequent to the total hip replacement were analyzed using two-way analysis of variance (ANOVA) for related measurements. Comparisons between groups were made using linear contrast methods.

Results

Strain Distribution of the Intact Femur

Figure 2 compares the strain distribution in the proximal femur before ($n = 5$) and after ($n = 4$) cementless femoral arthroplasty for reference points along the medial and lateral borders. Along the medial wall, two of the five intact femurs tested had a strain magnitude at the calcar greater than that at the subtrochanteric region; three of the five femurs held the reverse relationship. Both conditions have been reported in the past [23,24,34–37]. The magnitude of the distal point was the lowest value on the medial aspect in four of the five femora tested. Laterally, the subtrochanteric region was fairly consistent in having the greatest magnitude of strain, whereas the trochanteric region displayed the lowest strain values.

Figure 3 presents data for reference points along the anterior and posterior borders under the same conditions as for Fig. 2. Along the anterior margin, three of the five specimens demonstrated maximal strain magnitudes at the most proximal point, while two specimens had maximal values at the distal point. For the posterior border, the largest values were seen at the point most distal.

The increase in strain magnitude in response to increasing load along the medial and lateral borders of the intact femur is illustrated in Fig. 4. To demonstrate the response to increasing loads at the calcar, Fig. 5 presents data for each femur tested.

Strain Distribution After Cementless Femoral Arthroplasty

As demonstrated for the intact femur, the points along the medial margin demonstrated the largest measurements for strain magnitude. In contrast, while the strain values of the anterior border were the smallest found for the intact femur, the lateral margin displayed the lowest values following arthroplasty. Furthermore, along the medial and anterior borders, there was a shift in the point of greatest magnitude distally. The relationships along the lateral and posterior borders were unchanged.

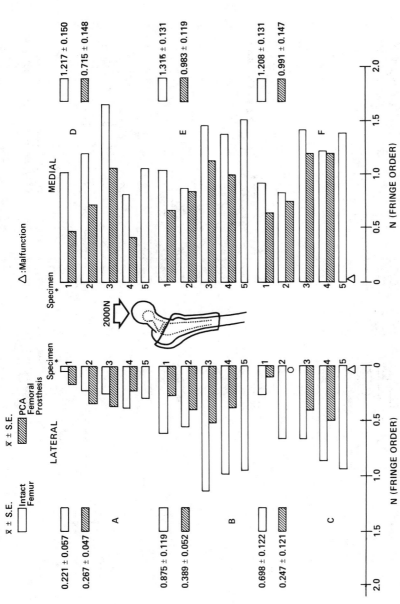

FIG. 2—*Bar graph illustrating the strain distribution in the intact femur and the strain distribution after cementless implantation of the femoral component. Values at various reference points along the medial and lateral borders are presented. Each femur served as its own control. The load was 2000 N. Note that, laterally, the subtrochanteric region has the greatest strain magnitude and that the trochanteric region is under tension. Medially, no consistent relationship exists between the strain magnitude at the calcar and that at the subtrochanteric region in the intact femur.*

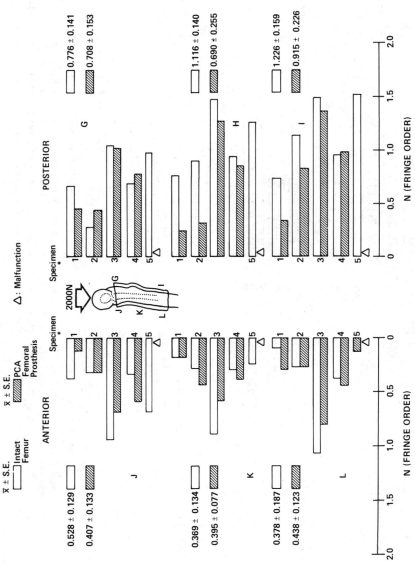

FIG. 3.—Bar graph illustrating the strain distribution in the intact femur and the strain distribution after cementless implantation of the femoral prosthesis. Values at various reference points along the anterior and posterior borders are presented. Each femur served as its own control. The load was 2000 N.

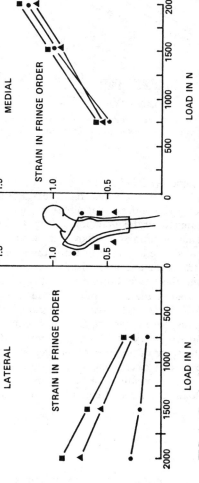

FIG. 4—Graph of averaged strain values for all intact femurs tested. The strain values at the various reference points along the medial and lateral borders appear to increase proportionately with the increase in load, with no abrupt changes noted up to 2000 N.

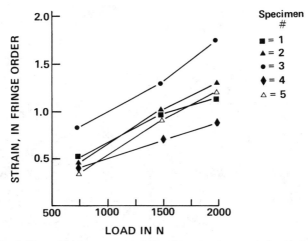

FIG. 5—*Graph illustrating the response of the calcar to an increase in load before cementless PCA arthroplasty. The strain magnitude at the calcar increases in proportion to the increase in load for the intact femur.*

As is the case with previous reports [36,37], there is a great deal of individual variation in strain magnitude and distribution from one femur to the next. For this reason, the changes in strain magnitude subsequent to cementless total hip arthroplasty (THA) are presented as a percentage of the values obtained for the intact femur (Fig. 1). The magnitude of strains exhibited by the loaded proximal femora were shown to decrease after cementless hip arthroplasty for most of the points tested. All points along the medial margins experienced lower strain magnitudes; the strains for the more distal points posteriorly and the subtrochanteric and distal points along the lateral border were also reduced. Three of the four implanted femora demonstrated an increase in strain subsequent to surgery at the greater trochanteric region. However, the level of change was not statistically significant.

Figure 6 shows that the strain response to increased loading remains approximately linear after cementless femoral arthroplasty for points along the medial and lateral margins. However, the change in response (the slope of the line) to the increased load is decreased. This is particularly apparent at the medial calcar. Figure 7 illustrates the strain response to increasing loads at the calcar after cementless THR. Again, although strain appears to increase proportionately with load, the level of increase, that is, the slope of the line, is diminished.

Discussion

A major concern over biological fixation is the proximal stress transfer from prosthesis to bone and the potential for both pathological stress transfer and significant stress shielding. A particular area of controversy in cemented THR has been bone resorption in the calcar region and its relationship to prosthesis loosening. It is known that the calcar frequently resorbs some years after femoral component implantation in up to 70% of the cases [42–46]. Although significant stress reduction has generally been considered the mechanism of resorption at the calcar [34,36,37,47], a possible decrease in the blood supply [42,45], thermal necrosis [48], and foreign-body reaction to nonpolymerized monomers of bone cement [49]

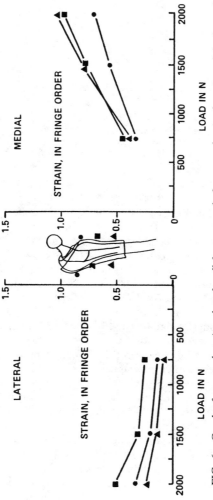

FIG. 6—*Graph of averaged strain values for all femurs tested after cementless implantation of the femoral component. Though the strain response to increasing load remains approximately linear, the rate of increase, or stress transfer, is diminished. The decrease is particularly noticeable at the calcar.*

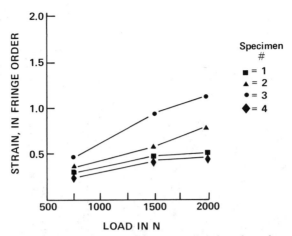

FIG. 7—*Graph showing the level of strain experienced at the calcar after total hip arthro-plasty, which increased as the femur was subjected to increasing load. However, the level of the increase in strain was of a lower magnitude than that seen for the intact femur.*

have been postulated as causes. Opinions differ as to whether this type of resorption is causally related to the loosening of prostheses [43,45,47,50].

As reported in previous laboratory studies, there was a significant decrease in the strain magnitude at the calcar following THR. Oh [36] and Oh and Harris [37] have reported greater than 90% reduction in strain values at the calcar after cemented THR (load = 2012 N). McBeath reported reductions greater than 75% (load = 445 N) [34], and Crowninshield reported up to 91% reduction for a collarless steel prosthesis and 50 to 60% reduction for a prosthesis with a collar design (load = 1000 N) [47]. Data from our study reveal a strain magnitude reduction of 34.7% at 750 N, 38.5% at 1500 N, and 43.7% at 2000 N at the calcar after cementless femoral arthroplasty with a press-fit, porous-surfaced prosthesis. These reductions are statistically significant (Fig. 1). Our findings indicate a substantially smaller reduction in strain than all previous reports for cemented THR. As our experimental technique measures the difference between the principal strains ($\epsilon_1 - \epsilon_2$), the significance of our results must be further discussed. Based on the principles of photo-elasticity, a decrease of the difference in principal strains in a region known to be subjected to compressive strains is a reflection of either (1) a decrease in the hoop strains, (2) a decrease in the level of axial strain, or (3) a decrease in the hoop strains concomitant with a decrease in axial strain. As the prosthesis used in this study was a press-fit design, we suggest that our findings for the calcar reflect a decrease in the axial strain [34,39]. However, additional studies using rosette strain gages are needed to validate our results further.

Radiographic evidence of endosteal bone formation near the distal portion of the stem, indicating significant stress transfer from implant to bone, has been reported for biologic fixation [20]. In our study, we found an increase in strain at the anterior margin and strain decreases at the other margins. The increase at the anterior point distally was not statistically significant. However, the decreases in strain at the medial (13.5%) and posterior (22.4%) points distally were significant ($P < 0.01$ at 2000 N). Oh [36] and Oh and Harris [37] have described a gradient of strain distribution along the proximal femur with the greatest strain values at the calcar, which decrease as one moves distally in the intact femur; after THR, the distribution is reversed. In our study, the region of the calcar demonstrated the largest strain values in three of the five intact specimens. Along the medial margin, the magnitude of strain decreased as the distal point was approached. After THR, this relationship was reversed in two of the three specimens. However, these relationships have not been a universal finding, according to other previous studies [23,24].

The cadaver specimens tested for this *in vitro* study were obviously without the benefit of bony ingrowth. Therefore, the relatively smaller magnitudes in strain change after THR are a reflection of the press-fit design and not necessarily a result of the porous-coated surface. Theoretically, while cemented THR achieves maximum stability at time of surgery, cementless THR with porous-surfaced implants will reach optimum stability over time. If strain values were obtained after the process of bony ingrowth had been in progress, one would suspect that the strain values would be different and that the proximal femur would share a greater portion of the applied load. This may further limit the significant decrease in the strain magnitude at the calcar and along the medial borders. Future studies evaluating specimens retrieved from deceased patients into which porous-coated prostheses have been implanted would give further insight into these processes.

Acknowledgments

We wish to express our gratitude to the Anatomy Board of Maryland for making available to us the cadaver specimens used in this study.

References

[1] Judet, R., Siquier, M., and Brumpt, B., *Clinical Orthopaedics and Related Research,* Vol. 137, 1978, pp. 76–84.

[2] Judet, R., "Rough-Surfaced Total Hip Prosthesis Without Cement," *Orthopaedic Transactions,* Vol. 5, 1981, pp. 382–383.

[3] Lord, G. and Bancel, P., "The Madreporic Cementless Total Hip Arthroplasty: New Experimental Data and a Seven-Year Clinical Follow-Up Study," *Clinical Orthopaedics and Related Research,* Vol. 176, 1983, pp. 67–76.

[4] Lord, G. A., "Bone Ingrowth and Uncemented Total Hip Replacement With Madreporic Stemmed Prostheses," *Orthopaedic Transactions,* Vol., 1981, pp. 383–384.

[5] Lord, G. A., Hardy, J. R., and Kummer, F. J., *Clinical Orthopaedics and Related Research,* Vol. 141, 1979, pp. 2–16.

[6] Cameron, H. U., *Clinical Orthopaedics and Related Research,* Vol. 165, 1982, pp. 188–190.

[7] Engh, C. A., *Clinical Orthopaedics and Related Research,* Vol. 176, 1983, pp. 52–66.

[8] Engh, C. A., Bobyn, J. D., and Gorski, J. M., *Orthopaedics,* Vol. 7, 1984, pp. 285–298.

[9] Hungerford, D. S., Hedley, A., Habermann, E., Borden, L., and Kenna, R. V., "Clinical Results with the PCA Hip," *Total Hip Arthroplasty: A New Approach,* University Park Press, Baltimore, 1984, Chapter 8, pp. 170–180.

[10] Ryan, G. M. and Brosher, A. P., *Orthopaedics,* Vol. 3, 1980, p. 660.

[11] Charnley, J., *Journal of Bone and Joint Surgery,* Vol. 47B, 1965, pp. 354–363.

[12] McKee, G. K., *Clinical Orthopaedics and Related Research,* Vol. 72, 1970, pp. 88–103.

[13] Bobyn, J. D., Cameron, H. U., Abdulla, D., Pilliar, R. M., and Weatherly, G. C., *Clinical Orthopaedics and Related Research,* Vol. 166, 1982, pp. 301–312.

[14] Galante, J. O., "Overview of Current Attempts to Eliminate Methylmethacrylate," *The Hip: Proceedings,* Eleventh Open Scientific Meeting of The Hip Society, Mosby, St. Louis, 1983, pp. 181–189.

[15] Hedley, A. K., Clarke, I. C., and Kozinn, S. C., *Clinical Orthopaedics and Related Research,* Vol. 163, 1982, pp. 300–311.

[16] Morscher, E. W., "European Experience with Cementless Hip Replacements," *The Hip: Proceedings,* Eleventh Open Scientific Meeting of The Hip Society, Mosby, St. Louis, 1983, pp. 190–203.

[17] Pilliar, R. M., *Journal of Orthopaedic Research,* Vol. 1, 1983, pp. 189–234.

[18] Pilliar, R. M., Cameron, H. U., Birmington, A. G., Szinek, J., and Macnah, I., *Journal of Biomedical Materials Research,* Vol. 13, 1979, pp. 799–810.

[*19*] Tonino, A. J., Davidson, C. L., Kloffer, P. J., and Linclan, C. A., *Journal of Bone and Joint Surgery,* Vol. 58B, pp. 107–113.

[*20*] Engh, C. A. and Bobyn, J. D., "Evaluation of Adaptive Femoral Bone Modeling," *The Hip: Proceedings,* Twelfth Open Scientific Meeting of The Hip Society, Mosby, St. Louis, 1984, pp. 110–132.

[*21*] Cook, S. D., Skinner, H. B., Weinstein, A. M., and Haddard, R. J., "Stress Distribution in the Proximal Femur After Surface Replacement: Effects of Prostheses and Surgical Techniques," *Biomaterials, Medical Devices, and Artificial Organs,* Vol. 10, 1982, pp. 85–102.

[*22*] Crowninshield, R. D., Brand, R. A., and Johnston, R. C., *Journal of Bone and Joint Surgery,* Vol. 62A, 1980, pp. 68–78.

[*23*] Koch, J. C., *American Journal of Anatomy,* Vol. 21, 1972, pp. 177–298.

[*24*] Rybicki, E. F., Siruonen, F. A., and Weis, E. B., Jr., *Journal of Biomechanics,* Vol. 5, 1972, pp. 203–215.

[*25*] Skinner, H. B., Cook, S. O., Weinstein, A. M., and Haddad, R. J., *Clinical Orthopaedics and Related Research,* Vol. 166, 1982, pp. 277–283.

[*26*] Svensson, N. C., Valliaffan, S., and Wood, R. D., *Journal of Biomechanics,* Vol. 10, 1977, pp. 581–588.

[*27*] Toridis, T. G., *Journal of Biomechanics,* Vol. 2, 1969, pp. 163–174.

[*28*] Evans, F. G. and Lisner, H. R., " 'Stress Coat' Deformation Studies of the Femur Under Static Vertical Loading," *Anatomy Research,* Vol. 100, 1948, pp. 159–190.

[*29*] Kalen, R., "Strain and Stresses in the Upper Femur Studied by the Stress Coat Method," *Acta Orthopaedica Scandinavica,* Vol. 31, 1961, pp. 103–113.

[*30*] Milch, H., *Journal of Bone and Joint Surgery,* Vol. 22, 1940, pp. 621–626.

[*31*] Pammels, F., "Uber die Bedentung der Bauprinzipien des Sturtzund Benwegrengsapparetes für die Beansprushung der Rohrenknochen," *Acta Anatomica,* Vol. 12, 1951, pp. 207–227.

[*32*] Caler, W. E., Carter, D. R., and Harris, W. H., *Journal of Biomechanics,* Vol. 14, 1981, pp. 503–507.

[*33*] Lanyon, L. E., Paul, I. L., and Rubin, C. T., *Journal of Bone and Joint Surgery,* Vol. 63A, 1981, pp. 989–1001.

[*34*] McBeath, A. A., Schopler, S. A., and Narechairia, R. G., *Clinical Orthopaedics and Related Research,* Vol. 150, 1980, pp. 301–305.

[*35*] Mizrahi, J., Livingstone, R. P., and Rofan, I. M., *Journal of Biomechanics,* Vol. 12, 1979, pp. 491–500.

[*36*] Oh, I., "Effect of Total Hip Replacement on the Distribution of Stress in the Proximal Femur: An *In Vitro* Study Comparing Stress Distribution in the Intact Femur with That After Insertion of Different Femoral Components," *The Hip: Proceedings,* Fifth Open Scientific Meeting of The Hip Society, Mosby, St. Louis, 1977, pp. 1–12.

[*37*] Oh, I. and Harris, W. H., *Journal of Bone and Joint Surgery,* Vol. 60A, No. 1, 1978, pp. 75–85.

[*38*] Jones, L. C. and Hungerford, D. S., "Measurements of Strain in the Fresh Human Femur Using the Photoelastic Coating Technique," *Transactions of the Society for Biomaterials,* Vol. 8, 1985, p. 199.

[*39*] Jones, L. C. and Hungerford, D. S., "The Photoelastic Coating Technique—It's Validation and Use," *Transactions of the Orthopaedic Research Society,* Vol. 12, 1987, p. 406.

[*40*] Sedlin, E. D. and Hirsch, C., *Clinical Orthopaedics and Related Research,* Vol. 37, 1966, pp. 29–48.

[*41*] Instruction Bulletins No. IB-233, IB-221, IB-228, and IB-223-A, Measurement Group, Inc., Photoelastic Division, Raleigh, NC, 1982.

[*42*] Blacker, G. and Charnley, J., *Clinical Orthopaedics and Related Research,* Vol. 137, 1978, pp. 15–23.

[*43*] Coventry, M. B. and Stauffer, R. N., "Long-Term Results of Total Hip Arthroplasty," *The Hip: Proceedings,* Tenth Open Scientific Meeting of The Hip Society, Mosby, St. Louis, 1982, pp. 34–41.

[*44*] Charnley, J. and Curic, Z., *Clinical Orthopaedics and Related Research,* Vol. 95, 1973, pp. 9–25.

[*45*] Charnley, J., *Clinical Orthopaedics and Related Research,* Vol. 111, 1975, pp. 105–120.

[*46*] Nicholson, O. R., *Clinical Orthopaedics and Related Research,* Vol. 95, 1973, pp. 217–223.

[*47*] Crowninshield, R. D., Brand, R. A., Johnston, R. C., and Pedersen, D. R., "An Analysis of Femoral Prosthesis Design: The Effects on Proximal Femur Loading," *The Hip: Proceedings,* Ninth Open Scientific Meeting of The Hip Society, Mosby, St. Louis, 1981, pp. 111–122.

[*48*] Andersson, G. B. J., Freeman, M. A. R., and Swanson, S. A. V., *Journal of Bone and Joint Surgery,* Vol. 54B, 1972, pp. 590–599.

[49] Willert, H. G., Ludwig, J., and Semlitsch, M., *Journal of Bone and Joint Surgery,* Vol. 56A, 1974, pp. 1368–1382.
[50] Moreland, J. R., Gruen, T. A., Mai, L., and Amstutz, H. C., "Aseptic Loosening of Total Hip Replacements: Incidence and Significance," *The Hip: Proceedings,* Eighth Open Scientific Meeting of The Hip Society, Mosby, 1980, pp. 281–291.

James A. Davidson,[1] *Michael Bushelow,*[2] *Andrew J. Gavens,*[1] *and Michael F. DeMane*[1]

Effect of Press Fit on Lateral Stem Stresses and the Integrity of Porous-Coated Femoral Prostheses: An *In Vitro* Strain Gage Study

REFERENCE: Davidson, J. A., Bushelow, M., Gavens, A. J., and DeMane, M. F., "**Effect of Press Fit on Lateral Stem Stresses and the Integrity of Porous-Coated Femoral Prostheses: An *In Vitro* Strain Gage Study,**" *Quantitative Characterization and Performance of Porous Implants for Hard Tissue Applications, ASTM STP 953,* J. E. Lemons, Ed., American Society for Testing and Materials, Philadelphia, 1987, pp. 264–275.

ABSTRACT: A strain gage analysis was performed to determine the stresses generated on the lateral surface of two types of porous-coated femoral prostheses. One type of prosthesis was made of wrought titanium (Ti-6Al-4V) alloy with a porous polysulfone coating; the other was a porous-coated alloy meeting the requirements of the ASTM Specification for Cast Cobalt-Chromium-Molybdenum Alloy for Surgical Implant Applications (F 75-82). Various degrees of fit were evaluated, ranging from a loose fit to one in which the stem was cemented into the femoral canal. The results showed that modeling pressfit porous-coated stems using cemented conditions can produce peak stresses on the lateral surface of the stem significantly lower than those for noncemented stems. Based on the fatigue strength of the stem material, the authors show that assessing stem integrity with the assumption that the cemented stem represents a complete bone ingrowth condition may lead to a design evaluation of the femoral stem that is overoptimistic for *in vivo* fatigue load conditions.

KEY WORDS: porous implants, porous-coated femoral prostheses, femoral stem stresses, fatigue integrity, press fit, titanium alloy, cobalt F 75 alloy

The use of porous materials for biological fixation of implant materials has been well documented in many animal studies conducted over the last two decades. Because of the success of these studies, the use of porous-coated femoral prostheses in humans has increased significantly in the last few years. Approximately 35% of all total hip arthroplasties performed in the United States today are done without the use of bone cement.

While biological fixation of a porous-coated femoral implant has shown to be advantageous, there is still concern about the fatigue integrity of these implants when a porous coating is present on the lateral, tension-loaded surface. Recent articles have demonstrated that a significant decrease in fatigue strength of both cobalt-based and titanium-based alloys occurs as a result of the porous coating sintering process [1,2]. Thus, it is critical that tensile stresses that can develop on the lateral surface of a femoral stem be well understood to ensure stem integrity.

A porous-coated femoral implant requires a very precise surgical procedure to ensure

[1] Materials research director, research engineer, and engineering manager of reconstructive products, respectively, Richards Medical Co., Memphis, TN 38116.
[2] Research engineer, Howmedica Corp., Rutherford NJ 07070.

that the implant is press fitted into the femur in an optimum fashion so that micromovement between the bone and the implant is minimal. Researchers have demonstrated that if the press fit is not optimal, and excess micromotion occurs, bone ingrowth may be inhibited. It has also been demonstrated recently, by the evaluation of two porous-coated implants, that the actual amount of biological fixation in humans is less than that observed in animal models [3]. Moreover, there was far less contact between bone and the implant than was indicated by X-ray review. For these reasons, the integrity of porous-coated stems when optimal fixation of these components is not achieved must be evaluated.

The integrity of porous-coated stems has been examined by modeling biological fixation by means of bone cement [4,5]. Unfortunately, there are several drawbacks to such a model: (1) the model does not enable immediate postoperative stem stress to be evaluated during a period in which no ingrowth has occurred; (2) the model does not account for the possibility of partial amounts of ingrowth; and (3) the model does not account for possible subsidence of the femoral prosthesis prior to ingrowth and the associated changes that may occur with respect to lateral stem stresses.

In this study, the integrity of a titanium alloy femoral component with a low-elastic-modulus porous polysulfone coating and a cobalt alloy femoral component with a high-elastic-modulus porous cobalt coating were tested by varying the degree of the press fit between the component and the femur. These tests modeled fixation ranging from a poor press fit with no bone ingrowth to an excellent press fit with complete bone ingrowth (that is, cemented). The results are compared with fatigue data characteristic of the particular stem material to assess the overall integrity of the stem.

Procedure

Two types of porous-coated femoral prostheses were evaluated to determine the lateral surface stress in relation to the degree of fit in the femoral canal of an embalmed cadaver femur.

Materials

One type of femoral stem used in this study was an intermediate-size (Size 3), collarless, porous polysulfone (PPSF) coated titanium alloy (Ti-6Al-4V) stem. The other stem was a small intermediate-size cast cobalt-chromium-molybdenum alloy stem (Moore-type endo-prosthesis) [ASTM Specification for Cast Cobalt-Chromium-Molybdenum Alloy for Surgical Implant Applications (F 75-82)], with a porous coating that was made of the same material and sintered to the stem surface. Unlike the titanium stem, this cobalt stem contained a collar. Photographs of both stems are shown in Fig. 1.

Strain Gage Analysis

Axial strain gages were attached to selected locations along the lateral surface of both stems, as illustrated in Fig. 2. The gages were applied to areas in which the porous coating had been removed and the substrate polished. Once the gages were secured, a thin protective layer of silicone rubber was placed over the gages and molded to fill the gap left by the removed porous coating layer. A strain gage was also applied to the calcar region along the medial side of the femur used to test the cobalt alloy stem.

For the PPSF stem, two embalmed adult human cadaver femurs were selected and roent-genograms were made of each to determine the degree of press fitting the prosthesis would have in the femur. Femur 1 was chosen so that a poor press fit would result. Templating of

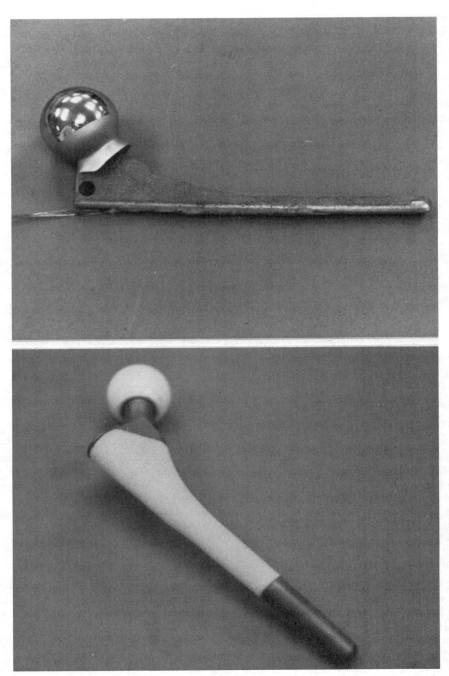

FIG. 1—*Photographs of the porous-coated cobalt alloy stem* (top) *and the PPSF titanium alloy stem* (bottom) *investigated in this study.*

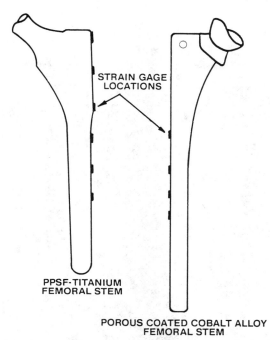

STRAIN GAGE
LOCATIONS

PPSF-TITANIUM
FEMORAL STEM

POROUS COATED COBALT ALLOY
FEMORAL STEM

FIG. 2—*Location of the strain gages along the lateral surface of the PPSF* (left) *and cobalt alloy* (right) *stems.*

the femur demonstrated that a prosthesis of a larger size than that used in this study would have been indicated clinically. The other femur, Femur 2, was chosen to effect a good press fit of the prosthesis to the bone. Implantation of the strain-gage-attached PPSF stem to effect a good press fit was performed using standard total hip arthroplasty techniques and appropriate instrumentation for the size of stem used in the study. To model the effect of bone ingrowth, the prosthesis was cemented into the second femur after the press fit test was completed.

For the porous-coated F 75 cobalt alloy stem, a third femur was used. For the first test, the femur was carefully reamed to effect a good press fit. To represent a poor fit, the same femur was overreamed. Once this test was completed, the stem was removed and cemented back into the femur to model full bone ingrowth conditions.

In all tests, the femoral condyles were removed from the cadaver femora, and the distal 7 cm of the femora were potted in bone cement in fixed tubes. The potted femora were mounted on a 9072-kg (20 000-lb)-capacity MTS closed-loop servo-hydraulic test machine at an angle of 22°. Loads were applied through a polyethylene cup potted in bone cement in 34-kg (75-lb) increments. A photograph of the test setup is shown in Fig. 3. For each test, a curve showing the applied load versus lateral stem strain was plotted. Comparison of these curves provided an indication of the effect of fixation on stem stresses for each type of stem.

Results

The lateral stem strains for the PPSF stem at 272 kg (600 lb) of applied load are shown for each degree of fit in Fig. 4. Each degree of fit is shown with the cemented condition

FIG. 3—*Photograph showing the test setup for testing the strain-gage-attached porous-coated femoral stems in a cadaver femur.*

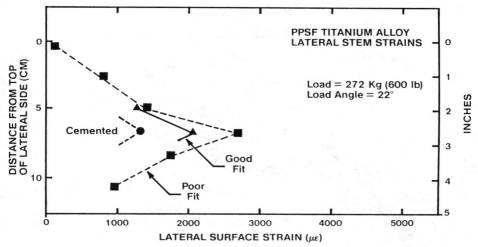

FIG. 4—*Lateral surface strain profiles for three conditions of fixation of the PPSF hip stem at a load of 272 kg (600 lb).*

showing the lowest peak strain. For the cemented and "good fit" conditions, only two gages were applied (in areas of peak strain) to minimize disruption of the lateral porous coating surface. For the cemented condition, one of the two gages failed prior to testing. However, the remaining functional gage was representative of the peak strain region. Similar plots are given in Fig. 5 representing the poor, good, and cemented conditions for the porous F 75 cobalt alloy stem.

In Figs. 6 and 7, the peak lateral stem stresses for the PPSF and the F 75 cobalt alloy stems, respectively, show the effect of fixation on stem stresses. In these figures, the peak strains are plotted as a function of applied load.

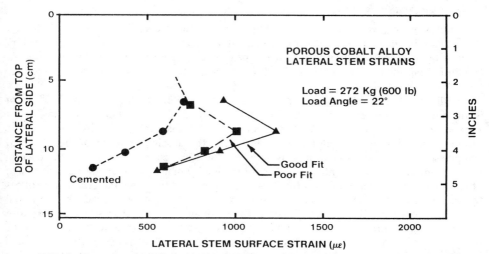

FIG. 5—*Lateral surface strain profiles for three conditons of fixation of the porous-coated cobalt alloy hip stem at a load of 272 kg (600 lb).*

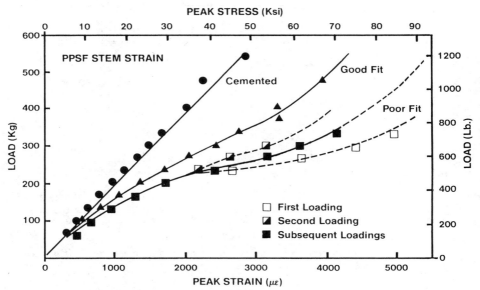

FIG. 6—*Peak lateral surface strains as a function of applied load and degree of fixation for the PPSF titanium alloy stem.*

Discussion

From the results presented in Figs. 6 and 7, it is clear that, for both types of porous-coated stems, the degree of fixation is important relative to the peak lateral stem stress. In general, as the fit is improved, the lateral stem strain decreases.

Cadaver Calibration Results

Comparison of Figs. 6 and 7 reveals two interesting differences. One difference is that the load-strain curve is very linear for the cobalt alloy stem and nonlinear for the titanium alloy stem. A second difference is that the peak strains are more severe for the good fit condition than for the poor fit condition for the cobalt alloy stem. The opposite is true for the PPSF stem.

The difference in load-strain linearity for the noncemented conditions of the PPSF stem is attributed to the combination of a low-elastic-modulus titanium alloy substrate and an even lower modulus coating. The elastic modulus of the titanium alloy is roughly half that of the cobalt alloy. On loading, more deflection in the proximal portion of the stem and compression of the porous coating can occur for the PPSF stem. Thus, as loading occurs, the bone gradually absorbs more and more of the load. Finally, the bone in the medial side of the femur establishes a uniform resistance to the force of the stem and gradually reduces the incremental increase in lateral stem strain with increasing applied load. As the initial press fit is improved, this process occurs at lower applied stem loads because of the more uniform contact initially, thereby producing lower stem strains, as indicated in Fig. 6 for the good fit condition. Cementing fixes the stem, and thus the lateral stem strain increases linearly, as is evident in both the PPSF and cobalt alloy results shown in Figs. 6 and 7. The higher modulus cobalt alloy stem is unable to deflect proximally as easily as the PPSF stem.

As a result, less load is distributed to the bone, and the load-strain curves behave in a more linear manner.

The reason the poor fit condition shows lower strains than the good fit condition in the cobalt alloy stem is attributed to the presence of the collar. Contact between the collar and the bone in the calcar region improves the stem support proximally and has been shown [6–9] to reduce the stem strain and increase compressive strain in the bone [9,10]. In the good fit condition, the cobalt alloy stem was tightly press fitted into the femur. Because of this, the stem resisted the applied load primarily through frictional resistance in the intramedullary canal. In the poor fit condition, less frictional resistance along the stem resulted, thus allowing the collar to load on the bone in the calcar region. Because of this increased proximal support by the collar, a smaller net moment and lower lateral stem strains resulted. Strain gages attached to the calcar region of the femur containing the cobalt alloy stem support this argument. The compressive strain in the collar, recorded at a 272-kg (600-lb) load for the good fit condition, was 280 µstrain. However, in the poor fit condition, this compressive strain increased several hundred percent to 1738 µstrain. In the cemented condition, this strain decreased back to a low level, reflecting load support by the cement.

One final observation regarding the lateral strain calibration tests is that the collarless PPSF stem tended to settle during testing of the poor fit condition. The curve shown in Fig. 6 with the open squares represents the recorded strain during the first loading cycle. As a result of settling, the second load cycle produced significantly lower strains, indicated by the partially closed squares. The stem finally settled on subsequent loading, producing strains indicated by the closed squares. All the other test results did not change after the application of subsequent loading cycles. Thus, for collarless stems, a good press fit results not only in lower stem strains, but also in less subsidence of the stem in the femoral canal.

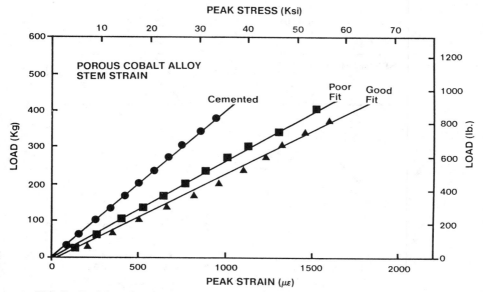

FIG. 7—*Peak lateral surface strains as a function of applied load and degree of fixation for the porous-coated cobalt alloy stem.*

Stem Fatigue Integrity

The stem strains determined from the cadaver calibration study are needed to determine the integrity of the stem under repetitive loading (fatigue) conditions encountered in service through normal daily activity of the patient. For best appreciation of this aspect of the study, the strain data were converted to stress values, using Young's modulus of the substrate material, and then normalized by dividing by the fatigue endurance limit of the material. In rotating bending fatigue tests, wrought titanium alloy exhibits a typical fatigue endurance limit in the range of 552 to 689 MPa (80 to 100 ksi) [1,11–15]. As a conservative approach, the peak lateral stem stresses for each press fit condition of the PPSF-coated titanium stem are normalized using a fatigue endurance limit of 552 MPa (80 ksi), as shown in Fig. 8.

In the cast and annealed condition, ASTM F 75 cobalt alloy shows a fatigue endurance [16–18] in rotating bending fatigue between 206 and 310 MPa (30 and 45 ksi). The R ratio (minimum cyclic stress divided by the maximum cyclic stress) is equal to −1 for rotating bending fatigue. Higher R ratio fatigue tests, such as that for cantilever bending (R = 0), show superior fatigue performance for a given maximum value of cyclic stress [19]. Femoral prostheses experience loading similar to the cantilever bending condition. For cast F 75 cobalt alloy in tension-tension fatigue, the typical fatigue endurance limit occurs at a peak cyclic stress of approximately 379 MPa (55 ksi) [20,21]. However, the cobalt alloy stem in this study contains a sintered coating. The combination of the sintering treatment and the notch effects from the presence of the beads on the lateral surface reduces the fatigue endurance limit in cantilever bending to a reported average value of 241 MPa (35 ksi) [5]. Thus, in Fig. 9 the peak lateral stem stress is normalized for a fatigue endurance limit of 241 MPa.

The typical patient's weight for the intermediate-size PPSF titanium stem is approximately 68 kg (150 lb). The typical patient's weight for the small/intermediate-size porous-coated cobalt alloy stem is approximately 57 kg (125 lb). In Figs. 8 and 9, a dashed line is shown representing a hip load equal to six times the body weight of a 68 and 57-kg person, respectively.

A factor of six times body weight (6 × BW) would be a conservative estimate of average

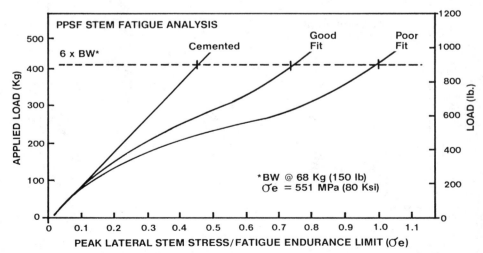

FIG. 8—*Peak lateral surface stresses, normalized with respect to the fatigue endurance limit, for the PPSF titanium alloy hip stem.*

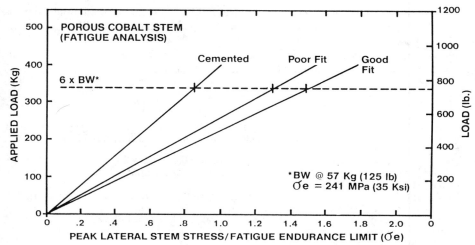

FIG. 9—*Peak lateral surface stresses, normalized with respect to the fatigue endurance limit, for the porous-coated cobalt alloy hip stem.*

hip loads occurring through normal daily activity [22–24]. Thus, it is desirable to show that the ratio of the peak stem stress to the fatigue endurance limit is less than 1.0, which suggests that the stem would perform indefinitely under these conditions. However, if this ratio is greater than 1.0, a finite fatigue life results and concern arises with respect to the stem integrity under these hip loading conditions.

In Fig. 8, one can see that the PPSF stem shows a ratio of less than 1.0 at 6 × BW for all press fit conditions. However, the porous-coated F 75 cobalt alloy stem (Fig. 9) shows a ratio below 1.0 only in the cemented condition. If this condition is, in fact, representative of a complete bone ingrowth condition, then the integrity of the stem under these loading conditions is sufficient. Because it is not clear when, or if, full bone ingrowth occurs, the stem may perform *in vivo* in a manner inferior to that represented by a stem tested in a completely cemented condition. The presence of bone or tissue ingrowth, which may occur prior to full weight bearing, however, will stiffen the system and result in peak stem stresses more characteristic of the fully cemented condition. Nevertheless, it remains unclear at the present time whether cemented conditions are, in fact, representative of complete ingrowth conditions.

It is obvious that the demands placed on a porous-coated press fit femoral component are not the same as those placed on a cemented stem, particularly in the early postoperative stages when bone or fibrous tissue ingrowth into the porous surface is occurring. Because of the unknowns associated with stem stresses on porous-coated stems, it is probably wise to assume, for design purposes, that the ideal cemented condition may never occur if, in fact, this condition is a reflection of full ingrowth. Thus, designing for an initial good press fit would provide an added margin of safety in determining whether a particular porous-coated femoral stem design can withstand the anticipated *in vivo* demands placed on it.

Conclusions

1. Maximum lateral stem stresses determined from the stem in the cemented condition were significantly lower than those for the noncemented condition.

2. Because of variations in fatigue performance of different orthopedic materials, a press fit femoral component design based on the assumption that a cemented condition represents full bone ingrowth may be unconservative.

3. For conservative design analysis of porous-coated femoral components, the use of stresses determined from a good initial press fit is recommended.

References

[1] Cook, S. D., Georgette, F. S., Skinner, H. B., and Haddad, R. J., Jr., "Fatigue Properties of Carbon and Porous-Coated Ti-6Al-4V Alloy," *Journal of Biomedical Materials Research,* Vol. 18, 1984, pp. 497–512.

[2] Georgette, F. S. and Davidson, J. A., "The Effect of HIP'ing on the Fatigue and Tensile Strength of a Cast, Porous-Coated Co-Cr-Mo Alloy," *Journal of Biomedical Materials Research,* Vol. 20, October 1986, pp. 1229–1248.

[3] Bobyn, J. D. and Engh, C. A., "Human Histology of the Bone–Porous Metal Implant Interface," *Orthopaedics,* Vol. 7, 1984, pp. 1410–1421.

[4] Pilliar, R. M. and Bratina, W. J., "Micromechanical Bonding at a Porous Surface/Structured Implant Interface—The Effect on Implant Stressing," *Journal of Biomedical Engineering,* Vol. 2, No. 1, 1980.

[5] "POROCOAT: A Technical Review of Porous-Coated Implants for Biological Fixation," Technical Report, DePuy, Inc., Warsaw, IN, 1984.

[6] Galante, J. O., Andriacchi, T., Rostoker, W., Schultz, A., and Belytschko, T., "Femoral Stem Failures in Total Hip Replacements," *THE HIP,* The Hip Society, Ed., C. V. Mosby Co., St. Louis, 1975, Chapter 15.

[7] Andriacchi, T. P., Galante, J. O., Belytschko, T. B., and Hampton, S., "A Stress Analysis of the Femoral Stem in Total Hip Prosthesis," *Journal of Bone and Joint Surgery,* Vol. 58-A, No. 5, July 1976.

[8] Weightman, B., "The Stress in Total Hip Prosthesis Femoral Stems: A Comparative Experimental Study," *Advances in Artificial Hip and Knee Joint Technology,* M. Schaldach, Ed., 1976, pp. 138–147.

[9] Lewis, J. L., Askew, M. J., Wixson, R. L., Kramer, G. M., and Tarr, R. R., "The Influence of Prosthetic Stem Stiffness and of a Calcar Collar on the Stresses in the Proximal End of the Femur with a Cemented Femoral Component," *Journal of Bone and Joint Surgery,* Vol. 66-A, No. 2, February 1984, pp. 280–286.

[10] Oh, I. and Harris, W. H., "Proximal Strain Distribution in the Loaded Femur," *Journal of Bone and Joint Surgery,* Vol. 60-A, No. 1, January 1978.

[11] Williams, D. F., "Titanium as a Metal for Implantation: I. Physical Properties," *Journal of Medical Engineering and Technology,* Vol. 3, 1977, p. 195.

[12] Stubbington, C. A. and Bower, A. W., "Improvements in Fatigue Microstructure Control," *Journal of Materials Science,* Vol. 9, 1974, pp. 941–947.

[13] Stubbington, C. A. and Bower, A. W., "The Effect of Section Size on the Fatigue Properties of Ti-6Al-4V Bars," *Titanium Science and Technology,* R. I. Jaffee and H. M. Burte, Eds., Plenum Press, New York, 1973.

[14] Zwicker, V., Bless, M., and Hofmann, U., "Effects of Ti-6Al-4V Metallurgical Structures on Fatigue Properties," *The Science and Application of Titanium,* R. I. Jaffee and W. E. Promisell, Eds., Pergamon Press, New York, 1970.

[15] Bartlo, T. J., "Effect of Microstructure on the Fatigue Properties of Ti-6Al-4V Bar," *Fatigue at High Temperature, ASTM STP 459,* American Society for Testing and Materials, Philadelphia, 1969, pp. 144–154.

[16] Lorenz, M., Semlitsch, M., Panic, B., Weber, H., and Wilbert, H. G., "Fatigue Strength of Cobalt-Base Alloys with High Corrosion Resistance for Artificial Hip Joints," *Sulzer Technical Review,* Sulzer Bros. Ltd., Winterthur, Switzerland, 1978.

[17] Ducheyne, P. and Hastings, G. W., *Metal and Ceramic Biomaterials,* Vol. 1, *Structure,* CRC Press, Boca Raton, FL 1984.

[18] Davidson, J. A. and Bardos, D. I., "Historical Need for High-Strength Biomedical Implant Materials," Paper No. 8501-001, *Proceedings,* Westec '85, American Society for Metals, Los Angeles, CA, 18–21 March 1985.

[19] Dieter, G. E., *Mechanical Metallurgy,* R. F. Mehl and M. B. Bever, Eds., McGraw Hill, New York, 1961.

[20] Miller, H. L., Rostoker, W., and Galante, J. O., "A Statistical Treatment of Fatigue of the Cast Co-Cr-Mo Prosthesis Alloy," *Biomedical Materials Research,* Vol. 10, 1976, pp. 399–412.
[21] Ducheyne, P., Wevers, M., and DeMeester, P., "Fatigue Properties of Implant Materials in Hip Prosthesis Form: A Standardized Test," *Journal of Biomedical Materials Research,* Vol. 17, 1983, pp. 45–57.
[22] Paul, J. P., "Loading on Normal Hip and Knee Joint Replacements," *Advances in Hip and Knee Joint Technology,* M. Schaldoch and D. Hoffmen, Eds., Springer-Verlag, Berlin, 1976, pp. 53–70.
[23] Crowninshield, R. D., Johnston, R. C., Brand, R. A., Pedersen, D. R., Wilson, J. A., and Tolbert, J. R., "An Engineering Analysis of Total Component Design, *Orthopaedic Review,* Vol. 12, No. 11, 1983.
[24] Zwicker, A. V. and Schmidt, H. J., "Mechanical Properties of Metallic Materials for Long-Term Use in Highly Stressed Locomotor Systems," *Advances in Artificial Hip and Knee Joint Technology,* Springer-Verlag, New York, 1976.

DISCUSSION

E. P. Lautenschlager[1] (*written discussion*)—In addition to filling up the empty spaces, I have always thought that bone cement may have a secondary role as a stress mitigator between the high-modulus metallic prosthesis and the low-modulus bone, much like placing tires between your metal automobile and the street. Therefore, although your current tests were quasi-static, might you consider keeping the same experimental setup and running it in a cyclic dynamic mode to observe any possible shifting in the strain patterns?

J. A. Davidson, M. Bushelow, A. J. Gavens, and M. F. DeMane (*authors' closure*)—Your point is well taken. Unlike the conditions for cemented implants, repetitive loading in a press-fit situation can result in a variation in stem stress patterns. Because of this potential effect, multiple loads were applied to the "poor" press-fit porous polysulfone hip stem. As expected, the stem stresses changed (decreased). However, after three loading cycles, the stem stresses stabilized. Probably further changes in stress may occur *in vivo* until bone and fibrous tissue stabilize the implant. It is unfortunately not possible to assess this situation accurately.

[1] Department of Biological Materials, Northwestern University Chicago, IL 60611.

Subrata Saha,[1] James A. Albright,[2] Michael E. Keating,[2] and Raghunath P. Misra[3]

A Biomechanical and Histological Examination of Different Surface Treatments of Titanium Implants for Total Joint Replacement

REFERENCE: Saha, S., Albright, J. A., Keating, M. E., and Misra, R. P., **"A Biomechanical and Histological Examination of Different Surface Treatments of Titanium Implants for Total Joint Replacement,"** *Quantitative Characterization and Performance of Porous Implants for Hard Tissue Applications, ASTM STP 953,* J. E. Lemons, Ed., American Society for Testing and Materials, Philadelphia, 1987, pp. 276–285.

ABSTRACT: The objective of this study was to evaluate biomechanically and histologically the effectiveness of different surfaces—specifically, smooth, textured, and plasma-sprayed porous surfaces—on titanium implants in achieving stable fixation for total joint replacements.

Hollow cylindrical metal plugs (5 to 6 mm in diameter) with different surfaces were implanted in the femora of two dogs. After five months of implantation, pull-out tests were carried out on these cylinders. The ultimate shear strengths were 11.54 ± 2.96, 3.17 ± 0.81, and 1.04 ± 0.32 N/m^2 for the plasma-sprayed, textured, and smooth surfaces, respectively. As expected, the smooth cylinders showed the minimum pull-out strengths and the porous-coated ones the maximum. Histological examination showed maximum bony ingrowth in the plasma-sprayed group and moderate growth in the textured group. Our results suggest that a textured surface or a porous coating (plasma spray) on titanium implants could significantly improve the long-term stability of uncemented total joint replacements.

KEY WORDS: porous implants, porous coating, titanium, shear strength, plasma-sprayed porous coating

Although total hip arthroplasty (THA), as developed by Charnley, has revolutionized the field of orthopedic surgery, most long-term follow-up studies have reported a high rate of component loosening or migration, leading to revision or failure of the joint replacement [1–3]. For instance, a review of Charnley total hip arthroplasties done at the Mayo Clinic showed an incidence of 24% femoral component and 6% acetabular component loosening within the five-year follow-up. This increased to 30% femoral component and 11% acetabular component loosening by ten years [4]. In younger patients, the risk of component loosening was shown to be even greater.

[1] Professor and coordinator of bioengineering, Department of Orthopaedic Surgery and Physiology, Louisiana State University Medical Center, Shreveport, LA 71130.

[2] Professor and chairman, and assistant professor, respectively, Department of Orthopaedic Surgery and Physiology, Louisiana State University Medical Center, Shreveport, LA 71130; Dr. Keating is currently affiliated with the Center for Hip and Knee Surgery, Indianapolis, IN.

[3] Professor, Department of Pathology, Louisiana State University Medical Center, Shreveport, LA 71130.

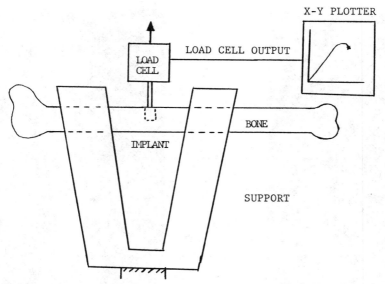

FIG. 1—*Diagram of the experimental setup for pull-out testing of the metal plugs.*

The exact mechanism of loosening is not well established. However, many investigators believe that acrylic bone cement is partly responsible for long-term aseptic loosening of the prosthetic components [5]. This has led to the increasing use of cementless fixation in total joint replacement. Several investigators have used porous materials that allow bony ingrowth as a method of fixation [6–8]. Various materials, including metals, polymers, and ceramics, have been used, but metals, chiefly cobalt and titanium alloys, are the most commonly used. Several types of porous surfaces are now commercially available for orthopedic implants; however, their role in achieving optimum fixation is not clear. The objective of this study was to evaluate biomechanically the effectiveness of different titanium alloy implants for total joint replacement. The bone/implant interfaces were also examined histologically to determine the nature of the bony ingrowth and its possible correlation with the ultimate shear strength.

Materials and Methods

Three groups of hollow cylindrical metal plugs (5 to 6 mm in outer diameter and 6 mm in length) of a titanium alloy (Ti-6Al-4V) with three different surfaces were obtained for implantation. These plugs had longitudinal threaded holes for ease of subsequent mechanical testing. All the implants used in the study were pressure washed, vapor degreased, and passivated. The plugs had the following surfaces:

(*a*) smooth—that is, a blast finish with a roughness of 1.27 to 1.52 μm (50 to 60 μin.) (this finish is typically found on orthopedic implants);
(*b*) textured—that is, a blast finish with a surface roughness of 3.81 μm (150 μin.); and
(*c*) a plasma-sprayed porous coating of pure titanium. (The pore size ranged from 1000 + μm at the outer surface to less than 200 μm at a point approximately 0.5 mm into the porous coating; the porous coating thickness was approximately 0.5 mm.)

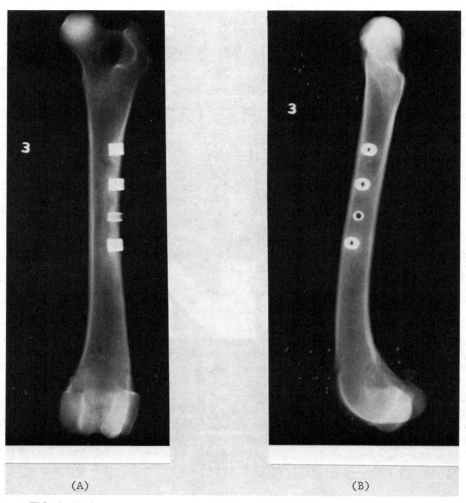

FIG. 2—*Radiographs of a canine femur showing the metal implants 5.25 months postoperatively:* (a) *anteroposterior view,* (b) *lateral view.*

Healthy adult mongrel dogs (approximately 30 kg in weight) were chosen as the experimental animal for implantation of the prostheses. The animals were maintained in accordance with the latest "Guide for the Care and Use of Laboratory Animals" established by the National Research Council of the National Academy of Sciences. Prior to surgical implantation, the animals were radiographically screened for normal bone morphology. Appropriate anesthetic agents were used during all surgical procedures and the principles of aseptic techniques were strictly followed. The implantation site for all the specimens was the mid-diaphyseal region of the femur. Holes with diameters slightly smaller than the implant diameter were drilled in the lateral cortex using an air-powered surgical drill. The drill diameter was 5.5 mm for the smooth and textured implants (6 mm in diameter) and 3.4 mm for the plasma-sprayed porous-coated implants (4.8 mm in diameter). The implants were then introduced into the drilled holes so that the top surface of the implant was level

TABLE 1—*Mean (±1 standard deviation) maximum shear stress and stiffness for the smooth, textured, and porous-coated (plasma-sprayed) implants.*

Surface	No. of Samples Tested	Pull-Out Stress, MN/m²	Stiffness, MN/mm
Smooth	5	1.042 ± 0.317	0.044 ± 0.012
Textured	4	3.170 ± 0.806	0.053 ± 0.022
Porous-coated (plasma-sprayed)	5	11.536 ± 2.959	0.250 ± 0.065

with the periosteal surface. The animals were allowed to recover normally and they showed no signs of limping during the follow-up period. There were no cases of infection and both animals remained healthy throughout the length of time of the protocol. However, one animal suffered a unilateral femur fracture through a drill hole, which healed normally. The animals were sacrificed at 5.25 months postoperatively.

After the animals were sacrificed, excess soft tissue was removed from the excised femurs

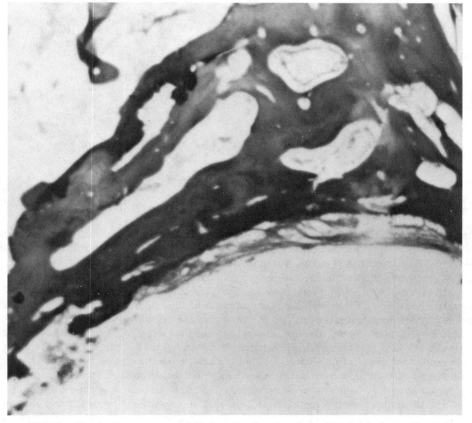

FIG. 3—*Histologic appearance of the bone/implant interface of implants after mechanical testing (decalcified thin sections, hematoxylin and eosin stain):* (a) *smooth surface (magnification, ×125).*

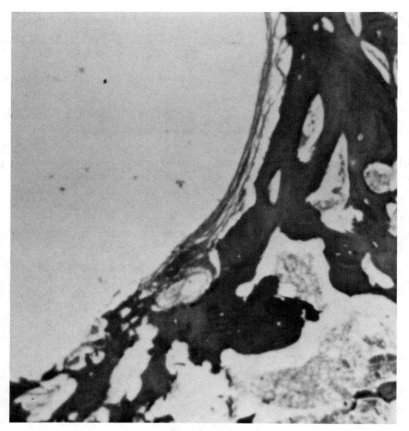

FIG. 3—*Continued:* (b) *textured surface* (*magnification,* ×200).

and the bones were radiographed. The bone covering the periosteal surface of the implants was then removed. Immediately afterwards, the bones were tested mechanically in an Instron servohydraulic mechanical testing machine. The bones were held horizontally in a special jig; a threaded rod was attached to the implant, and a tensile force was applied using a crosshead speed of 25.4 mm/min (Fig. 1). The load-deformation curve was recorded automatically on an X-Y plotter. The shear strength of the implant/bone interface was calculated by dividing the maximum load by the area of the cylindrical surface of the implant.

After mechanical testing, the bones were sectioned transversely, fixed in 10% buffered formalin solution, and decalcified in ethylenediaminetetraacetic acid (EDTA). They were then embedded in paraffin, prepared in thin sections, stained with hematoxylin and eosin, and examined with light microscopy. One femur that had fractured and healed was not tested mechanically. Cubes of bone containing the implants were first cut on a band saw and embedded in polymethyl methacrylate, and then thin sections showing the implant/bone interface were cut with a diamond saw (Beuhler). These sections were also stained and examined histologically using a light microscope (Leitz).

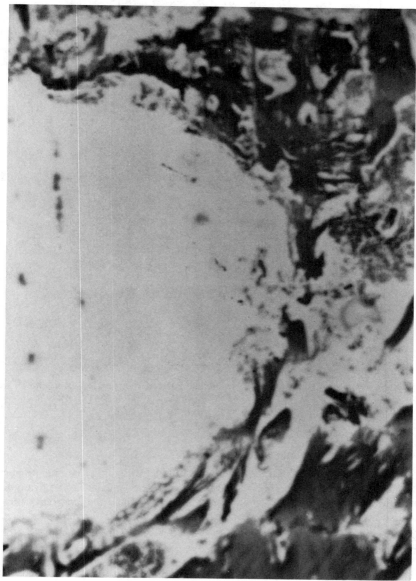

FIG. 3—*Continued:* (c) *porous-coated (plasma-sprayed) surface (magnification, ×125).*

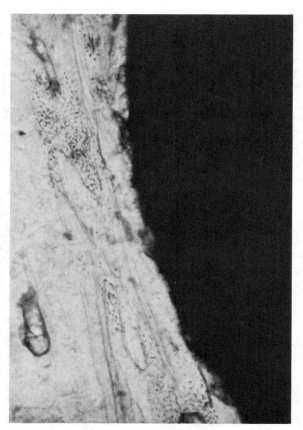

FIG. 4a—*Photomicrographs of the bone/implant interface showing close apposition of the bone to the implant (black) for a textured implant (undecalcified thin section with the implant in place).*

Results and Discussion

The animals appeared to tolerate the implants well and were able to ambulate normally. On gross examination of the excised bones, all the cylinders appeared to be firmly fixed in the bone. In most of the cylinders, bone had also partially filled the inner cavity from the endosteal surface, and in some cases, bone had also grown over the periosteal surface of the implants. Radiographs of the bones containing the metal cylinders did not show any radiolucent lines between the bone and the implant (Fig. 2). This suggests excellent biocompatibility of the implant material.

The interface shear strengths for the three groups of implants, as determined from the mechanical pull-out tests, are shown in Table 1. As expected, the smooth cylinders showed the minimum strength and the plasma-sprayed porous-coated cylinders the maximum. Analysis of variance applied to the strength data indicated that the differences in shear strength between each group were statistically significant ($P < 0.05$). Table 1 also shows that the interfacial stiffness was significantly higher for the porous surface than for the smooth and textured surfaces.

Our results agreed in general with those reported by Hahn and Palich [9], who also found significant improvement in the shear strength of porous (plasma-sprayed) titanium implants after 14 and 26 weeks of implantation, in comparison with uncoated control specimens.

Figure 3, *a* through *c*, shows examples of histological sections of the bone interfaces, after mechanical testing, for the three different implant surfaces examined. All the surfaces showed some areas in which a fibrous tissue layer was interposed between the implant and the surrounding mature bone, but this layer was smaller for the textured surface than for the smooth surface, and it was much thinner and even absent in most areas for the plasma-sprayed porous surface (Fig. 3). The plasma-sprayed surface also showed fragmentation of bony trabeculae as an evidence of mature bone ingrowth into the pores of the implant (Fig. 3*c*).

Figures 4*a* and 4*b* are photomicrographs showing the bone/implant interfaces of a textured and a plasma-sprayed porous coated implant. These specimens were not mechanically tested, thus the interfaces were intact. Mature bone was found to be in intimate contact with the surface and no evidence of any soft tissue interface region was found. As these specimens were obtained from the fractured bone, the implant was perhaps protected from high cyclic

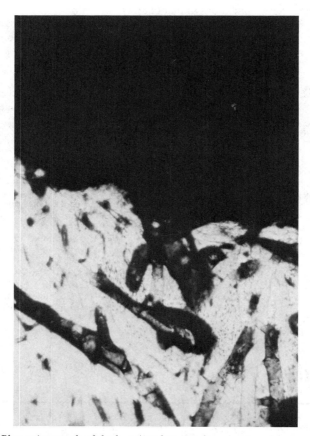

FIG. 4*b*—*Photomicrograph of the bone/implant interface showing close apposition of the bone to the implant (black) for a porous-coated (plasma sprayed) implant (undecalcified thin section with the implant in place).*

loading during locomotion. This might have allowed the formation of uninterrupted bone ingrowth, and thus no collagenous membrane was formed.

As the metal plugs used in this study were in the diaphysis of the femur, the implants were only subjected to a load during deformation of bone. The loading characteristics would be different if the implants had been fixed in the cancellous bone and directly loaded, as in the case of a total hip replacement. Therefore, further studies should be carried out to determine the bone/implant interfacial shear strength for different surfaces of an implant when fixed in cancellous bone. The authors plan to conduct such a study in the future.

Conclusions

The following conclusions can be drawn from this study:

1. This study shows the highly biocompatible nature of the titanium alloy Ti-6Al-4V for orthopedic implant applications.

2. A plasma-sprayed porous surface promoted direct adaptation of mature bone. This resulted in significant improvement in the shear strength of the bone/implant interface, in comparison with textured or smooth implant surfaces.

3. The use of titanium alloy with a plasma-sprayed porous surface or a textured surface in total joint replacements may improve the long-term stability of total joint replacements.

Acknowledgments

The authors would like to thank Biomet Inc. for supplying the metal implants and for financial support for this project. We acknowledge the technical assistance of J. A. Engelhardt in the mechanical testing of the implants.

References

[1] Sutherland, C. J., Wilde, A. H., Borden, L. S., and Marls, K. E., *Journal of Bone and Joint Surgery*, Vol. 64A, 1982, pp. 970–982.
[2] Ling, R. S. M., *Complications of Total Hip Replacement*, Churchill Livingstone, Edinburgh, 1984.
[3] Beckenbaugh, R. D. and Ilstrup, D. M., *Journal of Bone and Joint Surgery*, Vol. 60A, April 1978, pp. 306-313.
[4] Stauffer, R. N., *Journal of Bone and Joint Surgery*, Vol. 64A, September 1982, pp. 983–998.
[5] Keating, E. M. and Albright, J. A., *Schumpert Medical Quarterly*, Vol. 3, August 1984, pp. 46–53.
[6] Bobyn, J. D., Pilliar, R. M., Cameron, H. U., Weatherly, G. C., and Kent, G. M., *Clinical Orthopaedics and Related Research*, No. 149, June 1980, pp. 291–298.
[7] Engh, C. A., *Clinical Orthopaedics and Related Research*, No. 176, June 1983, pp. 52–66.
[8] Hungerford, D. S. and Kenna, R. V., *Clinical Orthopaedics and Related Research*, No. 176, June 1983, pp. 95–107.
[9] Hahn, H. and Palich, W., *Journal of Biomedical Materials Research*, Vol. 4, 1970, pp. 571–577.

DISCUSSION

J. B. Park[1] (*written discussion*)—I do not think that the static loading condition in the femoral shaft is a good test for porous ingrowth, although many use it for its cost and

[1]University of Iowa, Department of Biomedical Engineering, Iowa City, IA 52242.

convenience. Any meaningful data will only come from using load-bearing test conditions—for example, in the intramedullary cavity or another suitable location—with a simulated or actual prosthesis in place. Can you provide information on the "actual surface area" of the specimens tested?

S. Saha, J. A. Albright, M. E. Keating, and R. P. Misra (authors' closure)—At the end of the Results and Discussion section of the paper, we also mention that the loading characteristics would be different if the implants were fixed in the cancellous bone and directly loaded, as in the case of total hip replacement. However, we feel that the results are still valid as a relative comparison of three different surfaces. We did not measure the actual surface area of the tested specimens.

Patrick Dallant,[1] Alain Meunier,[1] Pascal Christel,[1]
Genevieve Guillemin,[1] and Laurent Sedel[1]

Quantitation of Bone Ingrowth into Porous Implants Submitted to Pulsed Electromagnetic Fields

REFERENCE: Dallant, P., Meunier, A., Christel, P., Guillemin, G., and Sedel, L., "**Quantitation of Bone Ingrowth into Porous Implants Submitted to Pulsed Electromagnetic Fields.**" *Quantitative Characterization and Performance of Porous Implants for Hard Tissue Applications, ASTM STP 953,* J. E. Lemons, Ed., American Society for Testing and Materials, Philadelphia, 1987, pp. 286–300.

ABSTRACT: This study has been conducted to evaluate the effect of electromagnetic fields of low frequency and low amplitude on bone growth into porous implants in order to consider a potential clinical application—that is, improvement of the anchorage in articular prostheses intended to be implanted without cement. The experiment has been conducted *in vivo* on sheep, using porous implants made of Ti-6Al-4V alloy implanted in the cortical bone. The electromagnetic stimulation was provided by external generators connected to Helmholtz coils placed on each side of the stimulated limb.

The results of this stimulation have been quantified by measuring the shear properties of the bone/implant interface and the bone regrowth around and into the porous implants in control and stimulated animals. These results were measured in a total of 16 animals sacrificed at three and five weeks after implantation. No improvement was shown in the mechanical properties of the bone/implant interface. On the other hand, the depth of bone tissue ingrowth and the ratio of bony ingrowth within the implant to the total new bone formation increased at five weeks (13 and 20%, respectively) when electromagnetic fields were used. However, this increase is not statistically significant.

KEY WORDS: porous implants, electromagnetic stimulation, bone growth, histomorphometry, push-out test

The most frequently used method for anchoring articular prostheses in bones is fixation by acrylic cement, which often deteriorates in the long-term range of implantation [1–3]. In order to overcome this problem, Judet [4] and Lord [5], in the early 1970's, proposed ensuring implant stability by using prostheses with irregular or porous surfaces. Bone growth into the irregularities or the pores of the material allows a "biological" anchorage, which is theoretically unlikely to deteriorate with time. Immediate failures are, however, occasionally observed [5]. These early failures can be related to the lack of bone growth within the implant's pores, the origin of which can be traced to one or more of the following deficiencies:

(*a*) insufficient adaptation of the prosthesis design to the bone geometry,
(*b*) poor bone stock with low osteogenic potential (that is, in the cases of revision), and

[1] Physician, research associate, surgeon, research associate, and professor, respectively, Laboratoire de Recherches Orthopédiques, Faculté de Médecine Lariboisière—Saint-Louis, 75010 Paris, France.

(c) the occurrence of shear micromotions at the bone/implant interface because of an inadequate postoperative program.

In order to avoid potential problems, patients are usually asked to put the operated limb on a non-weight-bearing regimen for several weeks, or to begin to load it progressively over a long period of time. Nevertheless, this program is difficult to achieve with elderly people, the patients most often concerned with joint replacement. This results in a prolonged stay in a wheelchair. Consequently, it would be advantageous to find a way to avoid initial non-anchorage of the prosthesis by stimulating bone ingrowth into porous implants, which would subsequently reduce or abolish the postsurgery constraints.

The possibility of speeding up osteogenesis by applying electrical stimulation [6–10] has been reported. Several researchers [11–16] have claimed positive results, with increased bone growth and shear strength of the bone/implant interface, when a direct electric current was delivered into the porous material. The prosthesis was connected to the cathode of the stimulator, while the anode was kept at a distance in the muscles.

However, stimulation through continuous current shows several drawbacks:

1. The bone growth observed was of variable significance, as it depends on the surface current density (in direct relation to the implant area in contact with the tissues) and the voltage of the current delivered by the stimulator, both of which parameters are difficult to manage in practice.

2. Although this method applied to noncemented prostheses has not yet reached the clinical stage, it is predictable that, when the method is used in humans, the stimulator or its electrodes will have to be removed after use (this is actually being done in treating nonunions with implanted electrical devices), requiring undesirable reoperations, with their potential risks of infection.

The osteogenic phenomena resulting from direct current delivered through implanted electrodes can be obtained with external use of pulsed electromagnetic induced currents (PEMIC) [10]. Their clinical use in the treatment of nonunions of long bones has become routine [7]. Therefore, it would be logical to infer that similar osteogenic potential would result from the use of PEMIC.

The previous considerations have led the authors of this paper to investigate primarily the initial stages of osteogenesis, which are critical for stabilization of the noncemented implant. The experiment described herein was performed to evaluate the effect of PEMIC, externally delivered using a pulse generator connected to Helmholtz coils, on early bone growth (at three and five weeks after implantation) into porous implants.

Materials And Methods

Implants

A titanium alloy (Ti-6Al-4V) was used for the implant. This material was chosen for its great resistance to corrosion. This nonmagnetic material spontaneously develops a titanium oxide passivation layer, which prevents any local electrolytic phenomena caused by induced currents in vivo.

The implants, manufactured by the Centre d'Etude Nucléaire of Grenoble, France, were cylindrical (4.8 mm in outside diameter) and 10 mm in length (Fig. 1). Hot pressing of Ti-6Al-4V spheres resulted in a 36 to 40% open porosity with pore sizes ranging from 200 to 300 μm (Fig. 2).

FIG. 1—*Photographic enlargement of the Ti-6Al-4V alloy porous implant (scale = 10 mm).*

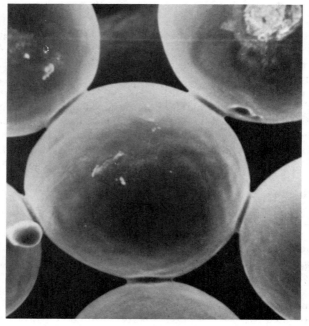

FIG. 2—*Electron microscope micrograph showing the implant's pores (×100).*

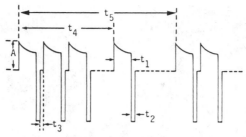

FIG. 3—*Schematic representation of the induced-current waveform employed in this study. The pulse burst waveform has the following characteristics:* t_1 = 250 μs, t_2 = 4 μs, t_3 = 5 μs, t_4 = 50 ms, *and* t_5 = 10.5 s; *amplitude* (A) = 1.6 mV. *The 1.6-mV amplitude utilized for calibration of the main polarity portion* t_1 *is an arbitrary unit referring to the probe coil described in the text.*

Electromagnetic Fields

Inductively coupled pulsating waveforms were supplied using facing and rectangular air-gap coils (70 by 120 mm) 70 mm apart. They were wound with 40 to 50 turns of different No. 16 B&S gage copper magnet wire. The coils were powered with voltage pulses timed to achieve the induced voltage configuration shown in Fig. 3. The coils were placed vertically external to a plastic cast, which maintained relatively constant geometry between the implants and the coils. The general characteristics of the signal utilized have been described elsewhere [17,18], and the electromagnetic equipment was supplied by Electro-Biology, Inc., Fairfield, NJ. The induced electric field was calibrated using a secondary coil probe, a 0.5-cm-diameter probe coil wound with 65 turns of a different, No. 36 B&S-gage, copper magnet wire [19].

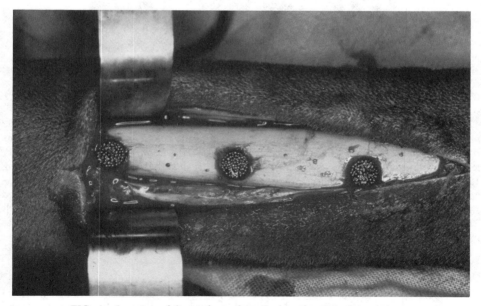

FIG. 4—*Location of the implants along the lateral aspect of a metatarsus.*

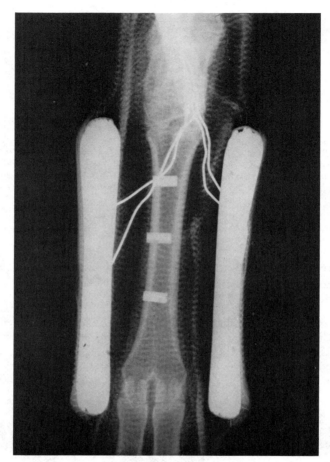

FIG. 5—*Anteroposterior X-ray of a sheep metatarsus with the three porous plugs, immobilized in a plastic cast carrying the Helmholtz coils.*

This coil was calibrated to correspond to an induced voltage field of 1.5 mV/cm *for the perimeter of the probe coil.* This voltage was related to the actual induced current by means of *in vivo* measurements using special electrodes [*19,20*]. Peak real-time current density levels were measured *in vivo* [*19*] and were between 10 and 50 μA/cm² for signals exhibiting 15 mV on the coil probe. The average magnetic field is 0.2 gauss, using the timing characteristics of the induced waveform shown in Fig. 3. The rationale for using such an induced signal was based on the results obtained from a previous fracture healing experiment [*9,21*]. This signal has demonstrated a significant speeding-up effect in the fracture healing process in rats.

Animal Experiments

Fourteen adult Prealpes sheep were used. Under general anesthesia, through a lateral approach, one of the two metatarsals on each sheep was exposed subperiosteally. Three

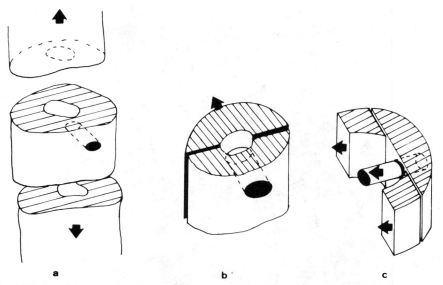

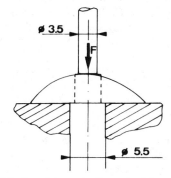

FIG. 6a—*Schematic diagram representing the different stages of sawing and push-out testing.*

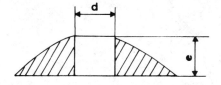

FIG. 6b—*Schematic diagram of the push-out test.*

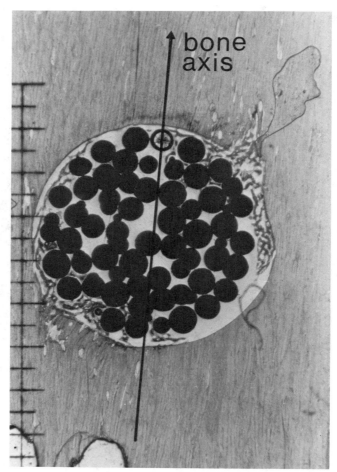

FIG. 7—*Photographic enlargement of a stained slice (3 weeks after implantation). Note the preferential ingrowth at 35° from the direction of the long bone axis.*

holes were slowly drilled under an irrigation of physiological saline in the lateral cortex with a 5-mm (outside diameter) drill (Fig. 4). Each of the three holes was then filled with a Ti-6Al-4V implant. The implant rested against the intact medial cortical bone. The incision was closed on two levels, and the distal extremity was put in a plastic cast with the ankle at 90° and the hoof immobilized in flexion.

On the outer and inner surfaces of the cast, the two rectangular coils were placed parallel and centered on the porous implants (Fig. 5). These coils were connected through a long cable to a pulse generator. The animals were stimulated for 12 h daily from 6 p.m. to 6 a.m. and were kept in restraining cages during the entire period of the experiment in order to avoid any damage to the electrical circuitry.

The stimulated animals were connected to active generators, while the controls, submitted to the same protocol, had inactive coils.

The animals were separated into two groups: one group was stimulated for three weeks and the other for five weeks.

After sacrifice, the metatarsals operated on were removed, rid of their soft tissues, and X-rayed. They were cut into three 2-cm sections, each section bearing an implant. The specimens were then immersed in a lactated Ringer's solution bath and kept frozen at $-40°C$ until mechanical tests were performed.

Mechanical Tests

Measurements of the mechanical properties of the bone/implant interface were performed using a push-out test. One of the major drawbacks of the method is the misalignment between the implant orientation and the compressive force direction applied to the specimen by the mechanical device used in the experiment.

In order to minimize this problem, a special holder was designed that allowed the specimens to be sawed in a direction perpendicular to the implant axis.

The specimens were sawed and cut in this special jig with a slow diamond saw lubricated with saline (Figs. 6*a* and 6*b*). The specimens thus obtained contained the cortical part of the implant. The cortical thickness was measured at four locations around the implant. Thirty-four specimens were available for mechanical testing.

The specimens were then placed in an Amsler vise-grip testing machine (Model 1 TZM 748). A plunger (3.5 mm in outside diameter) pushed out the implant with a velocity of 1 mm/min. The force applied and the implant displacement were recorded on an X-Y plotter. The load at rupture, the displacement at rupture and the average slope of the force-displacement curve were measured. From these values, the shear stress to failure, the deformation at rupture, and the mean rigidity were calculated.

Morphometric Analysis

After mechanical testing, the 34 available specimens were fixed in buffered formalin, then dehydrated and embedded in methyl methacrylate.

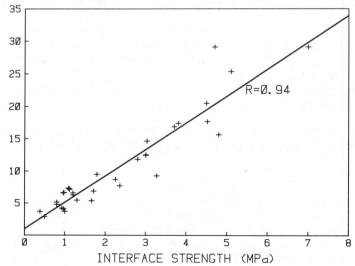

FIG. 8—*Stiffness versus strength of the bone/implant interface.*

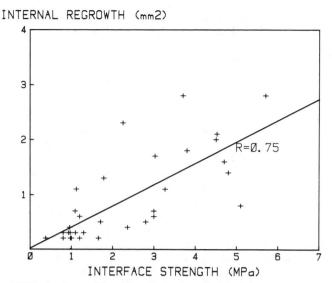

INTERNAL REGROWTH (mm2)

INTERFACE STRENGTH (MPa)

FIG. 9—*Internal bone regrowth area versus interface strength.*

Thin slices (500 μm) were cut perpendicular to the long axis of the implant with a slow diamond saw. Bearing in mind the working hypothesis (a better anchorage of the prosthesis in bone), the cut was done at the endosteal level, the seat of the bone/prosthesis interface in the clinical situation. The slices were ground down to 80 μm and then polished with a 0.5-μm alumina slurry. Photographic enlargements, using a transmission microscope, were obtained from the specimens stained with a modified Paragon (Fig. 7).

The photographs were analyzed with the semiautomatic technique developed in the authors' laboratory [22]. The system, based on a computer-driven graphic tablet, allows quantification of the following parameters from a contour line digitization of the picture:

(a) the "total" growth, that is, the total surface of new mineralized tissue;

(b) the "inner" growth, that is, the surface of the newly formed bone within the implant;

(c) the "outer" growth, that is, the surface of the bone outside the implant (between the drilled hole and the implant); and

(d) the overall characteristics of the implant (average porosity, mean sphere diameter, and other characteristics).

The external contour of each implant was defined as the smallest circle tangent to the most external beads separating the "inner" and "outer" bone growth.

Results

Fourteen animals were operated on by two different surgeons. Seven animals, paired by age and weight, were given electrical stimulation, and the remaining seven were controls. Six animals were sacrificed at three weeks and eight were sacrificed at five weeks.

In all of the specimens at an early stage of bone ingrowth, the main direction of bone formation was between 35 and 45° (Fig. 7) from the bone long axis. The sheep metatarsus is mostly subjected to anteroposterior bending. Therefore, the lateral aspect, where the implants are located, is mainly subjected to shear stresses. This can explain the typical ingrowth pattern found in this experiment.

After an initial elastic load deformation stage, the curve slowly plateaus and, thus, the true rupture phase is not detectable with sufficient accuracy. For this reason, the deformation at failure is not correlated with any of the other mechanical parameters (rigidity and stress to failure), and this parameter has not been taken into consideration in the rest of the study.

The linear correlation between the strength and stiffness of the interface was calculated. A coefficient of correlation (R) of 0.94 (Fig. 8) indicated that no gross misalignment or malpositioning of the specimens occurred during the mechanical tests.

Among all the histomorphometric parameters, the internal bony ingrowth area exhibited the best coefficient of correlation with the interface strength (Fig. 9). Although this coefficient ($R = 0.75$) is highly significant ($P = 0.01$), it was low enough to indicate that another parameter is involved in the interface mechanical strength. The gap existing between the edges of the drilled hole and the implant's outer surface was certainly responsible for this low coefficient of correlation.

Mechanical Results

The calculated values of the interface strength and stiffness are shown in Figs. 10 and 11, respectively. Both parameters double in two weeks. No difference appears between the stimulated and control animals at the two treatment periods subjected to these mechanical tests.

Histomorphometric Results

Most of the histomorphometric parameters did not show any difference between the control and stimulated groups at either three or five weeks. The internal regrowth expressed

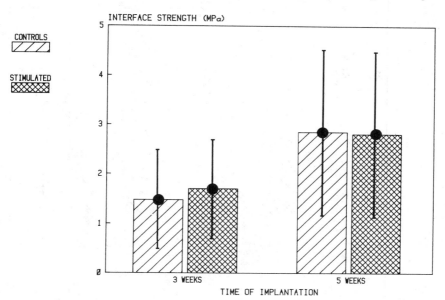

FIG. 10—*Bone/implant interface strength versus the time of implantation for the control and stimulated groups.*

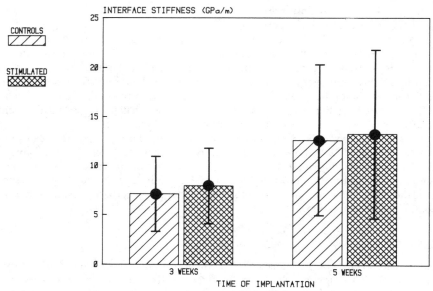

FIG. 11—*Interface stiffness versus the time of implantation.*

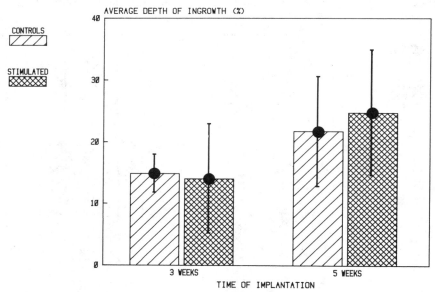

FIG. 12—*Average depth of ingrowth (in percentage of the plug's radius) versus the time of implantation.*

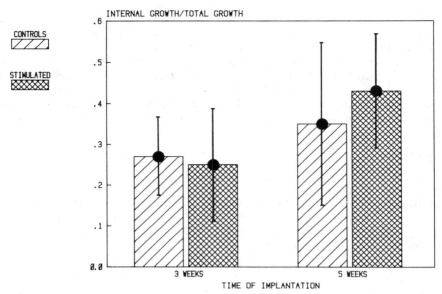

FIG. 13—*The internal-growth/ total-measured-growth ratio versus the time of implantation.*

in percentage of the implant's free area shows a threefold increase between three and five weeks (6 and 18% respectively).

The average depth of bony ingrowth (expressed in percentage of the implant radius) and the ratio of internal growth to the total growth (Figs. 12 and 13) exhibit a similar increase (13 and 20%, respectively) at five weeks for the stimulated group when compared with the control one. However, this difference was not statistically significant.

Discussion

Electromagnetic stimulation did not induce any significant differences in the mechanical properties of the bone/implant interface.

A study of the mechanical correlations showed that the distribution of the stimulated and control specimens around the regression line was completely random. From the fact that no variation in the resistance of the interface was found, one may conclude that the control and stimulated specimens present identical mechanical anchorage. Therefore, the authors assume that, if the stimulation has any effect, it can be found only in the quantification of new tissue and not in its mechanical quality.

Low-amplitude, low-frequency PEMIC tends to improve the depth of bone ingrowth, as well as the relative amount of bone within the implant after five weeks of stimulation. One may ask if this statistically insignificant effect should have any relevance or not.

Moreover, in contrast to the treatment of nonunions, pulsed electromagnetic induced currents may not be adapted to this type of application. A metallic porous implant certainly induces a large disturbance in the electromagnetic field pattern within and around the material, resulting in unpredictable electrical currents.

Conclusions

The hypothesis developed in the introductory part of this paper has not been demonstrated. The pulsed electromagnetic induced currents used in this experiment did not significantly improve the strength of the bone/implant interface. A large spreading of the data, as well as a possibly nonadapted PEMIC waveform may be responsible for these negative results.

Only further investigations using different waveform configurations specially designed for this type of application could give a definite answer concerning the potential use of PEMIC in postoperative management of noncemented implants.

Acknowledgments

The implants were fabricated under the direction of G. Moreau, chief of service of fabrication of materials of the Center of Nuclear Studies of Grenoble, France. The electromagnetic stimulators were graciously provided by Electro-Biology Inc., of New Jersey. This study was supported by a grant from the Medical and Scientific Research of the Caisse Regionale d'Assurance Maladie de l'Ile de France. The technical assistance of Y. Abols and T. Schmelck is gratefully acknowledged.

References

[1] Salvati, E. A., Wilson, P. D., Jr., Jolley, M. N., Vakili, F., Aglietti, P., and Brown, G. C., *Journal of Bone and Joint Surgery*, Vol. 63A, 1981, pp. 753–764.
[2] Stauffer, R. N., *Journal of Bone and Joint Surgery*, Vol. 64A, 1982, pp. 983–993.
[3] Sutherland, C. J., Wilde, A. M., Borden, L. S., and Marks, K. E., *Journal of Bone and Joint Surgery*, Vol. 64A, 1982, pp. 970–982.
[4] Judet, R., Siguier, M., Brumpt, B., and Judet, T., *Clinical Orthopaedics*, Vol. 137, 1978, p. 76.
[5] Lord, G. A., Hardy, J. R., and Kummer, F. J., *Clinical Orthopaedics*, Vol. 141, 1979, p. 2.
[6] Bassett, C. A. L., "Biophysical Principles Affecting Bone Structure," *The Biochemistry and Physiology of Bone*, Vol. 2, G. H. Bourne, Ed., Academic Press, New York, 1971, Chapter 3.
[7] Bassett, C. A. L., Mitchell, S. N., and Gaston, S. R., *Journal of the American Medical Association*, Vol. 247, 1982, pp. 623–628.
[8] Becker, R. O., Bassett, C. A. L., and Backman, C. H. in *Bone Biodynamics*, H. M. Frost, Ed., Little, Brown, Boston, pp. 209–232.
[9] Christel, P., Cerf, G., and Pilla, A., *Mechanisms of Growth Control*, R. O. Becker, Ed., Charles C Thomas, Springfield, MA, 1981, p. 237.
[10] Sedel, F. and Christel, P., *Revue de Chirurgie Orthopedique*, Vol. 64, 1978, pp. 107–122.
[11] Buch, F., Albrektsson, T., and Herbst, E., *Biomaterials*, Vol. 5, 1984, pp. 341–346.
[12] Collela, S. M., Miller, A. G., Stang, R. G., Stoebe, T. G., and Spengler, D. M., *Journal of Biomedical Materials Research*, Vol. 15, 1981, pp. 37–46.
[13] Park, J. B. and Kenner, C. H., *Biomaterials, Medical Devices, and Artificial Organs*, Vol. 3, 1975, pp. 233–243.
[14] Park, J. B., Young, S. O., Kenner, G. H., Von Recum, A. F., Myers, B. R., and Moore, R. R., *Biomaterials, Medical Devices, and Artificial Organs*, Vol. 6, 1978, pp. 291–303.
[15] Salman, N. N., and Park, J. B., *Biomaterials*, Vol. 1, 1980, pp. 209–213.
[16] Weinstein, A. M., Klawitter, J. J., Cleveland, T. W., and Amoss, D. C., *Journal of Biomedical Materials Research*, Vol. 10, 1976, pp. 231–247.
[17] Pilla, A. A., *Annals of the New York Academy of Sciences*, Vol. 238, 1974, pp. 149–170.
[18] Pilla, A. A. in *Bioelectrochemistry*, M. Keyzer and F. Gutman, Eds., Plenum Press, New York, 1980, p. 353.
[19] Pilla, A. A., Sechaud, P., and McLeod, B., *Journal of Biological Physics*, Vol. 11, 1983, p. 51.
[20] McLeod, B., Pilla, A. A., and Samspel, M. W., *Bioelectromagnetics*, Vol. 4, 1983, p. 357.
[21] Christel, P. and Pilla, A. A. in *Proceedings*, First Annual Meeting of the Biological Repair and Growth Society, Philadelphia, PA, 9–11 Nov. 1981.
[22] Dallant, P., Meunier, A., Christel, P., and Sedel, L. in *Biomaterials and Biomechanics 1983*, P. Ducheyne, G. Van der Perre, and A. E. Aubert, Eds., Elsevier Science Publishers, Amsterdam, 1984, pp. 295–300.

DISCUSSION

U. Gross[1] (*written discussion*)—Did you perform microradiography to detect the density of bone mineralization?

P. Dallant, A. Meunier, P. Christel, G. Guillemin, and L. Sedel (*authors' closure*)—We performed microradiography on each tested specimen. We have found that the mineral content of the newly formed bone was lower than normal. However, we do not have the equipment necessary to perform accurate densitometry determinations. For this reason, it was not possible to assess any effect of PEMIC on bone mineralization.

Z. R. Glaser[2] (*written discussion*)—It appears from one of your slides that new bone growth occurs at approximately 30° on both sides of the long bone at the location of the coils. A pair of spaces on both sides at 90° relative to the bone growth appears in the slide. Does this indicate loss of bone, resorption, or demineralization?

If the coils are rotated around the long bone so that they are over (on top of) the operated site, is there any change in bone growth?

Have you used more than one pair of coils? Is there any increase in the rate or amount of bone growth?

Is there any indication of altered immunologic parameters in the animals after the electromagnetic field exposure?

P. Dallant, A. Meunier, P. Christel, G. Guillemin, and L. Sedel (*authors' closure*)—The space found at 90° relative to the bone growth has a very regular border and is not related to a resorption process or demineralization but to the initial drilling, which, as performed, was not perfectly circular. However, some resorption cavities are found in the intact cortical bone located in the vicinity of the new bone apposition.

We did not perform the experiment proposed. The rotation of the coils would certainly modify the electromagnetic field, as well as the current induced in the implanted site, and might change the results of this experiment as well. Nevertheless, we do not think that a change in the field orientation would result in a dramatic change in the bone growth process.

We always used a single pair of coils. An increased number of coils would result in a modified electromagnetic field pattern, in a change in the intensity of the induced current, or in both. Considering that almost all our results tend to show that this very type of PEMIC has an insignificant effect, we do not expect to find large modifications of the data by changing only these two parameters.

No immunological evaluation was performed on these animals. The PEMIC investigated in this experiment have been used extensively in clinical situations, and we have found no reports in the literature on altered immunological parameters with this type of PEMIC. We did not observe, during this study, any infection either in the control or in the stimulated group.

J. B. Park[3] (*written discussion*)—It seems to me that the study is not designed correctly since there are no good controls (which should have an equal amount of electrical power as the stimulated group, for obvious reasons) and since it is not known how much induced current is generated in the specimen or adjacent tissues.

---

[1] Institute of Pathology, Free University of Berlin, Berlin, West Germany.
[2] Center for Devices and Radiological Health/U.S. Food and Drug Administration, Rockville, MD.
[3] Department of Biomedical Engineering, University of Iowa, Iowa City, IA.

I suggest also that the lateral side of the animal should serve as the control side rather than other animals since the animal-to-animal variations are much greater than those within an animal. It is also profitable to have a sham-operated animal as the starting point to verify the effect of PEMIC first before even trying to use any porous implants.

P. Dallant, A. Meunier, P. Christel, G. Guillemin, and L. Sedel (authors' closure)—The purpose of the experiment was not to determine what the parameters of the electromagnetic field are that result in bone formation (more fundamental research is certainly needed in this case) but to evaluate the effect of an already developed PEMIC on the early bone growth within a porous implant. The use of controls having an equal amount of electrical power within the operated site would not have allowed a true comparison between an electromagnetically stimulated and an actual clinical situation.

The contralateral metatarsus was not used as a control bone for two reasons:

1. Considering the short distance between the two legs of the animals, it might be possible that some residual electromagnetic fields would stimulate the contralateral side.

2. In order to obtain a stable location of the electromagnetic fields, the metatarsus had to be immobilized with a cast. A sheep that has both legs immobilized usually lies down and never gets up. This could certainly affect the experiment's results.

The PEMIC used in the experiment has proved to be efficient in previously published fracture healing experiments in rats.

For practical reasons, it was not possible to repeat a similar experiment on a larger number of sheep.

Systems for Future Applications

Paul Ducheyne[1] and John M. Cuckler[2]

Flexible Porous Titanium for Revision Surgery: Concept and Initial Data

REFERENCE: Ducheyne, P. and Cuckler, J. M., **"Flexible Porous Titanium for Revision Surgery: Concept and Initial Data,"** *Quantitative Characterization and Performance of Porous Implants for Hard Tissue Applications, ASTM STP 953,* J. E. Lemons, Ed., American Society for Testing and Materials, Philadelphia, 1987, pp. 303–314.

ABSTRACT: Bone stock deficiency is a major challenge encountered in the revision of failed hip arthroplasties. Loosening of the prosthetic components under aseptic conditions results in the loss of trabecular and cortical bone. At revision the surgeon is frequently faced with a thin smooth cortical tube with an almost complete absence of endosteal trabecular bone. Consequently, fixation of the revision component is seriously compromised. The objective of the present paper is to address this issue specifically by discussing a new concept in materials created to alleviate this major problem of deficient bone stock. The authors propose using a porous flexible titanium material to fill void spaces in the deficient bone structure. The material can easily be shaped at the time of surgery and has a pore size in the optimum range for bone ingrowth (100 to 600 μm). In addition, it is possible to infiltrate it with bone cement. Furthermore, the elastic properties of this material closely correspond to those of trabecular bone.

The objective in the design of this new material was to improve the current practice of revision surgery in three problem areas: first, by avoiding brittle fracture due to excess cement thickness; second, by decreasing the risk of disuse atrophy caused by large prosthesis stem sizes; and third, by filling the excess cavity with a material with elastic properties matching those of the missing tissue.

KEY WORDS: porous implants, titanium, elastic properties, bony ingrowth, cement penetration, metal fibers

Bone stock deficiency is a major challenge encountered in the revision of failed hip arthroplasties [*1*]. Loosening of prosthetic components under aseptic conditions results in a loss of trabecular bone caused by micromotion or macromotion between the cement and host bone. Atrophy of cortical bone may result from pistoning of the cement-prosthesis unit, stress shielding of bone, or bone resorption resulting from inflammatory response from polymeric wear particles of acrylic or polyethylene [*2*]. The cumulative result of these factors presents the surgeon with a thin smooth cortical tube, not infrequently accompanied by cortical defects, with an almost complete absence of endosteal trabecular bone. Consequently, fixation of the revision component is seriously compromised [*3*], resulting in failure rates of total hip revisions reportedly as high as 20% within eight years of follow-up [*4*]. Even higher rates of radiographic demarcation are observed, suggesting catastrophically high failure rates of revision hip arthroplasties with longer use [*5*].

Several surgical techniques have been proposed to solve various problems, but basically,

---

[1] Associate professor of biomedical engineering, Department of Bioengineering, School of Engineering and Applied Science, University of Pennsylvania, Philadelphia, PA 19104.

[2] Assistant professor, Department of Orthopaedic Surgery, University of Pennsylvania Medical School, University of Pennsylvania, Philadelphia, PA 19104.

as was pointed out by Harris [6], the problem to solve is the bony stock. To make up for deficient bone stock, alternatives such as allografts; foil, shell, and plate metal; additional polymethyl methacrylate (PMMA) cement; and large-section implants have been used [6–9].

Suboptimal fixation of revision components resulting from absence of trabecular bone occurs because of the inability of acrylic to form a "micro interlock" with the smooth cortical surface of the revision femur. Furthermore, the heat of polymerization of thick acrylic cement mantles surrounding a revision prosthesis may increase thermal necrosis of bone. The mechanical properties of polymethyl methacrylate may also be compromised when thicknesses greater than 3 to 4 mm are exceeded [10]. The use of larger prostheses to avoid thickness in cement mantles may worsen stress-shielding effects, resulting in further atrophy of the proximal femur.

In our work, we hypothesize that a flexible synthetic trabecular bone provides excellent fixation with methacrylate cement through the microinterlock provided by the intrusion of cement into the material and, as a result of bony ingrowth into the opposite side of the material, yields successful long-term anchorage of the prosthesis. The objective of this paper is to discuss the properties of a new, flexible porous material and to provide data corroborating the concept of synthetic trabecular bone proposed in our hypothesis. Specifically, we will discuss its porosity and pore structure, mechanical properties, and tissue ingrowth behavior.

Materials: Titanium Felt Synthesis

Throughout this study a porous titanium felt made specifically for application in joint revision surgery was used. The synthesis of this material represented the first objective of

FIG. 1—*Macrophotograph of the titanium felt used in this study.*

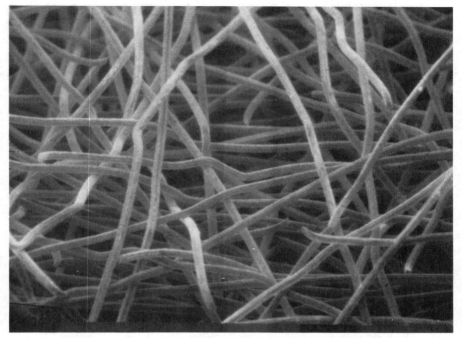

FIG. 2—*Scanning electron micrograph of the surface of the titanium felt* ($\times 60$).

the study. It was developed based on a set of minimum requirements formulated for the envisioned material:

(*a*) biocompatibility,
(*b*) sufficient flexibility that it can easily be shaped during surgery,
(*c*) elasticity of the same order of magnitude as that of trabecular bone,
(*d*) adequate coherency so that it can be handled during surgery, and
(*e*) a pore size in a range that allows both bony ingrowth and cement infiltration.

The material we typically used was composed of straight, 50-μm-diameter commercially pure titanium wire, compacted and sintered to a volumetric density ranging from 5 to 15%. This material is fundamentally different from previously developed porous wire systems [*11,12*] since the materials used in those systems have a totally different density and are therefore not as flexible as the felt.

Figures 1 and 2 are a macrophotograph and a scanning electron micrograph, respectively, of the surface of this feltlike material. Figure 3 shows a cross section of the material. These photographs clearly illustrate the abundant porosity available for tissue ingrowth.

It is of value to note that a pore size cannot easily be determined for this material. The usual techniques for pore size determination are not applicable [*13*]. The significance of porosity measurement when the pore volume is in the 85 to 95% range is highly questionable. With stereometry, the most versatile technique for pore determination, the main difficulty is that transformation of two-dimensional measurements to three-dimensional pore sizes is not possible at these high pore volumes. In addition, stereometry measurements frequently

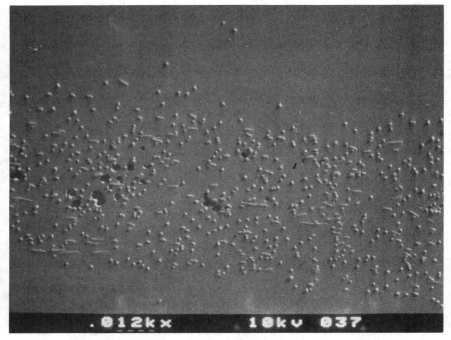

FIG. 3—*Cross-sectional micrograph of the titanium felt* (×12).

forego the physical meaning of the pore. As an example, observe Fig. 4, in which meaningless measurements are indicated. The length of Pore A is covered by a fiber. Thus, this pore is not available for ingrowth and does not have to be counted. The length of Pore B is, in reality, not available for ingrowth over its full length since it is partially covered by fibers; therefore, this measured length is too large in comparison to the size available for ingrowth.

The pore size of this material can best be described by a pore range, which, considering Figs. 2, 3, and 4 is about 100 to 600 μm. Such a pore size substantially coincides with the optimum pore size for ingrowth into the porous Co-Cr bead system [14].

Mechanical Properties and Tissue Ingrowth: Experimental Procedure

Four-point bending tests were performed on titanium felt specimens measuring 120 by 12 by 3.4 mm. The density was 12 ± 1%. The support points and loading points were 90 and 50 mm apart, respectively. Tests were conducted at a constant crosshead rate of 5 mm/min, and all the specimens were loaded and subsequently unloaded. The data were tabulated as the peak test loads reached, the maximum stress in the titanium felt at that time, the concurrent peak displacement, and the concurrent residual displacement at unloading. Apparent moduli of elasticity were obtained from the slope of the loading portion of the graph.

Titanium felt specimens were implanted bilaterally in femora of adult New Zealand white rabbits, using standard surgical techniques. A cortical window was cut in the proximal anterolateral femur after routine subperiosteal exposure. After curettage of the medial endosteum, a 3 by 5-mm section of the felt was implanted against the inner surface of the

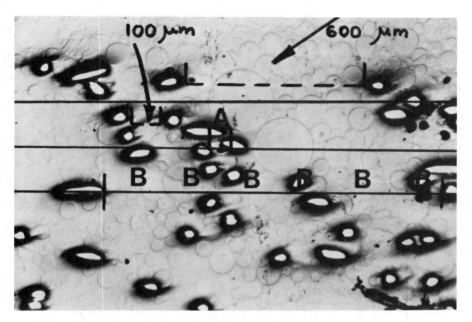

FIG. 4—*Microradiograph of a section of titanium felt, showing two meaningless pore measurements indicated by A and B (×60).*

femur, after which PMMA cement was packed in the canal to the level of the cortical defect. In the contralateral femur, the cortical window was filled with the titanium felt after filling the medullary space with PMMA cement. An autogenous cortical cancellous graft was then applied to the felt, followed by closure of the periosteum and soft tissue. The femora were harvested at 0, 6, 12, and 24 weeks and processed for microradiography and thin-section light microscopy.

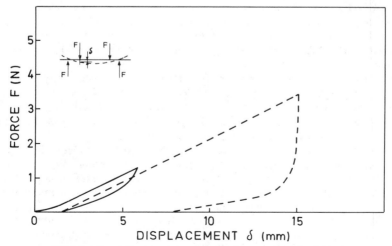

FIG. 5—*Load-displacement curve for the tested titanium felt.*

FIG. 6—*Three titanium felt specimens after testing. Note the permanent deformation.*

Results

Figure 5 shows a typical load-displacement curve. The loads and displacements are as defined in the inset. Figure 6 shows some specimens after testing. Because of the geometry of the test, the apparent modulus of elasticity describes the deformation behavior in the plane of the felt material. With an average apparent modulus of elasticity of 84 MPa, the corresponding stress and displacement characteristics are as shown in Table 1. The peak test load is not the ultimate load the material can support until fracture, but it represents the load during the loading portion of the test at which the loading was reversed. If the loading had been continued, continuously higher peak test loads would have been obtained, as can be deduced from Fig. 5. This figure shows two load-displacement curves on the same specimen: after full reversal of the first loading a second increase of load produces the same deformation pattern.

The loads that are needed to deform the 3.4-mm-thick felts permanently are very small; the order of magnitude is 1 to 5 N. Thus, the load required to shape the porous material can be manually applied during surgery. The data further show that the apparent elasticity, as defined, varies, ranging between 80 and 100 MPa. These values are of the same order of magnitude as the moduli of elasticity of trabecular bone [15].

The *in vivo* analysis of this material uses two models: a rabbit model with nonfunctional loading conditions and a dog model with functional loading conditions. We only report here on the rabbit model. This model was conceived to assess the ease of apposition of the titanium felt to the contoured bone structures, the possibility of cement impregnation, and the occurrence of osseous tissue ingrowth.

Microradiographic and histologic analyses show that at all times during implantation, and with both the endosteal and periosteal placement, there was an efficient adaptation between the felt and the concave or convex cortical surface. Figure 7 is a microradiograph substan-

TABLE 1—*Four-point bend test data documenting the relatively low loads, the permanent deformation, and the elasticity of titanium felt.*

Peak Test Load F, N	Maximum Stress at F, MPa	Displacement at F, mm	Permanent Displacement, mm	Apparent Modulus of Elasticity, MPa
1.40	0.9	3.7	0.5	101
1.40	1.1	6	1.5	77
2.50	1.6	6	1	90
3.30	2.8	13.7	6	86
3.60	2.8	13.8	6.3	80

tiating this point. In addition, with the presently used unelaborate cementing technique, some cement intrusion was present. What was most striking at zero weeks of implantation was the complete infiltration of the felt porosity by marrow tissue cells (Fig. 8). Bony ingrowth was persistently present in all areas where bone had been present prior to implantation; there was no osseous tissue ingrowth in those specimens or in those areas of specimens in which there is no bony tissue under normal physiological conditions. Figures 9 and 10 are photomicrographs of stained thin sections showing an endosteal and a periosteal specimen after six weeks. In the endosteal specimen there is limited ingrowth since these specimens were implanted in areas where there is normally no bone. Yet, bony trabeculae grew unobstructed through the material if they had been present before implantation (Fig. 9). In the periosteal specimens, osseous tissue was abundant since the implants had been, by and large, inserted in the surgically created cortical defect.

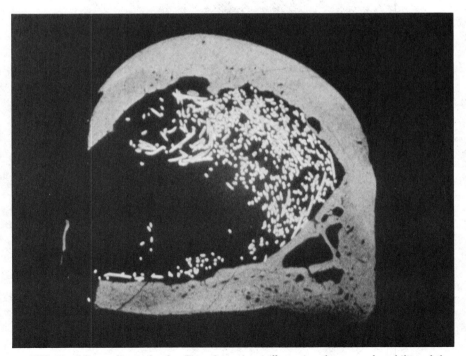

FIG. 7—*Microradiograph of a Time 0 specimen illustrating the easy adaptability of the titanium felt to the endosteal cavity of the femur.*

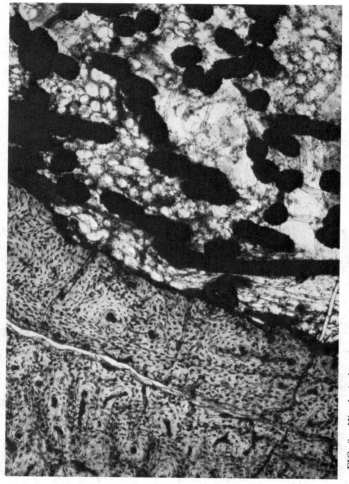

FIG. 8—*Histological section at zero weeks showing cement infiltration by bone marrow cells. The black dots are the fibers of the titanium felt ($\times 115$).*

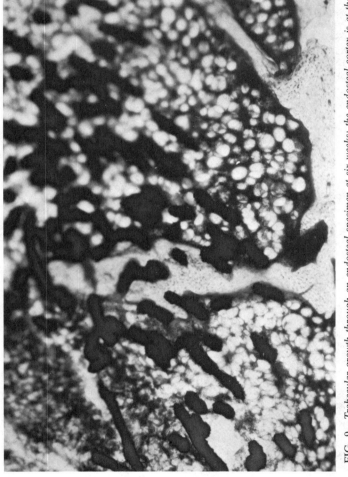

FIG. 9—*Trabecular growth through an endosteal specimen at six weeks; the endosteal cortex is at the bottom of the photomicrograph (×70).*

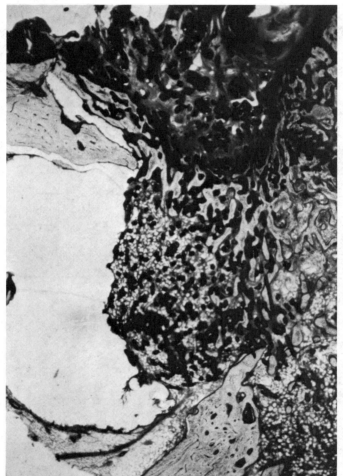

FIG. 10—*Cortical bone growth through a periosteal specimen at six weeks; the titanium felt extends across four fifths of the width of the photograph. There is a major regrowth of bone throughout the area of surgically removed cortex* (×23).

Discussion

The increasing rates of failure for revision hip arthroplasties indicate the need for solutions that address the fundamental problems of revision surgery. The underlying premise to our work is that long-term solutions should be based on the use of an ideal cement thickness, so that problems of overheating or brittle fracture of the cement are avoided. In addition, a normal stem thickness should be used, since this decreases the possible risk of disuse atrophy of the cortex. The excess cavity in the bone should then be filled with a material which is compliant, thus yielding a close apposition with the bone, and which, furthermore, has an elasticity averaging that of trabecular bone tissue.

We first directed our studies towards the synthesis of a shapable biocompatible material with an elasticity approaching that of trabecular bone. The titanium felt on which we have reported here may provide such a material. Subsequently, we engaged in the biological testing of this material.

The present *in vivo* experiments do not comprehensively document the behavior of this material in a fully functional situation, as is required to prove the working hypothesis. Such experiments are under way as a second phase of our biological evaluation program. The first effort, which is reported here, is a simple rabbit model and represents our initial biological testing in a species different from the one used for the functional loading study.

In the present study, implantation periods of 0, 6, 12, and 24 weeks were chosen to evaluate gradually increasing time intervals: 0 weeks to assess the surgical apposition, marrow cell infiltration, and cement penetration; 6 weeks to analyze the initial bone tissue ingrowth; and 12 and 24 weeks to assess the temporal course of bone ingrowth. The data of the rabbit model provide substantial evidence of easy surgical adaptation, abundant marrow cell infiltration, cement impregnation, and unobstructed bone tissue ingrowth.

Conclusions

The present *in vitro* and *in vivo* data substantiate the validity of major concepts developed at the origin of the synthesis of this new material: these include elasticity approaching that of bone, easy surgical insertion, limited cement penetration, and calcified tissue ingrowth. These data are now being expanded. A study closely simulating the clinical use and emphasizing the relationships between bone ingrowth, time-dependent stability, and bone remodeling is presently under way.

References

[1] Sim, F. H. and Chao, E. Y. S., "Hip Salvage by Proximal Femoral Replacement," *Journal of Bone and Joint Surgery,* Vol. 63A, 1981, p. 1228.

[2] Lee, A. J. C. and Ling, R. S. M., "Loosening," *Complications of Total Hip Replacement,* R. S. M. Ling, Ed., Churchill Livingston, London, 1984.

[3] Swanson, S. A. V., "Mechanical Aspects of Fixation," *The Scientific Basis of Joint Replacement,* S. A. V. Swanson and R. S. M. Ling, Eds., Pitman Medical, London, 1977.

[4] Pellicci, P. M., Wilson, P. D., Jr., Sledge, C. B., Salvati, E. A., Ranawat, C. S., Poss, R., and Callaghan, J. J., "Long-Term Results of Revision Total Hip Replacement," *Journal of Bone and Joint Surgery,* Vol. 67A, 1985, p. 513.

[5] Kavanaugh, B. F., Ilstrup, D. M., and Fitzgerald, R. H., "Revision Total Hip Arthroplasty," *Journal of Bone and Joint Surgery,* 67A, 1985, p. 517.

[6] Harris, W. H., "Allografting in Total Hip Arthroplasty," *Clinical Orthopaedics,* Vol. 162, 1982, pp. 150–164.

[7] Amstutz, H. C., Ma, S. M., Jinnah, R. H., and Mai, L., "Revision of Aseptic Loose Total Hip Arthroplasties," *Clinical Orthopaedics,* Vol. 170, 1982, pp. 21–33.

[8] Harris, W. H., "Revision Surgery for Failed Non-Septic Total Hip Arthroplasty—The Femoral Side," *Clinical Orthopaedics,* Vol. 170, 1982, pp. 8–20.

[9] Voorhoeve, A., "Austausch operationen von Huftgelenksendoprothesen unter Verwendung von Knochenzement—Metall Konstruktionen," *Chirurgische Praxis,* Vol. 27, 1980, pp. 275–290.

[10] Elson, R. A., "Revision Arthroplasties," *Complications of Total Hip Replacement,* R. S. M. Ling, Ed., Churchill Livingston, London, 1984.

[11] Rostoker, W., Galante, J. O., and Shen, C., "Some Mechanical Properties of Sintered Fiber Metal Composites," *Journal of Testing and Evaluation,* Vol. 2, 1974, pp. 107–112.

[12] Ducheyne, P., Martens, M., Aernoudt, E., Mulier, J., and De Meester, P., "Skeletal Fixation by Metal Fiber Coating of the Implant," *Acta Orthopaedica Belgica,* Vol. 40, 1974, pp. 799–805.

[13] De Hoff, R. T. and Rhines, F. N., *Quantitative Microscopy,* McGraw-Hill, New York, 1968.

[14] Bobyn, J. D., Pilliar, R. M., Cameron, H. V., and Weatherly, C. C., "The Optimum Pore Size for the Fixation of Porous Surfaced Metal Implants by the Ingrowth of Bone," *Clinical Orthopaedics,* Vol. 150, 1980, pp. 263–270.

[15] Carter, D. R. and Hayes, W. C., "The Compressive Behavior of Bone as a Two-Phase Porous Structure," *Journal of Bone and Joint Surgery,* Vol. 59A, 1977, p. 954.

Michael DeMane,[1] *Neil B. Beals,*[1] *David L. McDowell,*[2] *Frederick S. Georgette,*[1] *and Myron Spector*[3]

Porous Polysulfone-Coated Femoral Stems

REFERENCE: DeMane, M., Beals, N. B., McDowell, D. L., Georgette, F. S., and Spector, M., **"Porous Polysulfone-Coated Femoral Stems,"** *Quantitative Characterization and Performance of Porous Implants for Hard Tissue Applications, ASTM STP 953,* J. E. Lemons, Ed., American Society for Testing and Materials, Philadelphia, 1987, pp. 315–329.

ABSTRACT: Concerns about the metal ion release, high stiffness, and reduction in stem strength that are associated with porous metal coatings have prompted investigations of a relatively high-strength porous polymeric coating, porous polysulfone (PPSF). The present study was initiated in order to characterize the mechanical properties of PPSF and PPSF-coated titanium alloy and to evaluate the performance of PPSF as a coating on titanium alloy femoral stems in dogs. PPSF with various porosities was fabricated by sintering particles of Union Carbide P-1700 medical-grade polysulfone. Chemical coupling of PPSF to Ti-6Al-4V alloy specimens was achieved using silyl-reactive polysulfone. The shear strength of the PPSF/titanium alloy interface determined from push-out tests on coated cylindrical rods and transverse sections of the coated canine femoral stems was found to exceed 16 MPa. Tension and punch shear tests performed on the PPSF yielded values of 13 and 21 MPa, respectively. The relationship between strength and porosity characteristics was determined. Fatigue tests performed on PPSF-coated titanium femoral stems revealed no failure of the coating up to 1 million cycles. No difference between PPSF-coated and uncoated titanium specimens was found in the fatigue behavior. PPSF-coated titanium alloy canine femoral stems (32% porosity, 246 μm in average pore size) were implanted in 25 dogs. Evidence of loosening was found only in those animals in which the canal/fill ratio was low. No gross radiographic changes were found in the implanted femurs within the two-year postoperative period. Histology showed bone ingrowth fixation of the prostheses. Results of this investigation indicate that PPSF-coated femoral stems warrant continued investigation.

KEY WORDS: porous implants, porous materials, porous polymers, polysulfone, titanium alloy, femoral stems, orthopedic prostheses

Porous-coated joint replacement prostheses are currently undergoing clinical investigation. Animal and human studies [1–5] have demonstrated that these devices represent a viable alternative to conventional joint arthroplasties using acrylic-based cements. However, there are several important considerations that need be addressed when employing porous coatings:

(*a*) the increased amounts of substances released from the implant to the host as a result of the increased surface area of the porous coating,

---

[1] Development manager of reconstructive products, research engineer, and advanced materials technology manager, respectively, Richards Medical Co., Memphis, TN 38116.
[2] Associate professor of mechanical engineering, Department of Mechanical Engineering, Georgia Institute of Technology, Atlanta, GA 30332.
[3] Professor, Department of Orthopaedics, Emory University School of Medicine, Atlanta, GA 30303.

(*b*) the strength of the porous coating and its bond to the prosthetic stem,
(*c*) the strength of the coated stem, and
(*d*) the stiffness of the coating and the underlying prosthetic stem.

Investigations of porous metals have suggested that problems of metal ion release [6] and reduction in stem strength [7,8] result from the application of porous metal coatings to cobalt and titanium alloy femoral stems. Furthermore, some investigators have proposed that the high stiffness of porous-metal-coated prostheses may lead to adverse bone remodeling as a result of stress-shielding effects [9]. The frequency and severity of these problems are not known since they generally become manifest only over extended intervals of implantation.

Porous polymeric fixation systems [10–13] that might reduce or eliminate the aforementioned difficulties with porous metals are under investigation. It was hoped that the application of polymeric coatings might isolate the substrate metal alloy from the fluid environment of the body, thereby reducing the potential for metal ion release. Furthermore, the sintering temperature required to fabricate a porous polymer coating is well below that required to induce adverse microstructural changes in the titanium alloy. The application of the coating, moreover, does not introduce potentially deleterious notches or surface defects in the surface of the substrate alloy. This is an important consideration when dealing with notch-sensitive alloys such as Ti-6Al-4V.

A lower modulus porous polymeric coating may make possible a more favorable remodeling of bone in and around the prosthesis [12]. Finite-element modeling techniques suggest that low-modulus porous coatings may be associated with a more uniform stress distribution in surrounding bone than can be achieved with rigid systems and may enhance loading of trabeculae within the coating porosity [14]. Moreover, low-modulus coatings may facilitate intraoperative press-fit stability since the coating can undergo slight deformation. However, the strength of the porous polymeric coating and its attachment to the prosthesis need to be higher than that of the surrounding bone in order for the highest strength of the system to be achieved.

The potential value of a low-modulus coating coupled with strength requirements initially prompted this investigation of a high-strength engineering thermoplastic, polysulfone [12,15]. The present study was initiated to characterize the physical and mechanical properties of porous polysulfone (PPSF) and PPSF-coated Ti-6Al-4V alloy and to evaluate the performance of polysulfone as a porous coating on femoral stems in dogs.

Procedures

Medical-grade polysulfone (Union Carbide Corp. P-1700 MG) with a molecular weight of about 25 000 was employed in this study. Porous polysulfone (Fig. 1) was fabricated by sintering particles of the polymer at a temperature above the glass transition temperature. Chemical coupling of PPSF to titanium alloy (Ti-6Al-4V) specimens was achieved using a nonporous silyl-reactive polysulfone film applied to the metallic substrate prior to sintering. The physical and mechanical properties of PPSF and the clinical performance of PPSF-coated canine femoral stems were evaluated.

Porosity Characterization

The average pore size and percentage of porosity of PPSF specimens were determined using a standard quantitative optical analysis technique employing a manual point counting

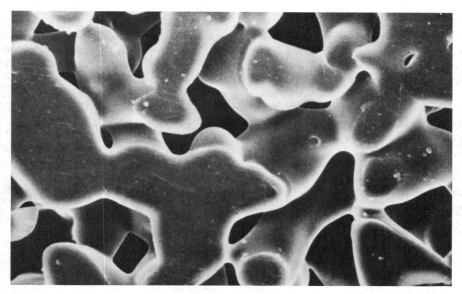

FIG. 1—*Scanning electron micrograph of a specimen of porous polysulfone ×300.*

procedure [*16*]. The pore characteristics of selected test specimens were also evaluated using a lineal analysis procedure performed with a computerized image analysis system. This method provided a distribution of pore sizes as well as the average pore size. Briefly, a transparency with parallel lines was superimposed in random fashion on the video image of a section of a porous specimen. The length of each grid line within a pore area was traced with the stylus of a digitized graphics tablet used in conjunction with the video display. The lineal measurements were used to calculate the percentage of porosity, average pore size, and pore size distribution.

Corrosion Behavior

Anodic polarization tests were conducted to assess the effect of a polysulfone coating on the corrosion behavior of Ti-6Al-4V alloy. A Princeton Applied Research corrosion cell (Model K0047) with a calomel reference electrode was utilized with a Princeton Applied Research potentiostat/galvanostat. A film of polysulfone was applied to one side of polished Ti-6Al-4V specimens (15.5 mm in diameter and 25 cm thick) using the coating procedure employed in the production of porous-polysulfone-coated joint replacement implants. Microscopic evaluation of metallic devices coated in this manner typically reveals a continuous film of solid polysulfone, 0.25 to 0.5-mm thick, adhering to the underlying substrate. The control specimens were tested in an uncoated condition. All the test specimens were secured in a Princeton Applied Research specimen holder (Model K0105), which exposed 1 cm² of the specimen to the test solution. Polarization curves starting at 50 mV negative to the equilibrum potential of the specimen were obtained using a scan rate of 10 mV/min. The scans were carried through the transpassive region of each specimen. Triplicate testing of specimens with and without a polysulfone coating was performed in Ringer's solution maintained at 37°C.

Mechanical Testing

A compressive shear test was used to determine the strength of the PPSF/Ti-6Al-4V alloy interface. Cylindrical rods of the alloy (6.4 mm in diameter) were coated with porous polysulfone and sectioned transversely to produce 5-mm-thick push-out specimens. The specimens were then positioned in a supporting frame (Fig. 2) in a mechanical test unit. The load required to produce failure was determined at a crosshead speed of 1 mm/min. The interfacial shear strength values were computed by dividing the load to failure by the calculated surface area over which the coating was attached.

Punch shear evaluation of porous polysulfone was performed in accordance with the ASTM Test for Shear Strength of Plastics by Punch Tool (D 732-84). Specimens with a mean porosity of about 30% and a 200-μm average pore size were mounted in a punch shear fixture (Fig. 3) and loaded at a rate of 1.25 mm/min until shear failure occurred. A value for shear strength was obtained by dividing the load required to shear the specimen by the calculated area of the sheared edge. Specimens were maintained in Ringer's solution at 37°C for periods of 1, 14, and 90 days prior to testing to assess the effect in *in vitro* aging on the punch shear strength.

Porous polysulfone tension test specimens (Fig. 4) with a range of porosities (13 to 44%) were fabricated by varying the sintering times in the fabrication process. Monotonic tension tests were performed using an applied stroke rate of 5 mm/min.

Titanium alloy femoral stems, coated with porous polysulfone, were potted in epoxy and mounted in a servohydraulic test machine in air (Fig. 5). The stems were subjected to sinusoidal loading from 0.67 to 4.67 kN at 2 Hz. After 1×10^6 cycles, the stems were

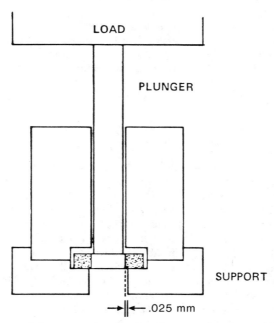

FIG. 2—*Diagram of the compressive shear test fixture employed to evaluate strength of the PPSF/titanium alloy interface.*

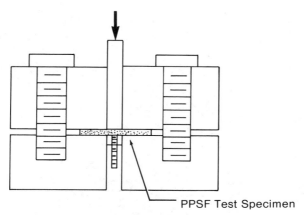

PPSF Test Specimen

FIG. 3—*Punch shear test fixture.*

sectioned transversely and polished for subsequent microscopy evaluation to assess the effect of cyclic loading on the integrity of the porous polysulfone/Ti-6Al-4V interface.

To determine the effect of the coating and coating procedure on the substrate mechanical properties, fatigue testing was performed using cylindrical test specimens of Ti-6Al-4V (Fig. 6) polished to a 600-grit finish. Bonded PPSF coatings were applied to eight specimens. Four uncoated test specimens were also evaluated to assess the effects of the coating on the fatigue properties of the underlying titanium alloy. The testing was performed on an R + R Moore fatigue apparatus at 167 Hz in air at room temperature. The tests were terminated after failure or when 2×10^7 cycles had accumulated to determine a fatigue endurance limit.

Animal Implantation Studies

Animal implantation studies were conducted to evaluate the clinical performance of the PPSF-coated titanium alloy system. A porous polysulfone coating (1 mm thick) was applied to 25 Ti-6Al-4V alloy canine femoral stems (Fig. 7). Pore analysis revealed that the coatings

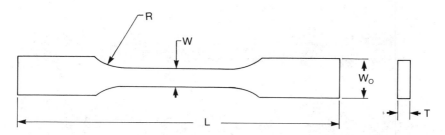

W = Width of gage length = 1.3 cm
W_O = Width of grip = 1.9 cm
L = Total length = 19 cm
T = Thickness = 3.2 mm or under
R = Radius of fillet = 5.1 cm

FIG. 4—*Dimensions of the tension test specimens.*

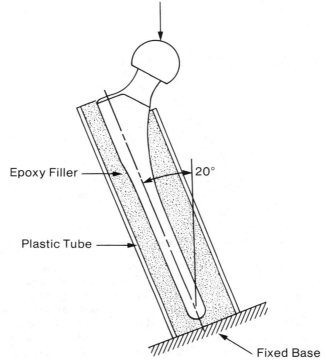

FIG. 5—*Configuration employed for fatigue testing of PPSF-coated femoral stems.*

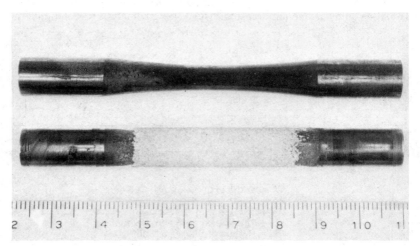

FIG. 6—*Uncoated and PPSF-coated titanium alloy specimens employed for fatigue testing (rotating bend test).*

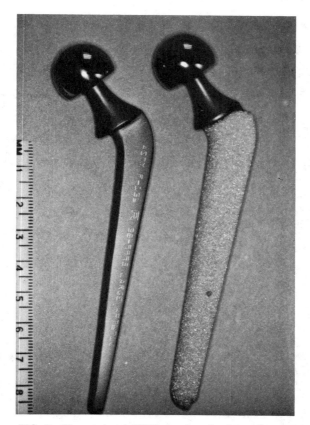

FIG. 7—*Uncoated and PPSF-coated canine femoral stems.*

were approximately 30% porous with an average pore size of about 250 μm. An additional coated stem was allocated for evaluation of the strength of the PPSF/metal interface. This specimen was potted in polymethyl methacrylate and sectioned transversely to produce four specimens approximately 1 cm thick. The specimens were set onto a metal ring larger in diameter than the PPSF coating dimension. A compressive load was applied to the metal stem section using stroke control at 1 mm/min. The direction of loading relative to the stem was from distal to proximal in order to push the section out of the slight taper of the stem. The load to failure was divided by the estimated surface area of failure.

The PPSF-coated canine femoral stems were inserted into 25 dogs for intervals up to two years. A polyethylene cup was implanted in the acetabulum with bone cement using conventional surgical procedures. Radiographs were obtained after selected time periods postoperatively. Radionuclide imaging with Tc^{99m} methylene diphosphonate (MDP) was performed on several animals. Postmortem evaluation included histology and microradiography.

Results

Physical and Mechanical Analyses

Potentiokinetic analysis of polysulfone-coated and uncoated Ti-6Al-4V alloy revealed that ion release, as measured by current density, was measurably reduced by the presence of

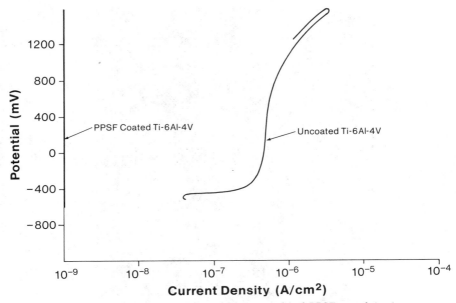

FIG. 8—*Anodic polarization curves for uncoated and PPSF-coated titanium.*

the polymeric coating. Figure 8 demonstrates that the uncoated Ti-6Al-4V alloy exhibited a current density of approximately 5×10^{-7} A/cm² through its passive region. This finding is in close agreement with published literature concerning the anodic polarization behavior of Ti-6Al-4V in Ringer's solution [17]. In contrast, the polysulfone-coated Ti-6Al-4V specimen exhibited no increase in current density with increasing potential. The current density exhibited by the coated specimens was below detection on the equipment utilized, and is therefore believed to be less than 1×10^{-11} A/cm² (Fig. 8).

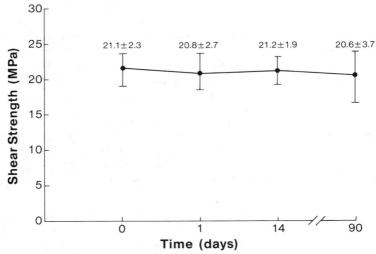

FIG. 9—*Punch shear data for PPSF immersed in Ringer's solution for selected time periods.*

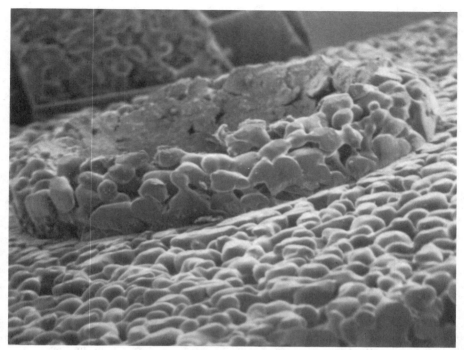

FIG. 10—*Scanning electron micrograph of the failed surface of a PPSF shear test specimen* (×*15*).

The interfacial shear strength of PPSF-coated cylindrical pins was 16.2 MPa as determined by compressive shear testing. Scanning electron microscope (SEM) analysis of the test specimens after failure revealed that in most instances failure occurred just within the PPSF coating. Shear strengths for the PPSF/canine femoral stem interface ranged from 19 MPa for the most proximal section to 34 MPa for the distal section. Failure was found to occur in the PPSF coating. These results suggest that the strength of attachment of the polysulfone to the underlying titanium alloy is at least as great as the shear strength of the PPSF layer.

Punch shear evaluation of PPSF yielded values of about 21 MPa. A plot of the punch shear strength versus the time in Ringer's solution at 37°C suggests that aging *in vitro* has no significant effect on the shear properties of the PPSF (Fig. 9). This is consistent with the hydrolytic stability of polysulfone. SEM inspection of a failed specimen (Fig. 10) revealed uniform failure within the PPSF network.

Monotonic tension tests of PPSF specimens (Fig. 11) resulted in a range of ultimate tensile strengths (6 to 26 MPa) and elastic moduli (0.45 to 1.48 GPa). Linear regression analysis of the relationship between the porosity and the ultimate tensile strength of porous polysulfone revealed that a good correlation ($r = -0.897$) existed and predicted that the ultimate tensile strength for a solid (that is, 0% porosity) specimen was 31.5 MPa (Fig. 11). The extrapolated value for the ultimate tensile strength of solid polysulfone differs significantly from the ultimate tensile strength (71 MPa) of solid injection-molded polysulfone (P-1700) subjected to the sintering conditions employed in the fabrication of the porous specimens. This difference may be accounted for, in part, by the effect of stress concentrations present in the porous structure and the triaxial stress pattern influencing yielding and plasticity.

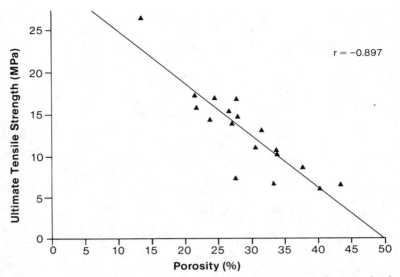

FIG. 11—*Graph of the ultimate tensile strength versus the percentage of porosity (each point represents the average strength for at least five specimens sintered using the same conditions).*

Different sintering conditions often produced specimens with similar porosity percentages but significantly different tensile strengths (Fig. 11). Examination of the morphology of these specimens revealed differences between the sintered particles in the thickness of the necks. The lower strength specimens had thinner necks between particles. Because of the difficulty in distinguishing neck regions with thicknesses approaching the particle size, a procedure was adopted which counted only those necks with thicknesses of 100 μm (ap-

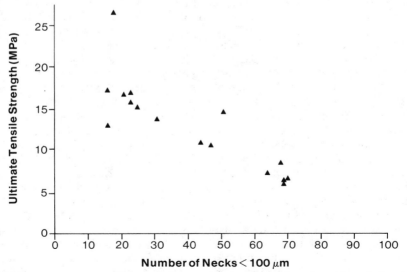

FIG. 12—*Graph of the ultimate tensile strength versus the number of particle necks less than 100 μm thick (in an area of approximately 14 mm²).*

proximately one fourth of the particle size) or less. Quantitative analysis revealed the correlation between the tensile strength and this parameter of neck thickness (Fig. 12). A correlation coefficient of -0.85 was obtained by linear regression analysis.

Lineal analysis with a computerized image analysis system revealed different pore size distributions for PPSF specimens with comparable average pore sizes.

Microscopic evaluation of transverse sections of PPSF-coated Ti-6Al-4V femoral stems exposed to 1×10^6 cycles (from 0.67 to 4.67 kN) revealed an intact PPSF/Ti-6Al-4V interface. Adhesion of the polymer to the underlying substrate was evidenced in all the sections evaluated, including the proximal lateral aspect of the embedded stem, where the tensile and shear stresses were greatest. The applicability of this method is limited since the epoxy grouting material models uniform and complete bone ingrowth fixation of the prosthesis.

Stress-life (S-N) plots of polysulfone-coated and uncoated Ti-6Al-4V rotating bend specimens (Fig. 13) indicate that the fatigue behavior of the titanium alloy is not affected by the application of the porous polysulfone coating. In addition, the microstructure of the Ti-6Al-4V alloy under the PPSF coating is a fine-grained, equiaxed microstructure, which is typical of high-fatigue-strength Ti-6Al-4V. It is not surprising that the application of a porous polysulfone coating has no measurable effect on the fatigue strength of the titanium alloy because the coating procedure is carried out at temperatures lower than 288°C (550°F), which is well below the temperature required to induce microstructural changes within the substrate.

Canine Femoral Stem Studies

The clinical results were categorized on the basis of the initial stabilization. Radiographs of animals with femoral canal diameters affording a high canal/fill ratio (that is, the stem diameter divided by the canal diameter) displayed little gross remodeling through two years of implantation. However, animals with large canals and insufficient initial stability displayed radiographic features of loosening after a few months. These features included a radiodense

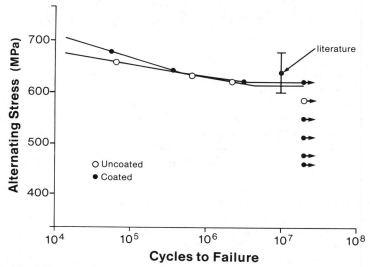

FIG. 13—*S/N curves for uncoated and PPSF-coated titanium alloy. The data points with arrows are runouts.*

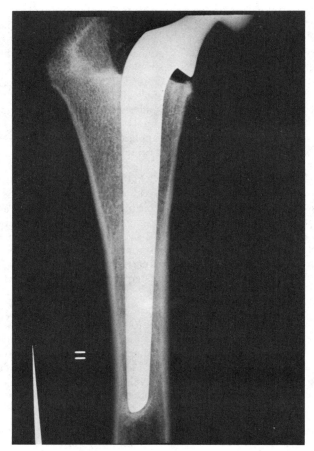

FIG. 14—*Radiograph of a PPSF-coated canine femoral stem after 14 months of implantation.*

line or area at some distance from the coating and increased bone density in the medullary canal distally. Evaluation of radiographs of the 13 animals surviving five months or longer revealed that all of the 4 dogs with large canals (greater than 13.5 mm in diameter, measured at the isthmus) displayed radiographic features of loosening. None of the 9 animals with acceptable canal/fill ratios displayed these radiographic features.

Radiographs of femurs in which the canal/fill ratio was high displayed focal bone loss of the medial cortex under the collar and a slight increase in bone density at the tip of the stem and proximally (Fig. 14).

In the immediate postoperative period, a faint radiodense line could often be seen at the margin of the polysulfone coating. This probably was produced by the presence of fragmented trabecular bone, crushed during the rasping procedure of surgery. Three months after implantation the radiodense line was often more clearly demarcated and appeared to encircle the prosthesis in several animals. Once formed, this radiodense line persisted indefinitely (Fig. 14). This radiographic feature was found histologically to be bone adapted to the porous coating.

Radionuclide imaging with technetium revealed increased bone turnover in the operated femur during the first few months postoperatively. This increased bone activity was due in large part to the periosteal reaction to surgery. Nuclear images of selected femurs that radiographically displayed features consistent with loosening showed a persistence of increased uptake distally. Femurs having stems with acceptable canal/fill ratios displayed uptake comparable to that seen in the unoperated limb, suggesting a near-normal bone turnover.

Histological evaluation revealed the presence of bone trabeculae along the surface of the porous polysulfone coating and within the porosity. As early as one month after implantation, bone was found bridging the PPSF/bone interface and filling the gap produced by the discrepancy between the shape of the broach and the coated stem (Fig. 15). At one month new bone was found around and distal to the tip of the stem in regions of the medullary canal where cancellous bone does not normally exist. The implants displaying evidence of loosening in radiographs and nuclear images had no bone ingrowth histologically; the porosity was filled with fibrous tissue.

The percentage of bone ingrowth in the coatings of prostheses implanted between one month and two years varied with location along the length of the stem. Generally, more bone was found within the pores proximally and in regions where the stem was in close proximity to the endosteal surface of the cortex.

Failure of the coating (that is, detachment from the metal stem or fracture through the porous polysulfone) was only found in one animal, which experienced a spiral fracture of the femur. In this case the fracture plane through the bone continued into the coating. While signs of failure of the coatings in the other animals were not found by gross or microscopic evaluation, specific investigations were not conducted to attempt to identify such a failure mechanism.

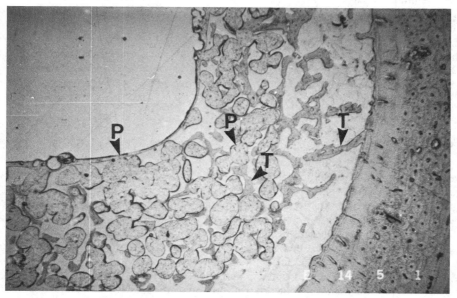

FIG. 15—*Ground section showing bone ingrowth into a PPSF-coated canine femoral stem after one month (Paragon stain): P = polysulfone, T = trabecular bone.*

Discussion

PPSF specimens with a wide range of porosity, percentages, average pore size, neck thicknesses, and pore size ditributions were produced using different sintering conditions. The results indicate that any one parameter, such as average pore size, alone may not be sufficient to predict the mechanical properties and clinical performance of a porous coating adequately.

The results suggest that PPSF coatings have several advantages over porous metal coatings. The solid polysulfone film underlying the PPSF coating reduces the surface area contributing to metal ion release and, in fact, impedes the electrochemical corrosion of the metallic stem. The potential required to overcome the ion diffusion barrier for conduction across the polysulfone film is not known. However, it is unlikely that potentials as high as those tested in this investigation will be generated *in vivo*. The stability and efficiency of the polysulfone film as a function of the duration of its immersion in physiological fluids is under investigation.

The bonded PPSF coating has no detrimental effect on the fatigue properties of the underlying metal substrate. The strengths of PPSF and its attachment to titanium alloy significantly exceed the shear strength of trabecular bone, approximately 2 MPa [18]. Previous investigations of the pull-out strength of porous titanium and PPSF specimens in dogs yielded interfacial shear strengths of 2 to 3.5 MPa [1,19]. Since failure occurred in the cancellous bone of the proximal femur, these values also represent the shear strength of trabecular bone. The tensile and shear strengths of the PPSF specimens and the PPSF/titanium alloy interface, as reported here, afford a substantial safety factor when compared with bone strengths. Testing is in progress to document that the fatigue strength of PPSF is sufficient to support load-bearing prostheses.

Results of the current study of PPSF-coated canine femoral stems confirm previous studies [15] showing the bone ingrowth stabilization of these load-bearing PPSF-coated implants. The benefit of the low modulus of the PPSF coating may be reflected in the near-normal remodeling around the coated canine femoral stem. Titanium and cobalt alloy canine femoral stems coated with porous metal resulted in gross adverse remodeling around many of the prostheses in previous animal studies [20,21]. The more favorable remodeling observed around PPSF-coated stems in the current study may be due to the reduced section modulus of the titanium stem/PPSF coating composite. However, the shape and size of the prosthesis, as well as factors related to implantation, may also have played a role.

Conclusions

1. The presence of a thin film of polysulfone on Ti-6Al-4V corrosion specimens inhibits the *in vitro* release of metal ions into solution in comparison with uncoated titanium alloy test specimens in short-term potentiokinetic analyses.

2. The fatigue behavior of titanium alloy rotating bend specimens is not adversely affected by the application of a porous polysulfone coating.

3. The mechanical properties of porous polysulfone and of the porous polysulfone/titanium interface, as described herein, are sufficient to maintain the integrity of a porous polysulfone prosthesis under anticipated *in vivo* loads.

4. Porous-polysulfone-coated canine femoral stems were stabilized by bone ingrowth for periods of up to two years in animals in which the stems adequately filled the femoral canal. No gross adverse nonanatomic adaptive remodeling of the femur occurs with use of this femoral stem in dogs.

Acknowledgments

This work was supported in part by DHHS-NIH Research Grant AM 28023. The technical assistance of S. L. Harmon, G. Stokes, and C. Cease is gratefully acknowledged. The authors are also grateful to Dr. K. M. Greenwood, for his assistance in implanting the PPSF-coated femoral stems in dogs.

References

[1] Galante, J., Rostoker, W., Lueck, R., and Ray, R. D., *Journal of Bone and Joint Surgery,* Vol. 53A, 1978, p. 101.
[2] Cameron, H. U., Macnab, I., and Pilliar, R. M., *International Journal of Artificial Organs,* Vol. 1, 1978, p. 104.
[3] Hungerford, D. S., Krackow, K. A., and Kenna, R. V., *Orthopedic Clinics of North America,* Vol. 13, 1982, p. 103.
[4] Engh, C. A., *Clinical Orthopaedics,* Vol. 176, 1983, p. 52.
[5] Galante, J. O., *Hip,* 1983, p. 181.
[6] Koegel, A. and Black, J., *Journal of Biomedical Materials Research,* Vol. 18, 1984, p. 53.
[7] Cook, S. D., Georgette, F. S., Skinner, H. B., and Haddad, R. J., *Journal of Biomedical Materials Research,* Vol. 18, 1984, p. 497.
[8] Yue, S., Pilliar, R. M., and Weatherly, G. C., *Journal of Biomedical Materials Research,* Vol. 18, 1984, p. 1043.
[9] Bobyn, J. D., Cameron, W. U., Abdulla, D., Pilliar, R. M., and Weatherly, G. C., *Clinical Orthopaedics,* Vol. 166, 1982, p. 301.
[10] Homsy, C. A., Cain, T. E., Kessler, F. B., Anderson, M. S., and King, J. W., *Clinical Orthopaedics,* Vol. 89, 1972, p. 220.
[11] Sauer, B. W., Lade, R. B., and Mercer, H., *Transactions of the Orthopaedic Research Society,* Vol. 5, 1980, p. 315.
[12] Spector, M., Michno, M. J., Smarook, W. H., and Kwiatkowski, G. T., *Journal of Biomedical Materials Research,* Vol. 12, 1978, p. 665.
[13] Blaha, J. D., "Polymers and Silicone Rubbers," *Transactions,* Seminar on Resources for Basic Science Educators, American Academy of Orthopaedic Surgeons, Williamsburg, VA, 1984.
[14] Anand, S. C., St. John, K. R., Moyle, D. D., and Williams, D. F., *Annals of Biomedical Engineering,* Vol. 5, 1977, p. 410.
[15] Spector, M., Davis, R. J., Lunceford, E. M., and Harmon, S. L., *Clinical Orthopaedics,* Vol. 176, 1983, p. 34.
[16] Dehl, R. E., "Characterization of Porosity in Porous Polymeric Implant Materials," NBS IR 83-2645 (FDA), National Bureau of Standards, Gaithersburg, MD, 1983.
[17] Lucas, L. C., Buchanan, R. A., and Lemons, J. E., *Journal of Biomedical Materials Research,* Vol. 15, 1981, p. 731.
[18] Galante, J., Rostoker, W., and Ray, R. D., *Calcified Tissue Research,* Vol. 15, 1970, p. 236.
[19] Spector, M., Harmon, S. L., Eldridge, J. T., and Davis, R. J., "Porous Polymer Coatings for Orthopaedic and Dental Prostheses," *Mechanical Properties of Biomaterials,* G. W. Hastings and D. F. Williams, Eds., Wiley, New York, 1980, p. 299.
[20] Chen, P.-Q., Turner, T. M., Ronningen, H., Galante, J., Urban, R., and Rostoker, W., *Clinical Orthopaedics,* Vol. 176, 1983, p. 24.
[21] Bobyn, J. D., Pilliar, R. M., Bennington, A. G., and Szivek, J. A., *Transactions,* Fifth European Conference on Biomaterials, Paris, France, 4–6 Sept. 1985, p. 193.

Ulrich Gross,[1] Wolfgang Roggendorf,[1] Hermann-Joseph Schmitz,[1] and Volker Strunz[2]

Biomechanical and Morphometric Testing Methods for Porous and Surface-Reactive Biomaterials

REFERENCE: Gross, U., Roggendorf, W., Schmitz, H.-J., and Strunz, V., **"Biomechanical and Morphometric Testing Methods for Porous and Surface-Reactive Biomaterials,"** *Quantitative Characterization and Performance of Porous Implants for Hard Tissue Applications, ASTM STP 953,* J. E. Lemons, Ed., American Society for Testing and Materials, Philadelphia, 1987, pp. 330–346.

ABSTRACT: Two implantation sites were evaluated in the rabbit for measurement of the tensile strength of bonding between bone and both surface-reactive and metallic biomaterials. In rabbits, the tensile strength of bonding was 2.4 N/mm² for glass-ceramic KG Cera rectangular solid implants that had been firmly clamped against cortical bone of the femur midshaft with champy plates for 56 days. The tensile strength of bonding was consistently less with cylindrical implants (flattened on one side) of the same material held against cortical bone of the femur midshaft with sutures or embedded into trabecular bone of the distal rabbit femur. The surface roughness of surface-reactive biomaterials affected the tensile strength of bonding only insignificantly.

KEY WORDS: porous implants, bonding surface-reactive biomaterials to bone, bone bonding tensile strength, animal models, tissue morphometry at implant interface, glass ceramic, porous biomaterials

The most common method of force transmission between an implant and bone is by the use of intramedullary stemmed implants attached to bone with bone cement. Other methods include the use of screws, holes, or attachment by porous ingrowth. A newer class of surface-reactive (bioactive) implant materials has been developed for attachment of implants to bone or for tooth replacement. The mechanism of this physicochemical bonding is not based on mechanical interlocking and is not fully understood [1].

The most widely accepted method for testing the efficacy of bonding or attachment between an implant and bone has been to measure the shear or torque force required to produce failure. This technique has also been used with surface-reactive biomaterials [2,3]. However, bond strength measurements obtained by shear or torque techniques will necessarily include forces required for failure of both mechanical interlocking and surface bonding. Thus, the strength of bonding between a surface-reactive biomaterial and bone would be measured more appropriately by tensile strength methods, in which the strength of attachment attributable to mechanical interlocking does not influence the test results [4].

---

[1] Professor of pathology, assistant professor, and scientific assistant, respectively, Institute of Pathology, Klinikum Steglitz, Free University of Berlin, 1000 Berlin 45, Federal Republic of Germany.
[2] Assistant professor, Department of Maxillo-Facial and Plastic Surgery, Klinikum Steglitz, Free University of Berlin, 1000 Berlin 45, Federal Republic of Germany.

Mechanical interlocking on a microscale must also be avoided; for example, the surface must be free from undercuts that could provide interlocking. The surface structure of the material being tested must be adequately characterized by morphological methods, including scanning electron microscope (SEM) and surface roughness measurements.

In measuring the tensile strength of the attachment between an implant and bone, failure typically occurs in the weakest of three locations: (1) entirely within the bone, (2) entirely within the material, or (3) at the interface of biomaterial and bone [5]. The fracture line can run through all three areas, and it is then practically impossible to attribute failure to any one.

The tensile strength of bonding between an implant and bone has been measured by several techniques [4–7] (Table 1), but no method has become standard. In most studies, the tensile strength of bonding has been determined by the area of the specimen and the force-load required to produce failure. However, bone seldom covers 100% of an implant interface. Tissues observed at the implant interface include bone; osteoid, chondroid, and soft tissue; and bone marrow [8]. It is assumed that bone is the main contributor to the tensile strength of bonding and that other tissues are less important. Collagen fibrils encased in the surface layer of Bioglass in a cell-free system have been reported by Hench [9]. It has not been determined conclusively whether this process is operative in vivo. The authors believe that the tensile strength of bonding must be determined by measuring the area of bone at the interface (not on the total area of the implant) for bone-specific tensile strength (in newtons per square millimetre). Thus, the percentage of the implant interface covered by bone and other tissues must be determined from control specimens that have not been subjected to tensile strength testing.

Recently, studies were done to find the implantation site that would result in maximum coverage of the implant surface area with bone [8,10]. In the rat femur, implant specimens were placed in proximal and distal trabecular bone and in midfemur cortical bone. In morphometric studies, the percentage of bone at the implant interface was measured, producing comparable results for midfemur cortical bone and distal trabecular bone [8,10]. However, because of the small size of the implants and specimens in these experiments, the rat was found to be an unsuitable animal model for measurement of tensile strength.

In the rabbit femur, cylindrical implants were inserted in holes made in trabecular bone of the distal femur and secured against flat midfemur cortical bone with sutures. Rectangular solid implants were clamped flat against midfemur cortical bone with champy plates. The use of rectangular solid implants has previously been reported by others [7]. When implanted, the specimens were without a biomechanical force-load, except for that associated with motion of the animals, an occurrence that cannot be avoided in animal studies.

The cylindrical specimen was preferred for testing as an insertion into a hole in trabecular bone, because it allowed the development of primary bone in the gap between the flattened area and the surrounding trabecular bone. This permitted assessment of the biomechanical quality of the newly developed bone.

The aim of this investigation was to determine some parameters that affect the tensile strength of surface-reactive biomaterials in rabbits.

Materials and Methods

The two specimen shapes evaluated were (1) cylinders 4 mm in diameter by 6 mm long and flattened on one side to a depth of 0.8 mm; and (2) rectangular solids 4 by 4 by 6 mm.

Four materials were evaluated in this study:

1. Glass ceramic KG Cera, containing 46.2% silicon dioxide (SiO_2), 25.5% calcium metaphosphate, [$Ca(PO_3)_2$], 4.8% sodium monoxide (Na_2O), 20.2% calcium oxide (CaO), 0.4%

TABLE 1—*Tensile strength of various implant materials.*

Material	Surface Condition	Species	Site	Type of Implant	Time, days	Tensile Strength, N/mm²	Reference
Titanium	sand-blasted, roughness 20 μm plasma-coated, roughness 20 μm	monkey (*Macaca speciosa*)	ulna	disk	210	1 to 4.1	Steinemann et al. [6]
Glass ceramic (KG Cera)	roughness below 10 μm	pig	lower jaw	square-shaped block	104	2.19 1.09	Bunte et al. [4]
Glass ceramic (KG Cera)	roughness below 10 μm	rabbit	femur midshaft	square-shaped block	56	2.4	Gross et al., this study
Hydroxyapatite	polished by sanded paper No. 400	dog	tibia	disk	56	1.32	Schmickal, [7]
Apatite-wollastonite-glass ceramic Hydroxyapatite Bioglass 45 S5	not defined	rabbit	tibia	blade, 15 by 10 by 2 mm	56	~1.3 1.0 0.45	Nakamura et al. [5]

potassium oxide (K_2O), and 2.9% magnesium oxide (MgO). Cylinders were evaluated with maximum surface roughnesses (RTs) of the flattened area (DIN German Standard 4762/1) [11] of 0.06, 10, 28, and 52 μm. The rectangular solid blocks evaluated had a maximum surface roughness (RT) of 10 μm. Scanning electron microscopy (SEM) of the surfaces at high magnification revealed only a very few scratches on the polished material (RT = 0.06 μm) (Fig. 1) and shell-like fractures of the surface with increasing surface roughness (Figs. 2–4).

2. Ti-6Al-4V alloy. Cylindrical specimens with a maximum surface roughness of 10 μm (Fig. 5) were evaluated.

3. Pure titanium. The surface roughness of the flattened area of titanium cylinders was essentially the same as that for the flattened areas of cylinders of Ti-6Al-4V alloy (RT = 10 μm) (Figs. 5 and 6). The surfaces of the metal specimens were different from the surfaces of the glass ceramic ones (Fig. 2) with comparable surface roughness.

4. Ti-5Al-2.5Fe alloy. The flattened area of the cylindrical specimens contained 33 holes 0.5 mm in diameter and with depths of 0.4, 0.8, and 1.2 mm (Fig. 7).

Prior to implantation, all the specimens were sterilized with ethylene oxide.

Rabbit Experiments

Female chinchilla rabbits (96 in number) with an average body weight of 3.0 to 3.5 kg were used as experimental animals. Two animals were kept in each cage (87 by 54 by 36

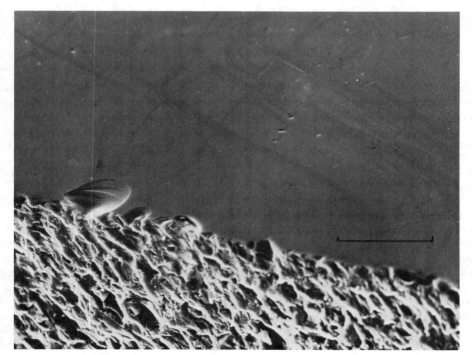

FIG. 1—*SEM photomicrograph of the flattened surface of a glass ceramic (KG Cera) cylinder implant with a surface roughness of 0.06 μm. The scale mark indicates 50 μm.*

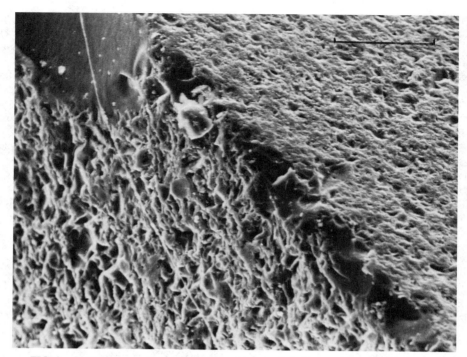

FIG. 2—*SEM photomicrograph of the flattened surface of a glass ceramic (KG Cera) cylinder implant with a surface roughness of below 10 μm. The scale mark indicates 50 μm.*

cm). The diet was standard, with water given *ad libitum*. The animals were allowed to run free at least 2 h each day for exercise before and some days after the operation, when they were observed to ambulate normally.

Anesthesia was induced by intramuscular injection of Ketanest [2-(*o*-chlorphenyl)-2-methylamino-cyclohexane-1-*on*-hydrochloride] and Rompun [5,6-dihydro-2-(2-(2,6-xylidino)-4*H*-1,3-thiazin], 25 and 5 mg per kilogram of body weight, respectively. When it was necessary to extend the operation time, anesthesia was continued by injections or inhalation of diethyl ether.

Both thighs of each animal were shaved and disinfected, and the skin and longitudinal fascia were incised laterally; a self-holding wound retractor was inserted between the extensor and flexor muscles, thus exposing the anterolateral aspect of the femur. With a diamond burr, the cortical bone of the midfemur was flattened, and the rectangular implant was attached by a small champy plate (Fig. 8) or the cylinder was ligated. The median approach was used to expose the patella sliding plane. A hole was drilled perpendicular to the long axis of the femur into the underlying trabecular bone with a burr (4 mm in diameter and providing inner irrigation with saline solution), and the cylinder was inserted (Figs. 9 and 10).

The rabbits were sacrificed by intramuscular anesthesia, as described before, and subsequent carbon dioxide inhalation. Bone segments with the attached implants were excised *en bloc*.

Technique for Tension Testing

Fresh, unfixed specimens were used for tension testing. The rectangular implants were carefully freed from the bone adhering to their lateral aspects, so that only the basal area remained attached to the underlying bone. The cylinders were completely freed from all bone at the circumference but not at the flattened area attached to the underlying bone. This preparation was done with a diamond burr drill and saline rinsing. Any dislodgement of the implant was carefully avoided. Because fixtures glued to the specimens to facilitate tension testing failed to bond, a clamp was used (Fig. 11). The testing machine (Strebel) was equipped with a U-shaped bar for retention of the bone and a flexible wire between the machine and the mechanical clamp on the specimen. The velocity for tensile loading was 1 mm/min. The force-load required to produce rupture was measured in newtons.

Histological Techniques

Specimens intended for histomorphological examination were fixed by immersion in buffered 4% formaldehyde solution (Lillie) at 4°C for 24 h. These specimens were subsequently processed by stepwise dehydration with ethanol, followed by immersion in liquid methyl methacrylate (MMA) for embedding. The MMA was then polymerized into polymethyl methacrylate (PMMA). Sections 50 μm thick were obtained by cutting with a diamond saw (Leitz). Surface staining was accomplished by the Giemsa technique [12].

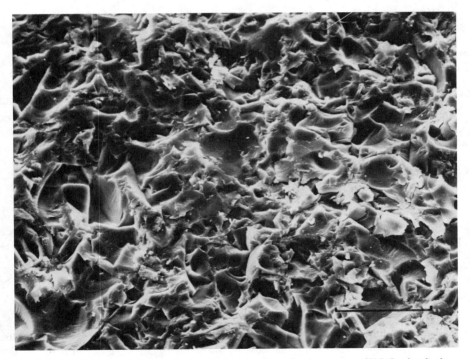

FIG. 3—*SEM photomicrograph of the flattened surface of a glass ceramic (KG Cera) cylinder implant with a surface roughness 28 μm. The scale mark indicates 50 μm.*

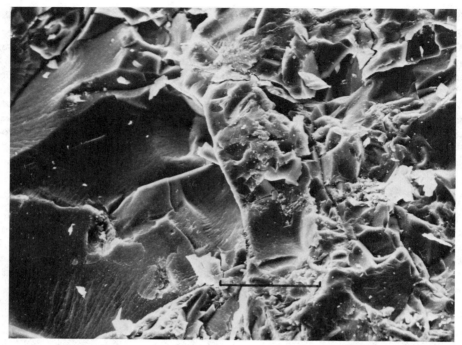

FIG. 4—*SEM photomicrograph of the flattened surface of a glass ceramic (KG Cera) cylinder implant with a surface roughness 52 μm. The scale mark indicates 50 μm.*

Calculation and Statistical Analysis

The percentage of bone, osteoid, chondroid, and fibrous tissue in contact with the specimens was determined by the morphometric studies done on the PMMA-embedded specimens discussed earlier.

Statistical analysis of the percentages of various types of tissue present were calculated for each implant specimen using morphometric data from four slices selected from each specimen.

Mean values were also calculated from all of the specimens in each test group.

Values from the tension testing were expressed in newtons per square millimetre of bone contact area. The mean values, standard deviations (SD), and standard error (SE) were calculated for each test group. The statistical significance was determined using the U-test (Mann-Whitney); data were accepted as significant if the *P* level was below 0.05.

Results and Discussion

At the flattened cortical bone of the femur shaft, glass ceramic KG Cera fixed under a champy plate displayed a higher tensile strength at 56 than at 84 days postoperatively (Table 2), which could be due to stress protection. Fixation of an implant to the flattened cortical bone with a suture results in higher tensile strength after 84 days than fixation under a champy plate (Table 2). Therefore, the mode of fixation and motion of an implant influence bone bonding during the healing period. Each model for testing tensile strength should therefore be investigated for maximum values. This accords with clinical experience, es-

FIG. 5—*SEM photomicrograph of the flattened surface of a Ti-6Al-4V alloy cylinder implant with a surface roughness below 10 μm. The scale mark indicates 10 μm.*

FIG. 6—*SEM photomicrograph of the flattened surface of a pure titanium cylinder implant with a surface roughness below 10 μm. The scale mark indicates 50 μm.*

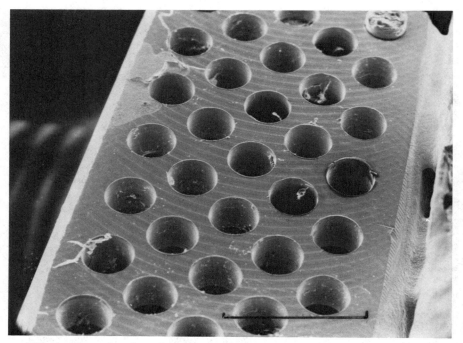

FIG. 7—*Flattened surface of a Ti-5Al-2.5Fe alloy cylinder implant with holes after tension testing. The scale mark indicates 1 mm.*

pecially with implants made of pure titanium and used as posts for dentures [6,13], which are kept without load or motion during wound healing and bone development in the implantation bed.

The tensile strength of a cylinder with a proximally oriented flattened area in the trabecular bone of the distal femur turned out to be lower at 84 days than at 56 days postoperatively (Table 2). This observation could be due to stress protection by the implant or related to bone remodeling at the interface. It can be speculated that bone remodeling also varies at the different implantation sites, which thus affects the mechanical qualities of the interface. The difference between orientations (distal or proximal) of the flattened area of a cylinder in trabecular bone does not affect the tensile strength after 84 days (Table 2). A distal orientation of the flattened area was chosen because it markedly decreased bleeding from the gap between the flattened area of the cylinder and the implantation bed. Furthermore, trabecular bone is denser towards the distal part of the femur than toward the proximal part, which is very close to the marrow cavity. Statistically, there is no difference in tensile strength between implants at the cortical bone midshaft and at the trabecular bone in the distal femur after 56 days. Therefore, tensile strength can be determined with equal results at these two sites after the given time.

Implants with increased surface roughness showed a tendency toward increased tensile strength (Table 3); however, there was no difference statistically. Morphometrically, there was no statistical difference in the percentage of bone bonding to the implant interface at the exposed flattened area for surface roughnesses of 0.06, 28, and 52 μm (Table 4). Also, at the circumferential area of the cylinders, bone bonding was statistically the same (Table 4 and Figs. 12 and 13). No chondroid tissue was found at the circumferential area of the cylinders, whereas there was some chondroid at the flattened area; this may be due to the different biomechanical situation in the small gap between implant and implantation bed.

Relating the amount of bone at the interface to the measured tensile strength 84 days postoperatively, the specific tensile strength of bone was 1.05, 1.1, and 1.4 N/mm^2 for the glass ceramic KG Cera materials, with surface roughnesses of 0.06, 28, and 52 μm, respectively. These values compare well with those given for other surface-reactive materials in Table 1. The tensile strength of trabecular bone was between 4.6 and 6.2 N/mm^2 in the distal femur of rabbits [10]. The tensile strength of the glass ceramic KG Cera and bone bonding is thus approximately 20 to 25% of that of the spongy bone in the distal rabbit femur.

Cylinders made of Ti-6Al-4V alloy and pure titanium (twelve implants each) with a surface roughness below 10 μm inserted in the distal femur were collected 84 days postoperatively, but they separated from the surrounding bone during preparation before the tension test could be performed. Four implants each of Ti-6Al-4V and pure titanium, implanted in the same manner for 168 days, also separated from bone during preparation. Therefore, the tensile strength as measured by Steinemann et al. [6], on the order of 1.5 N/mm^2 for plasma-coated pure titanium implants and 2.5 to 3.5 N/mm^2 for sand-blasted pure titanium implants between 125 and 210 days after implantation in the ulnar bone of monkeys (*Macaca speciosa*), must be related to the considerably higher surface roughness. The surfaces of our implants were obviously too smooth for bone attachment.

Cylinders made of Ti-5Al-2.5Fe alloy with 33 holes [0.5 mm in diameter and with depths of 0.4 mm ($n = 2$), 0.8 mm ($n = 4$), and 1.2 mm ($n = 4$)] implanted in the trabecular bone and oriented distally (Fig. 7) displayed tensile strengths of 2.2, 2.2, and 2.84 N/mm^2,

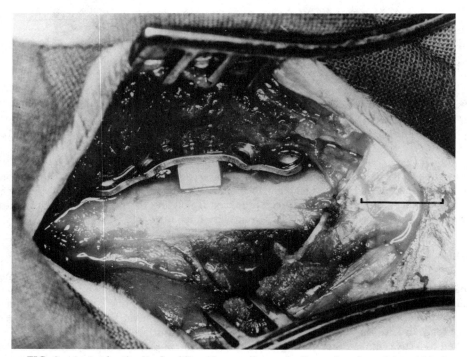

FIG. 8—*An implant* in situ *fixed by a champy plate at the flattened surface of the midshaft of a rabbit femur. The scale mark indicates 12 mm.*

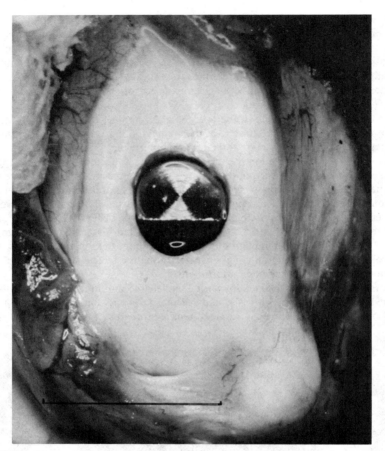

FIG. 9—*Flattened cylinder implant inserted in the distal femur below the patella sliding plane. The flattened area is oriented distally. The scale mark indicates 8 mm.*

respectively, 84 days postoperatively. The surface roughness of the holes in which bone trabeculae had developed (attachment to their walls was observed histologically) and the vacuum effect, as well as shear forces produced by pulling the bone and implant apart, seems to be the source of tensile strength in these implants.

Conclusions

Testing of a surface-reactive biomaterial can be done at the cortical bone of the femur shaft as well as in the trabecular bone of the distal femur in rabbits. The tensile strength can be measured best at the cortical bone of the femur midshaft but can also be determined in the trabecular bone of the distal femur 56 days postoperatively. There is no statistical difference in tensile strength between polished and rough surfaces of surface-reactive or so-called bioactive implant materials. Morphometry of the implant interface is recommended for determination of bone-specific tensile strength. There was no tensile strength in pure titanium and Ti-6Al-4V implants with surface roughness below 10 μm.

TABLE 2—*Tensile strength of glass ceramic (KG Cera) implants with a surface roughness of 10 μm.*

Location of Implant	Type of Implant	Days	Tensile Strength, N/mm^2	SE	nS^a	nA^b	Significance[c]
Femur shaft	rectangular implants	56	2.39	0.2	11	6	a
		84	1.59	0.2	6	6	b
	cylinders ligated	28	1.17	0.4	6	3	c
		84	1.95	0.5	9	5	d
Distal femur	cylinders oriented distally	28	1.04	0.2	11	6	e
		84	0.97	0.08	10	5	f
	cylinders oriented proximally	28	0.74	0.3	5	3	g
		56	1.78	0.4	7	5	h
		84	0.62	0.1	4	2	i

[a] nS = number of specimens.
[b] nA = number of animals.
[c] There is a significant difference ($P < 0.05$) in the U-test between a and b, b and d, and b and i. There is no significant difference ($P > 0.05$) in the U-test between d and f or a and h.

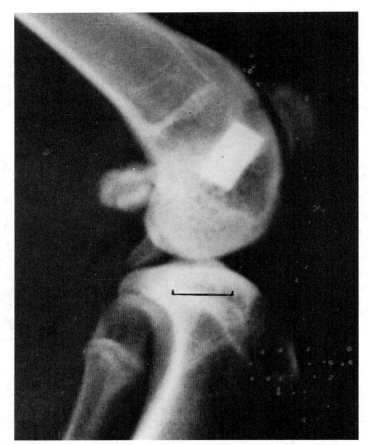

FIG. 10—*X-ray of a cylinder implant inserted in the distal femur below the patella sliding plane. The scale mark indicates 6 mm.*

TABLE 3—*Tensile strength of cylinder (KG Cera) implants with different surface roughnesses of the flattened area oriented distally in the distal femur of rabbits 84 days postoperatively.*

Surface Roughness, μm	Tensile Strength, N/mm²	SE	nS[a]	nA[b]
0.06	0.82	0.08	12	6
10	0.97	0.08	10	5
28	0.96	0.07	9	6
52	1.11	0.11	12	6

[a] nS = number of specimens.
[b] nA = number of animals.

TABLE 4—*Percentage of tissues (±SE) at the bone/implant interface of cylinders (KG Cera) with different surface roughnesses of the flattened area, oriented distally, and the circumferential area in the distal femur of rabbits 84 days postoperatively.*

Area of Implant	Surface Roughness, μm	Type of Tissue				nS^a	nA^b
		Bone	Osteoid	Chondroid	Soft		
Flattened area	0.06	78 ± 3	0.3	4 ± 2	17 ± 2	8	4
	28	86 ± 2	0	2.3 ± 1.6	12 ± 2	9	6
	52	79 ± 5	0	5 ± 3.5	16 ± 3	4	2
Circumferential area	0.06	86 ± 2	0	0	14 ± 2		
	28	88 ± 1	0	0	12 ± 1		
	52	83 ± 2	0	0	17 ± 2		

[a] nS = number of specimens.
[b] nA = number of animals.

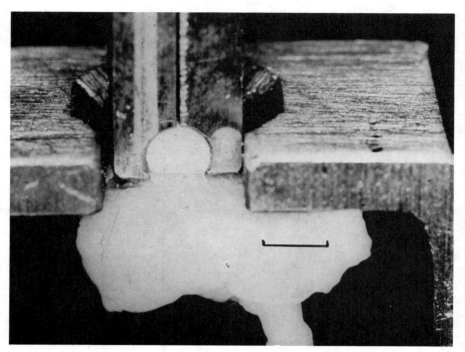

FIG. 11—*Cylinder implant bonded to bone before tension testing. The scale mark indicates* *4 mm.*

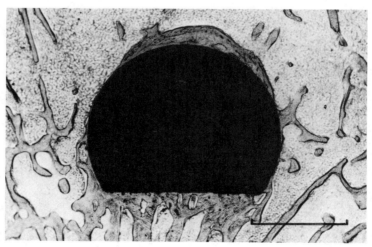

FIG. 12—*Photomicrograph of a section sawed through a flattened glass ceramic (KG Cera) cylinder implant with a surface roughness below 0.06 μm, 84 days after implantation in the distal femur (Giemsa staining). The scale mark indicates 2 mm.*

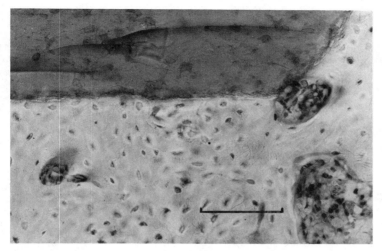

FIG. 13—*Photomicrograph of bone bonding to the flattened surface of a glass ceramic (KG Cera) cylinder implant with a surface roughness below 0.06 μm, showing a haversian canal at the edge of the flattened area (Giemsa staining). The scale mark indicates 100 μm.*

Acknowledgments

This study was supported by Grant 01 ZQ 110 of the Ministry for Research and Technology, Bonn, Federal Republic of Germany. Dr. G. Fuhrmann, Bundesanstalt für Materialprüfung, Berlin, assisted in the experiments. The materials were supplied by E. Leitz Wetzlar, Wetzlar, and by Mecron, Berlin, Federal Republic of Germany.

References

[1] Hench, L. L., Splinter, R. J., Allen, W. C., and Greenlee, T. K., *Journal of Biomedical Materials Research Symposium,* Vol. 2, 1971, pp. 117–141.
[2] Piotrowski, G., Hench, L. L, Allen, W. C., and Miller, G. L., *Journal of Biomedical Materials Research,* Vol. 6, 1975, pp. 47–61.
[3] Strunz, V., Männer, K., Gross, U. M., Hirsch, A., Deutscher, K., and Brömer, H. in *Dental Implants,* G. Heimke, Ed., Hanser, Munich, 1980, pp. 27–34.
[4] Bunte, M., Strunz, V., Gross, U. M., Brömer, H., and Deutscher, K., *Deutsche Zahnärztliche Zeitschrift,* Vol. 32, 1977, pp. 825–828.
[5] Nakamura, T., Yamamuro, T., Higashi, S., and Kokubo, T., *Journal of Biomedical Materials Research,* Vol. 19, 1985, pp. 685–698.
[6] Steinemann, S. G., Eulenberger, J., Maeusli, P.-A., and Schroeder, A. in *Biological and Biomechanical Performance of Biomaterials,* P. Christel, A. Meunier, and A. J. C. Lee, Eds., Elsevier Science, Amsterdam, 1986, pp. 409–414.
[7] Schmickal, T., "Vergleichende Untersuchungen über die Knochenhaftung verschiedener Implantatwerkstoffe mit Hilfe enossaler Implantate," inaugural dissertation thesis, University of Cologne, Cologne, Federal Republic of Germany, 1984.
[8] Gross, U. and Strunz, V., *Journal of Biomedical Materials Research,* Vol. 19, 1985, pp. 251–271.
[9] Hench, L. L., *Transactions of the Society for Biomaterials,* Vol. 8, 1985, p. 25.
[10] Gross, U., Roggendorf, W., Schmitz, H.-J., and Strunz, V. in *Biological and Biomechanical Performance of Biomaterials,* P. Christel, A. Meunier, and A. J. C. Lee, Eds., Elsevier Science, Amsterdam, 1986, pp. 367–372.

[11] "Oberflächenrauheit; Begriffe," (E DIN 4762/1) Deutsches Institut für Normung e.V., Berlin, May 1978.
[12] Gross, U. M. and Strunz, V., *Stain Technology,* Vol. 52, 1977, pp. 217–219.
[13] Braenemark, P. I., Hansson, B. O., Adell, R., Breine, U., Lindström, J., Hallen, O., and Öhman, A., *Scandinavian Journal of Plastic and Reconstructive Surgery, Supplement,* Vol. 16, 1977, pp. 1–132.

Edwin C. Shors,[1] Eugene W. White,[1] and Robert M. Edwards[2]

A Method for Quantitative Characterization of Porous Biomaterials Using Automated Image Analysis

REFERENCE: Shors, E. C., White, E. W., and Edwards, R. M., "**A Method for Quantitative Characterization of Porous Biomaterials Using Automated Image Analysis,**" *Quantitative Characterization and Performance of Porous Implants for Hard Tissue Applications, ASTM STP 953*, J. E. Lemons, Ed., American Society for Testing and Materials, Philadelphia, 1987, pp. 347–358.

ABSTRACT: An automated image analysis method is described for the quantitative characterization of porous biomaterials. The method consists of impregnating porous materials with epoxy embedding medium and polishing planar sections. Contrast was optimized under a scanning electron microscope using backscattered electron images or under a light microscope using reflected light. The images were analyzed using an image analysis system and software developed for this study. The parameters measured on each image feature of the solid and void fractions, frame by frame, were the area fraction, internal perimeter, minimum crossing distance, number of features per area, equivalent circular diameter, perimeter squared per area, and distribution of equivalent circular diameter squared.

The usefulness of this method was validated by comparing the median values of these parameters and their statistical distributions on Replamineform porous biomaterials derived from echinoid and coral skeletons. We believe that this method for quantitative characterization of the three-dimensional structure of biomaterials is applicable to a broad range of materials and is readily amenable to standardization.

KEY WORDS: porous implants, porous biomaterials, replamineform biomaterials, quantitative image analysis

In recent years, many new porous biomaterials have been developed for hard and soft tissue ingrowth. Tissue ingrowth into porous implants is becoming increasingly more important in orthopedics, dentistry, and cardiovascular surgery. It aids in the repair, regeneration, and reconstruction of tissues damaged by disease, trauma, or congenital anomalies. For example, fused beads or wire, attached to metallic hips and knees, facilitates biological fixation through the ingrowth of bone [1]. Similarly, porous hydroxyapatite encourages regeneration of bone in oral surgery for treatment of severely resorbed alveolar ridges [2] and in orthopedic surgery for repair of traumatic fractures [3]. In vascular prostheses, porosity promotes healing, infection resistance, and endothelialization [4].

Although porosity is recognized as essential for many of these new devices, there is considerable debate about the optimal porosity required. Primarily, this is because there is no accepted method for characterizing porous materials. Current methods require either certain material properties or certain configurations. For example, mercury porosimetry is

[1] Vice president, Research and Development, and consultant, respectively, Interpore International, Irvine, CA 92714.

[2] Manager of software development, LeMont Scientific, Inc., State College, PA 16801.

limited to certain sizes and shapes of material: the specimens must be larger than 1 cm³ and have pores smaller than 300 μm. In addition, its use for compliant materials, such as many polymers, is controversial [5]. Other methods for determining porosity, such as water permeability and gravimetric density, are under evaluation by medical device standards committees, but these methods are also restricted by the size and shape of the material. Alternatively, biomaterials can be evaluated directly under a stereoptic light microscope or scanning electron microscope. Unfortunately, these techniques are subjective, tedious, and descriptive.

Unquestionably, there is a need for a new technique for quantitatively evaluating macroporous biomaterials, similar to those developed for characterizing particulates [6]. This new technique would be valuable in comparing materials made by different manufacturing processes. It would also be valuable in determining the consistency of a material, thus serving as a quality assurance tool. Lastly, the technique could assist in conceptualizing the physical properties of materials, such as pore diameter and interconnectivity.

This study was undertaken to determine the usefulness and limitations of automated image analysis from two-dimensional (planar) sections for quantitative characterization of porous biomaterials. Essentially, this analysis involved making true planar sections of biomaterials after they had been impregnated with an embedding medium, cut, and polished. From the planar sections, certain features were quantitated. Replamineform biomaterials and their precursors were chosen for this preliminary evaluation because of their highly interconnected and uniform porosity over a range exceeding 1.5 orders of magnitude [7].

Methods and Materials

Replamineform biomaterials are derived from a process that replicates the calcium carbonate skeletons of certain porous marine invertebrates [7]. In this study, the precursor calcium carbonate skeletons were evaluated directly. In addition, porous hydroxyapatite (Interpore 500 porous hydroxyapatite, Interpore International, Irvine, CA) produced from *Goniopora* coral was evaluated. The porous structures were impregnated with an epoxy resin (Spurr low-viscosity embedding medium, Polysciences, Inc., Warrington, PA) under vacuum or by centrifuge. In most cases, a pigment (opaque resin color, Gerisch Products, Torrance, CA) was incorporated into the epoxy resin to make it opaque. This increased the contrast between the translucent solid and the epoxy resin when viewed under the light microscope. All specimens were ground using abrasive paper (400-grid Wet Paper, 3M Corp., St. Paul, MN) and polished on a rotating lap wheel using polishing cloth (Texmet, Buehler Ltd., Lake Bluff, IL) and polish (alumina 0.3 and 0.05 μm, Union Carbide Corp., Danbury, CT).

The specimens were observed under either a light microscope or a scanning electron microscope (SEM). Black and white micrographs were obtained from a metallographic light microscope using dark-field illumination. After the calcium salts had been dissolved away by acid, the contrast was enhanced with bright-field illumination. The light microscope with 7× to 25× magnification was satisfactory for evaluating *Porites* coral, *Goniopora* coral, and Interpore 500 hydroxyapatite (Figs. 1a and 1b). For the smaller pore structures, the *Protoreaster* sea star and *Heterocentrotus mammallatus* sea urchin spine, a scanning electron microscope using a backscattered electron detector was required because the pigment did not give adequate contrast under the light microscope (Figs. 2a and 2b). Backscattered electrons can distinguish between the high-density calcium salts and the lower-density embedding medium.

Image analysis was performed using an automated optical analysis system (Oasys, LeMont Scientific, Inc., State College, PA). The images were processed either directly from the microscope or from photographs. The optical analysis system was developed primarily for characterizing complex particulate materials. The analysis is operator independent except for the setting of the gray-scale thresholds. Operator-to-operator variation in setting the thresholds of two-phase structures is less than 0.5%. By using a high-resolution video camera in conjunction with a minicomputer, approximately 250 000 pixels were resolved in each frame, with gray levels of 0 to 250 points. In addition, the system has provisions to correct for background shading resulting from uneven illumination. This was sometimes required for light microscopic pictures taken at low magnification. Data smoothing with a filter (3 × 3 pixel averaging) and enhancement routines were also optionally used on grainy images.

The library of software programs with the system generated all the usual stereological measurements (point count, lineal analysis, and line scan). In the line scan mode, each contiguous feature was recognized and characterized regardless of shape complexity.

The parameters distilled from the wide selection of available routines and used to characterize the biomaterials were the area fraction, internal perimeter, minimum crossing distance, number of features per area, equivalent circular diameter, distribution of equivalent circular diameter squared, and perimeter squared per area.

Area Fraction

The area fraction was determined simultaneously for both the porous and the solid phases. The volume fraction (percentage) can be directly derived by recognizing the equivalence of area to volume when specimen preparation (true planar surfaces) and sampling are considered.

Internal Perimeter

The internal perimeter was computed directly from the total interface perimeter between the solid and porous phases. It was computed as a ratio of the interface perimeter length (in millimetres) to the image area (in square millimetres). Like the volume percentage, the internal surface area can be directly derived from the internal perimeter by using adequate sampling considerations.

Minimum Crossing Distance

The minimum crossing distance (MCD) was derived from a 32 by 30 point grid on the image. At each point, the local image area was scanned in a "star burst" pattern (Fig. 3). Four test lines radiated through a point at 45° increments until the adjacent phase was encountered. The minimum distance through the point across the feature represents the minimum crossing distance.

Number of Features per Area

For the number of features per area, each contiguous feature in the field was counted, regardless of shape complexity. This parameter provided a measure of the scale and the relative interconnectedness of the two phases.

Equivalent Circular Diameter

The equivalent circular diameter (ECD) was determined by computing the diameter of a circle that contains the area of each feature. From the classified data, the median and

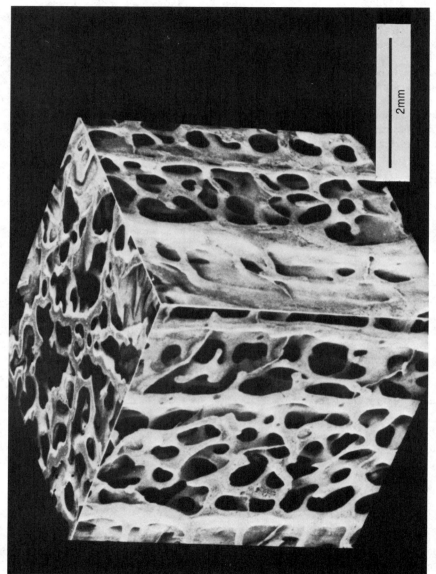

2mm

FIG. 1a—Macroporosity of Goniopora coral as viewed three-dimensionally from composite scanning electron micrographs.

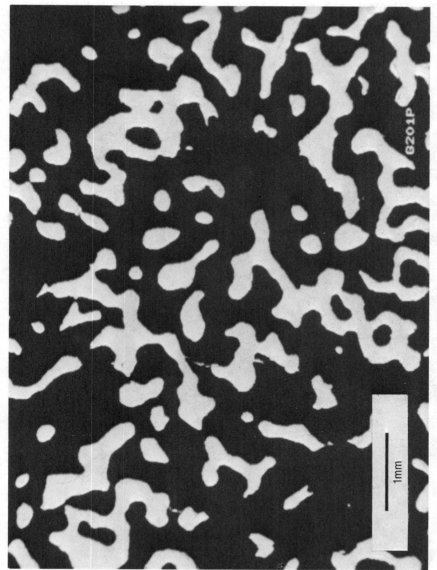

FIG. 1b—*Macroporosity of Goniopora coral as viewed two-dimensionally from epoxy-embedded planar sections under a light microscope using dark-field illumination.*

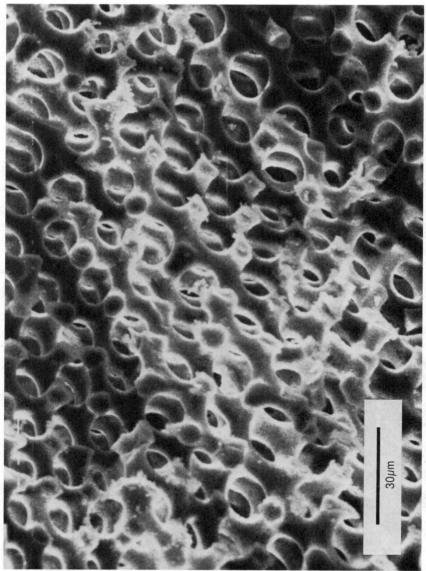

FIG. 2a—*Macroporosity of the Protoreaster sea star as viewed three-dimensionally on a fractured surface under an SEM.*

30μm

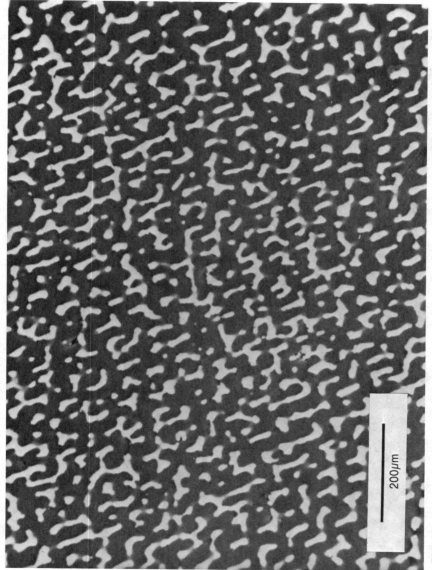

FIG. 2b—*Macroporosity of the Protoreaster sea star as viewed two-dimensionally from epoxy-embedded planar sections under an SEM using backscattered electrons.*

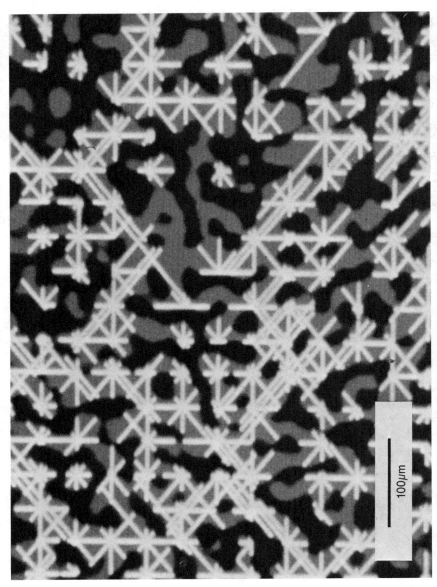

FIG. 3—*Macroporosity of an H. mammallatus sea urchin spine embedded in epoxy showing the "star burst" quantitative image analysis routine used to determine the minimum crossing distance of the solid phase.*

100μm

maximum values were derived. A tendency toward large contiguous features of porosity indicates a highly interconnected structure.

Distribution of Equivalent Circular Diameter Squared

The distribution of the equivalent circular diameter squared (ECD^2) (nd^2) was computed using the classified ECD data by multiplying the number of features in each class interval by the ECD^2 of the class midpoint. In order to evaluate the influence of the largest contiguous features, the values in percentage for the largest (last) and the next to the largest (next) feature were determined.

Perimeter Squared per Area

The value of the perimeter squared of each feature in relation to its area (P^2/A) is an index sensitive to the shape complexity. The median and maximum values were derived from histograms.

Results

This study was designed to develop methods for characterizing porous biomaterials by using quantitative, two-dimensional image analysis. Developing procedures for sample preparation and observation was a prerequisite to the data collection. It also became apparent that a myriad of parameters could be evaluated, and the challenge was to select the most important.

The Replamineform biomaterials and their precursors are structurally anisotropic. However, we found that it was practical to orient only the materials with larger pores (*Porites* coral, *Goniopora* coral, and Interpore 500 porous hydroxyapatite). The smaller-pore echinoid structures (*Protoreaster* sea star and *Heterocentrotus mammallatus* sea urchin spine) were not oriented at a microscopic level. This dichotomy was required for viewing the images through the scanning electron microscope and the light microscope. The echinoid structures with their smaller features required the SEM because they lacked contrast with the embedding medium when viewed at magnifications greater than about 50. The problem was reversed for the coral structures. It was difficult to get significant numbers of image features per frame using the SEM because it cannot achieve magnifications less than 15.

Table 1 lists the parameters evaluated in this study for the representative Replamineform biomaterials and their precursors. This table is not meant to provide definitive information about the consistency of the materials; instead, it provides a suggestion of the magnitude and the interrelationships of the parameters for typical structures.

Two unambiguous parameters are the area fractions (percentages) for the void and solid phases and the internal perimeter. With a sufficient sampling design, these parameters will precisely describe the physical, three-dimensional properties of the material; that is, the volume fraction and internal surface area. In addition, they will correlate with the density derived using pycnometry, and the surface area derived from nitrogen absorption or other methods.

The other parameters measured in this study are more ambiguous and conceptual. The parameter that seems to correlate well with pore diameters, estimated from visual analysis of three-dimensional scanning electron micrographs, is the minimum crossing distance (MCD).

The interconnectivity or permeability of the structure is described by the number of features per unit area and the equivalent circular diameter (ECD). For all of the Replam-

TABLE 1—*Quantitative image analysis of Replamineform biomaterials and precursors.*

Material	Phase	Area Fraction	Internal Perimeter, mm	Minimum Crossing Distance, μm	Features per Area, No./cm²	Equivalent Circular Diameter, μm		nd^2, %		Perimeter Squared per Area, Maximum
						Median	Maximum	Largest	Next Largest	
Protoreaster sea star	pore	0.672	80.4	36.0	238	17.9	794	99.9	0.1	1 000
	solid	0.328		20.1	62 962	21.5	100	3.1	.9	158
H. mammallatus sea urchin spine	pore	0.761	52.8	52.4	2 330	25.7	1 000	96.7	3.3	5 010
	solid	0.239		29.0	23 670	26.5	158	4.5	2.8	316
Porites coral, transverse	pore	0.588	7.4	319	166	140	6 309	84.9	2.1	5 011
	solid	0.412		193	302	150	2 511	17.4	17.4	1 000
Goniopora coral, longitudinal	pore	0.659	5.8	533	75	174	6 310	95.4	1.0	1 995
	solid	0.341		297	590	30	1 995	20.1	12.7	501
Interpore 500 hydroxyapatite, transverse	pore	0.625	4.2	627	33	189	12 600	91.6	5.8	3 981
	solid	0.375		414	155	269	3 981	14.4	9.1	1 000

ineform materials, it is apparent that the number of features per unit area is greater for the solid than for the void fraction. Conversely, the maximum value for the ECD is greater for the void than for the solid fraction. This suggests that the void fraction is highly interconnected, giving high permeability to the Replamineform materials. More sampling will be required to determine whether the median value for the ECD consistently correlates with the pore diameter, because the relationship is not obvious.

Several of the parameters identified for this study related the perimeter of an individual feature to its surface area. Although the median value of the perimeter squared per area (P^2/A) ratio is about the same for all specimens, the maximum value for the void fraction is consistently greater than the value for the solid fraction. This association is consistent with the nd^2 parameter (number $\times$ ECD^2).

Discussion

The purpose for this study was to determine the usefulness and limitations of using automated image analysis from two-dimensional, planar sections for quantitative characterization of porous biomaterials. We believe that the development of the methodology will be iterative, undergoing many iterations over the next few years. Although porous biomaterials have been in use now for over 20 years, unquestionably their development has accelerated over the last five years. This is true in orthopedics, dentistry, and vascular surgery, as well as other specialties [1–4].

We are reluctant at this point to define definitively either the most appropriate parameters or their values for selected biomaterials. Instead, we believe that this technology must be refined more thoroughly before statements about accuracy and precision can be made. This will require large numbers of specimens with careful attention to the sampling procedures. As techniques of specimen preparation and observation are refined, more definitive data will emerge. For example, the contrast between the void and solid fractions when using bright-field light microscopy must be improved for translucent nonmetallic materials. In addition, techniques must be improved for viewing large features at magnifications less than about 15× and small features at greater than about 50×.

As the practical problems are resolved, the theoretical and conceptual meaning of the data become more important. With the exception of the area fraction (volume fraction) and internal perimeter (internal surface area), the variables derived from two-dimensional planar sections do not translate directly into three dimensions. Thus in our opinion, the quantitative image analysis described here cannot claim to give true three-dimensional characterization. However, it does complement techniques and subjective observation from three-dimensional micrographs, such as those shown in Figs. 1a and 2a. It is an excellent tool for quality control of implant materials and for comparing specimens prepared by different processes.

The hope for the distant future is based on computer simulation. Lined and whole-feature measurements can be converted to three dimensions by using specific three-dimensional model algorithms. Preliminary studies indicate that geometric, repeating units of model porous structures translate from two-dimensional images to three-dimensional images [8].

Immediate efforts are directed towards improving the coatings process for viewing specimens under the light microscope using bright-field illumination. This is particularly necessary where feature definition is distinct but contrast is poor. Concomitantly, methods are under development for quantitatively characterizing composite porous structures. This requires the capability of distinguishing gray levels from three distinct phases. It will be used for quantitating biomaterials that are ingrown with biological tissue, such as bone, and composites of two or more biomaterials with interconnected porosity [9].

References

[1] Ducheyne, P. in *Functional Behavior of Orthopedic Biomaterials,* Vol. II, P. Ducheyne and G. W. Hastings, Eds., CRC Press, Boca Raton, FL, 1984, pp. 163–199.
[2] Piecuch, J. F., Topazian, R. G., Skaly, S., and Wolfe, S., *Journal of Dental Research,* Vol. 62, 1983, pp. 148–154.
[3] Holmes, R., Mooney, V., Bucholz, R., and Tencer, A., *Clinical Orthopaedics and Related Research,* Vol. 188, 1984, pp. 252–262.
[4] White, R. A., *Vascular Grafting,* C. B. Wright and J. Wright, Eds., P.S.G. Inc., Littleton, MA, 1983, pp. 315–325.
[5] Dehl, R. E., *Journal of Biomedical Materials Research,* Vol. 18, 1984, pp. 715–719.
[6] Underwood, E. E. in *Quantitative Stereology,* Adison-Wesley, London, 1970, p. 274.
[7] White, R. A., Weber, J. N., and White, E. W., *Science,* Vol. 176, 1972, pp. 922–924.
[8] Dinger, D. R. and White, E. W., *Scanning Electron Microscopy,* Vol. 3, 1976, pp. 409–415.
[9] Holmes, R. E., *Plastic and Reconstructive Surgery,* Vol. 63, 1979, pp. 626–633.

DISCUSSION

D. S. Metsger[1] (*written discussion*)—How extensively did you edit the binary images and how much did this influence the data? It appeared from your micrographs that the viewer must subjectively define pores and pore boundaries and then differentiate them from other structures. Would you please include with your answer a discussion on how the experimenter's decisions affect the reported values and also suggest objective criteria for evaluating micrographs and editing their corresponding binary images?

E. C. Shors, E. W. White, and R. M. Edwards (*authors' closure*)—No binary images were edited. A 3 by 3 smoothing filter was used on grainy structures. The variation due to operator-to-operator subjectivity was 0.5%.

[1] Miter, Inc., Columbus, OH 43229.

Philip J. Boyne[1]

Comparison of Porous and Nonporous Hydroxyapatite and Anorganic Xenografts in the Restoration of Alveolar Ridges

REFERENCE: Boyne, P. J., "**Comparison of Porous and Nonporous Hydroxyapatite and Anorganic Xenografts in the Restoration of Alveolar Ridges,**" *Quantitative Characterization and Performance of Porous Implants for Hard Tissue Applications, ASTM STP 953,* J. E. Lemons, Ed., American Society for Testing and Materials, Philadelphia, 1987, pp. 359–369.

ABSTRACT: This paper reports the results of surgical investigation of three types of hydroxyapatite implanted in ten adult *Macaca fascicularis* monkeys. The hydroxyapatite was placed in edentulous *alveolar ridges* of the animals. Three types of implants were evaluated—porous hydroxyapatite, nonporous granular hydroxyapatite, and xenografts of bovine bone. The results indicated that all three implants increased the thickness and height of the alveolar ridge. The anorganic xenogeneic bone was more effective in being replaced by host bone than either the porous or nonporous hydroxyapatite. Bony ingrowth into the hydroxyapatite in the porous implants was more extensive than bony growth around nonporous particles.

KEY WORDS: porous implants, hydroxyapatite, anorganic xenografts, alveolar ridge restoration

The clinical problem of containment of particulate nonporous hydroxyapatite (HA) implants when placed on deficient edentulous alveolar ridges has resulted in many clinical empirical attempts to resolve the difficulty. In addition to containment of the particles of the HA implant to prevent the so-called "migration" of the material into the surrounding soft tissue and neurovascular tissue, restoring and maintaining optimal ridge contour and topography are of paramount importance to the successful functional prosthetic reconstruction of the mandible. One approach to this problem is to utilize particles of *porous* rather than nonporous hydroxyapatite so that revascularization of the particles and connective tissue ingrowth from the host bed can consolidate the implant at the surgical host site more quickly. By using porous hydroxyapatite (PHA) it might be possible to avoid migration of the particles, and other implant displacement phenomena, and thus to effect a more optimal ridge form and contour for prosthetic rehabilitation, as well as allow the development of more optimal clinical application of this material [1,2].

In order to study the host tissue response to porous granular hydroxyapatite (PHA) and to compare this response histologically with that for nonporous granular hydroxyapatite particles (NPHA), a study was undertaken of implants in adult male cynomolgus *Macaca fascicularis* monkeys. The object of this research was to characterize the implant/tissue interface of PHA particles and NPHA particles and to determine if there was osseous or connective ingrowth into and around particles of either material and, further, to determine

---

[1] Chief, Oral-Maxillofacial Surgery, Department of Surgery, Loma Linda University Medical Center, Loma Linda, CA 92350.

the degree of incorporation of particles into the bone of the ridge. A third implant composed of xenografts of cortical bovine bone (XB) particles, rendered completely devoid of organic material (anorganic), was utilized to serve as a biologic control to the moderately porous hydroxyapatite.

Methods

Ten fully adult male *Macaca fascicularis* monkeys were rendered completely edentulous from the cuspid tooth posteriorly on all four quadrants, and a markedly severe alveolectomy was performed on each alveolar ridge, reducing the overall height of the ridge approximately 10 mm. The edentulous ridges were allowed to heal for 60 days, at the end of which time three of the alveolar ridges of each animal were augmented with the following materials, one for each quadrant:

(a) porous hydroxyapatite (PHA) particles, 20/40 U.S. Standard Mesh (Interpore International 200),
(b) nonporous granular hydroxyapatite (NPHA) particles, 20/40 U.S.Standard Mesh (Cooke-Waite Corp.) [1–4], and
(c) xenogenous bovine bone (XB) treated with ethylenediamine in 20/40 particulate form (Bio-oss—Geistlich-Pharma) [5].

One quadrant in each animal was allowed to remain unoperated and served as a control (Tables 1 and 2).
The alveolar ridges were augmented by a tunnel procedure commonly used clinically in

TABLE 1—*Types of implants placed in each animal by quadrant.*

	Mandible		Maxilla	
Animal No.	Right	Left	Right	Left
	AT 100 DAYS			
1	PHA	NPHA	XB	C
2	C[a]	PHA	NPHA	XB
3	XB	C	PHA	NPHA
4	NPHA	XB	C	PHA
5	PHA	NPHA	XB	C
	AT 130 DAYS			
6	C	PHA	NPHA	XB
7	XB	C	PHA	NPHA
8	NPHA	XB	C	PHA
9	PHA	NPHA	XB	C
10	C	PHA	NPHA	XB

[a] Key to abbreviations:
PHA = porous hydroxyapatite.
NPHA = nonporous hydroxyapatite.
XB = xenograft bone.
C = control, unimplanted.

TABLE 2—*Total number of implants of each type placed in each quadrant site.*

Implant Location	Implant Material			
	PHA	NPHA	XB	Control
Mandible				
Right	3	2	2	3
Left	3	3	2	2
Maxilla				
Right	2	3	3	2
Left	2	2	3	3

human patients [1,4]. A vertical incision was made in the area of the mental foramina, and approximately 4 g of implant material was placed in each quadrant.

The surgical areas were closed and the animals were placed on a soft diet and labeled with tetracycline (given 250 mg intramuscularly) at two months and one week prior to sacrifice.

The animals were sacrificed at the end of 100 and 130 days and the alveolar ridges were retrieved for study. These specimens were processed into stained decalcified sections for light microscopy and unstained undecalcified sections for ultraviolet microscopy. The specimens were evaluated for the following characteristics:

(*a*) the nature of the host tissue acceptance of the implant,
(*b*) the degree, if any, of osseous ingrowth both in and around the implant particles,
(*c*) the degree of soft tissue ingrowth in and around the particles, and
(*d*) the degree of periosteal bone formation over the outer surface of the implant mass and the overall recontouring and remodeling of the ridge as a result of the implant.

Results

The results of the review of the routine light microscopy of decalcified sections and ultraviolet microscopy of undecalcified sections at the end of 100 days indicated that in the sites containing the porous hydroxyapatite (PHA) and the XB, new bone formation had proceeded from the previous host alveolar ridge to surround the implant particles to a depth measuring approximately three layers of particles in the case of the Interpore implants, and as much as five to six particle layers in the case of the Bio-oss implants (Figs. 1 through 5).

In the case of the nonporous granular implants, new bone formation was seen only at the first layer of the implant particles. Only rarely did reparative bone occur deeper than the first layer into the implant material in these implants (Figs. 6 and 7).

The 130-day specimens did not demonstrate any appreciable increase in bone formation in the NPHA grafts, whereas new bone formation was observed to be extending deeper into the implant mass in both the PHA and XB implants.

Periosteal bone formation surrounding the outer aspect of the implant mass was seen only in the case of the XB implants in the 130-day specimens.

Very few particles of PHA and XB were found in ectopic sites displaced from the bony ridge proper. The ridges with PHA and XB gave the appearance of being consolidated, and the contour and the ridge crest were wider buccolingually and more appropriate for prosthetic function than the NPHA-restored ridges.

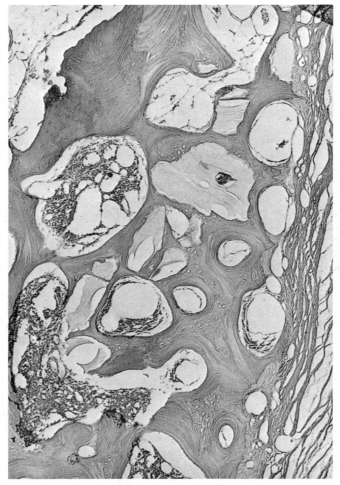

FIG. 1—*Low-power view of an alveolar ridge implanted with the xenograft bone (XB) particles 100 days after implantation, showing bone forming around several layers of implant particles.*

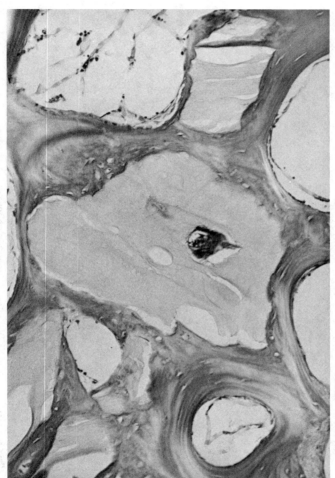

FIG. 2—*Higher-power view showing bone formation around particles of the xenogeneic bone (XB) approximately three to four particle layers deep into the implanted mass.*

FIG. 3.— High-power view showing a particle of xenogeneic bone with excellent bone formation surrounding the implant and, in the center of the implant particle, indicating revascularization of the old haversian canal of the cortical bone and formation of new bone in the center of the revascularized canal, all of which show excellent acceptance of the material and good apposition of bone growth.

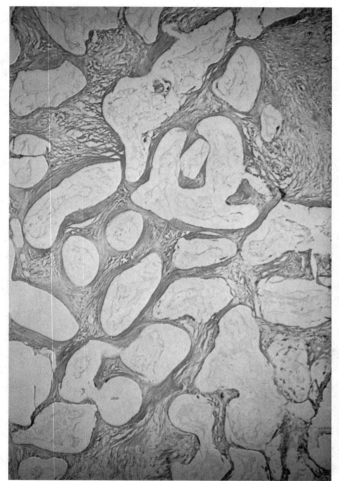

FIG. 4—*Low-power view of an Interpore 200 PHA grafted ridge. The porous particles are engulfed in connective tissue, and bone is growing through the pores of the implant, as is shown in Fig. 5.*

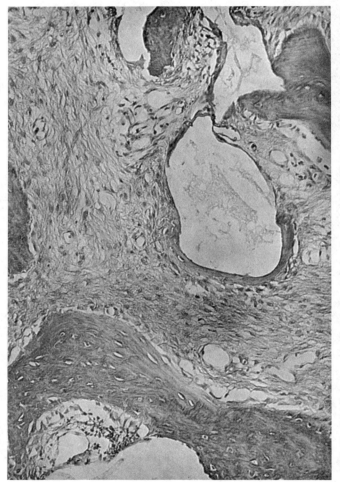

FIG. 5—A high-power view showing bone formation around PHA Interpore implant particles and within the center of the particles after revascularization of the pores.

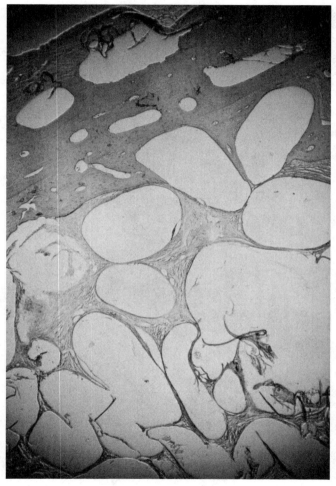

FIG. 6—Low-power view of a ridge after having been implanted with nonporous hydroxyapatite, showing mostly connective tissue around the particles with very little bone formation. Bone is forming only on those particles actually touching the host bony wall.

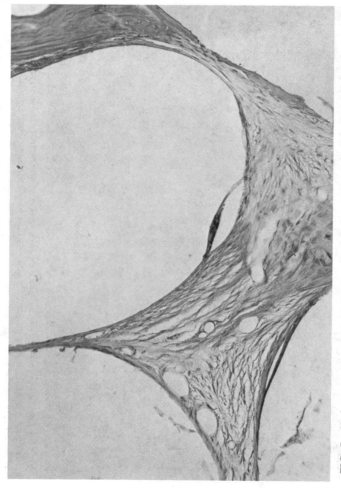

FIG. 7—High-power view of Fig. 6 showing connective tissue surrounding the NPHA implant particles in a capsulation response.

Conclusions

1. The overall clinical conformation of the alveolar ridges containing either the porous hydroxyapatite (PHA) and the xenograft (XB) was superior to that of the ridges containing solid particulate implants.

2. Results of this study appear to indicate that the XB implants of completely despeciated bone may have an advantage over solid hydroxyapatite particles in being gradually replaced by the host bone of the ridge, during which time a lamellated bone structure is maintained.

3. Porous hydroxyapatite particles have an advantage in being replaced very slowly by bone and inviting the ingrowth of bone tissue, which leads to a more complete and rapid stabilization of the implant particles at the recipient site.

References

[1] Boyne, P. J., "Current Clinical Experience with Durapatite," *Proceedings,* Symposium on Alveolar Ridge Augmentation in Edentulous Patients, *Communications Media for Education,* Princeton, NJ, Vol. 3, 1985, p. 311.

[2] Boyne, P. J., "Performance of Bone Grafts in Reconstructive Surgery," *Biocompatibility of Natural Tissues and their Synthetic Analogues,* D. F. Williams, Ed., CRC Press, Liverpool, 1984.

[3] Boyne, P. J., Rothstein, S. S., Gymaer, K. I., and Drobeck, H. P., Long-Term Study of Hydroxylapatite Implants in Canine Alveolar Bone," *Journal of Oral and Maxillofacial Surgery,* Vol. 42, 1984, pp. 589–594.

[4] Boyne, P. J., O'Leary, T. J., and Cox, C. F., "Hydroxylapatite, Beta Tricalcium Phosphate, and Autogenous and Allogeneic Bone for Filling Periodontal Defects, Alveolar Ridge Augmentation, and Pulp Capping," *JADA,* Vol. 108, May 1984, pp. 822–831.

[5] Boyne, P. J., Stringer, D. E., and Shafqat, J. P., "Comparison of Host Tissue Respone to Implants of Synthetic Hydroxylapatite and Ethylene Diamine Extracted Bone," *Proceedings,* Second World Congress on Biomaterials, Washington, DC, 27 April–1 May 1984.

*Itzhak Binderman,[1] M. Goldstein,[1] I. Horowitz,[1] N. Fine,[1]
S. Taicher,[2] Arthur Ashman,[3] and A. Shteyer[2]*

Grafts of HTR Polymer Versus Kiel Bone in Experimental Long Bone Defects in Rats

REFERENCE: Binderman, I., Goldstein, M., Horowitz, I., Fine, N., Taicher, S., Ashman, A., and Shteyer, A., **"Grafts of HTR Polymer Versus Kiel Bone in Experimental Long Bone Defects in Rats,"** *Quantitative Characterization and Performance of Porous Implants for Hard Tissue Applications, ASTM STP 953,* J. E. Lemons, Ed., American Society for Testing and Materials, Philadelphia, 1987, pp. 370–376.

ABSTRACT: Gaps were prepared surgically in femora of 42 albino rats, and were kept apart by a looped wire. Grafts of Kiel bone (KB) and HTR™ polymer in different environments were implanted into those gaps. The control groups showed a nonbony healing 36 days after the osteotomy. KB or HTR polymer alone failed to induce bone bridging, while KB with autologous bone marrow induced bone bridging in the gaps. The most significant finding in this study was the ability of calcium hydroxide [$Ca(OH)_2$] grafted with either KB or HTR polymer to cause differentiation of bone matrix in the surgical gaps. The combination of HTR polymer and $Ca(OH)_2$ induced formation of trabecular bone bridging to a greater extent than all the other material combinations.

KEY WORDS: porous implants, experimental bone gap healing, Kiel bone, HTR polymer, calcium hydroxide effect on grafts

Delayed union and nonunion of fractures have been a therapeutic challenge to the orthopedic surgeon. Bone grafting has traditionally been recommended for an established nonunion [1], but the specific technique, the type of graft, and the need for internal fixation remain controversial. Recently, electrical signals have been used to induce healing of nonunion fractures [2,3]. During the past two decades, bone grafting has been extensively employed for the bridging and replacement of bone defects, but autografts have not always been practical for this purpose. Alloplasts [4,5] or heterografts [6] might substitute for autografts; therefore, the techniques for these approaches should be developed. Xenografts like Kiel bone [6,7] and synthetic grafts like HTR™ (hard tissue replacement) polymer [4,5] might be effective materials for bridging the posttraumatic gap in the long bone. Large bone gaps in animals may serve for research purposes to compare various techniques for treating nonunions.

The purpose of this study was to compare the healing achieved in large gaps in the long bones of rats both when untreated (controls) and when packed with heterograft (Kiel bone)

[1] Head, associate professor, professor of surgery, and laboratory assistant, respectively, Hard Tissue Laboratory, Ichilov Medical Center, Sackler Medical School, Tel Aviv, Israel.

[2] Professor and chief, respectively, Department of Oral and Maxillofacial Surgery, Hebrew University—Hadassah Medical Center, Jerusalem, Israel.

[3] Former head of dental research, Mount Sinai Hospital, New York, NY 10019, and associate professor, Department of Biomaterials, New York University College of Dentistry, New York, NY, 10210.

or synthetic graft (HTR polymer) materials, with and without added bone marrow or calcium hydroxide [Ca(OH)$_2$].

Materials and Methods

Materials

Kiel bone (KB) was purchased from Braun Milsungen A.G., Milsungen, West Germany. This Kiel bone is a commercially available product prepared from fresh calf bone with proteins partly extracted, thus giving it lower immunogenic potential [7].

HTR™ (hard tissue replacement) polymer was donated by the MBS Co., New York [4,5]. It is a nonresorbable, microporous bone grafting material, a result of a patented technology that essentially combines polymethyl methacrylate (PMMA) and polyhydroxyl methacrylate (PHMA) in a process that results in a biocompatible composite without the addition of catalysts, inducers, or impurities [4,5]. It is available as beads of 20 to 24 mesh.

The bone marrow was accumulated from femora of an inbred strain of rats.

The calcium hydroxide [Ca(OH)$_2$] was prepared as a supersaturated solution by dissolving 0.5 g of Ca(OH)$_2$ in 100 mL of distilled water. The alkaline solution was filtered through a 0.45-μm filter in a sterile filtration unit. At this stage the pH of the Ca(OH)$_2$ reached 12 to 13. In some experimental groups (Group 4) the alkalinity was titrated with phosphoric acid (H$_3$PO$_4$) to a pH of 7.4, most probably forming calcium phosphate salts. The alkaline form of Ca(OH)$_2$ was used in Groups 3 and 7.

Methods

In 42 albino rats, weighing 120 to 150 g, the right femur was exposed surgically. With a low-speed dental drill using a Carborundum disk, an ostectomy of 3 to 4 mm was performed in the midshaft of the bone. Stainless steel wire was shaped to an "omega" form and its free ends were inserted into the shafts of the two fragments. This wire had two main functions—fixation of the fracture and maintaining a gap between the two fragments. The gap was immediately filled with one or a combination of bone-grafting materials in seven different groups, as indicated in Table 1. The muscles were then closed and the skin was sutured with 3–0 chromic catgut sutures.

The Kiel bone was immersed in the Ca(OH)$_2$ for 15 to 20 min for the alkaline (Groups 3 and 7) and neutral forms (Group 4). HTR polymer powder of a 20 to 24 mesh size was immersed in the alkaline form of Ca(OH)$_2$ solution for 15 to 20 min. The grafted materials were then filtered and dried under vacuum to produce a thin layer of HTR-Ca(OH)$_2$ or KB-Ca(OH)$_2$. Allogenic bone marrow collected from rat femora was used in Group 5.

The rats were sacrificed 36 days after the implantation and the bones were prepared for histological examination. Microscopically, serial sections were evaluated for the development of fibrous connective tissue, cartilage, and trabecular bone.

Results

In the present study, the authors have compared several grafts and combinations of grafts implanted in surgically prepared circumferential gaps in the midshafts of young rat femora. Results showed that when the two fragments were kept apart by the aid of an omega-shaped wire, a fibrous type of tissue developed, producing a nonunion healing of the bone (Fig. 1a and b). This group served as a control. Table 1 shows that, in the control group (Group 1), bone bridging did not occur within 36 days after implantation in any of the rats operated

TABLE 1—*Summary of results.*[a,b]

Group No.	Bone-Grafting Material	Fibrous Tissue	Cartilage	Trabecular Bone	Figure
1	control	L++	++	−	1a, 1b
2	KB	D+++	+	−	1c, 1d
3	KB + Ca(OH)$_2$ (pH = 12 to 13)	D+	+++	+	1e, 1f
4	KB + titrated Ca(OH)$_2$ (pH = 7.4)	D++	++	−	
5	KB + marrow	−	+++	++	1g, 1h
6	HTR polymer	D+++	+	−	2d, 2e
7	HTR + Ca(OH)$_2$ (pH = 12 to 13)	−	++	+++A	2a, 2b

[a] Key to abbreviations:
  L = loose connective tissue.
  D = dense connective tissue.
[b] Key to symbols:
  − = none.
  + = little.
  ++ = moderate amount.
  +++ = greatest amount.
  A = trabecular bone bridging.

on. In all the rats of this group, only loose fibrous tissue developed in the gap between the two bony fragments. On some occasions, loci of cartilage were observed (Fig. 1b).

Implantation of Kiel Bone

When particles of Kiel bone filled the gap, a more dense fibrous tissue surrounded the KB particles (Table 1, Group 2). Many macrophages and fewer multinucleated cells (osteoclast-like) were seen in the near vicinity of the KB (Fig. 1c and d). It seems that some new bone was formed at the bone fragments (Fig. 1d). KB which was treated with Ca(OH)$_2$ (Table 1, Group 3) prior to implantation stimulated the formation of a rigid callus with components of cartilage and bone (Fig. 1e and f). Definitely, this group showed a rigid bridge that filled the gap between the fractured fragments. Some spaces were still filled with dense connective tissue, which would most probably be remodeled into bone with time. Interestingly, when the high-pH solution of Ca(OH)$_2$ was titrated with concentrated phosphoric acid to reach a neutral pH, the induction ability of the graft was inhibited (Table 1, Group 4). Kiel bone soaked in allogenous bone marrow (Table 1, Group 5) developed into cartilage with new bone bridging the artificially prepared bone gap (Fig. 1g, h, and i). From all these combinations, the authors found that the most solid and mature bone bridge was formed by KB with marrow and by KB with the alkaline solution of Ca(OH)$_2$.

Implantation of HTR Polymer

When HTR polymer alone was grafted into the gap, a dense connective tissue response was created (Table 1, Group 6). The HTR material induced a very mild cellular response with almost no macrophage cells in the neighborhood of the HTR particulate. On the other hand, HTR polymer treated with the Ca(OH)$_2$ solution produced a graft material that induced bone formation in the gap and resulted in a solid support (Fig. 2a). Cartilage and

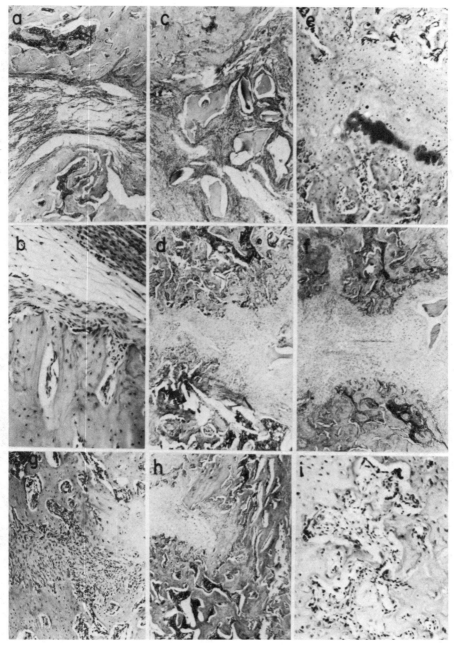

FIG. 1—*Photomicrographs of bone gaps in Groups 1 through 5:* (a) *and* (b) *control groups, in which fibrous connective tissue dominates;* (c) *and* (d) *KB grafts, in which dense connective tissue dominates;* (e) *and* (f) *KB with* $Ca(OH)_2$ *grafts, in which cartilage is the dominating tissue and there is some evidence of trabecular bone;* (g), (h), *and* (i) *KB with bone marrow grafts, which show cartilage with an increase in trabecular bone (hematoxylin and eosin* $\times 40$ *magnification).*

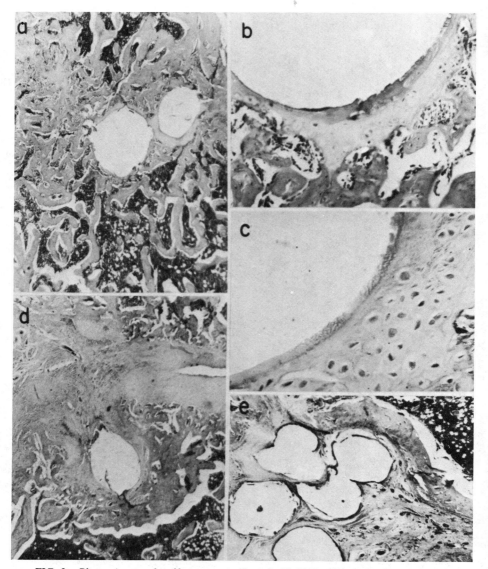

FIG. 2—*Photomicrographs of bone gaps implanted with HTR polymer. (a–c) HTR polymer with Ca(OH)₂: (a) ×80, (b) ×120, and (c) ×200. The HTR particles are surrounded with trabecular bone and cartilage; the HTR particle is encapsulated by a thin layer of connective tissue and cartilage separating it from the trabecular bone. (d,e) HTR polymer grafts: the HTR particles in (d) the gap (×80) and (e) the periosteum (×120) are surrounded by dense connective tissue.*

bone cells grew very close to the HTR polymer granules with no intermediate layers of fibroblasts (Fig. 2b and c). In comparison with the Kiel bone treated with Ca(OH)₂, the HTR polymer in combination with Ca(OH)₂ seemed to be superior, since no fibroblast cellular interphase was seen between bone or cartilage. While KB induced a significant giant cell and macrophage response, the HTR polymer did not induce any similar cellular response.

Discussion

A major identifiable factor which contributes to nonunion has been shown to be a fracture gap maintained by plates or external fixation [8]. Bridging of this gap by alloplasts or xenografts at an early stage is suggested as a method for preventing nonunion healing. The induction of osteogenesis is the main purpose of any bone grafting procedure, and fresh autogenous iliac cancellous bone is considered to have the highest potential of inducing osteogenesis [8]. The advantage of autogenous bone grafts lies in the presence of a small number of surviving osteoblasts, in their possibly better bone-inducing properties, and in their lack of antigenic properties. Since autografts are not always a practical procedure, searching for other means is desirable. Both grafting materials used in this study have low antigenic properties [4,6,7,9]. Kiel bone with autogenous red marrow has been successfully used in patients [7]. Both Kiel bone and HTR polymer appear in molded shapes, they are nonresorbable, and, thus, they have the advantage of forming a stable scaffold for new conductive bone growth. They also can be combined with bone marrow or different solutions that might promote osteogenicity.

The present study showed that, in early stages of bone healing, combined grafts of Kiel bone and autologous marrow induced bone bridging of surgically prepared gaps in the diaphyses of femora, which otherwise would have maintained nonunion after 36 days of healing. KB bone or HTR polymer alone failed to induce bone bridging and produced a dense fibrous tissue at the interface between the two fragments. The significant finding of this investigation is the ability of $Ca(OH)_2$, grafted on either KB or HTR polymer, to induce differentiation of bone in the fractured femur gaps.

The combination of KB with $Ca(OH)_2$ produced more cartilage and less trabecular bone and remnants of dense fibrous tissue, while the combination of HTR polymer treated with $Ca(OH)_2$ solution induced formation of trabecular bone bridging, less cartilage, and almost no fibrous tissue. Although it seems that HTR polymer in combination with $Ca(OH)_2$ could replace the cellular component of the combined graft of KB and marrow, further investigations are necessary.

Calcium hydroxide is being used in dentistry to induce differentiation of tooth pulp cells to produce a mineralized collagenous matrix resembling dentin and bone [10–12]. Spangberg [13] has found in guinea pigs that implantation of $Ca(OH)_2$ in mandibular bone caused less irritation than many compounds used in dentistry and that it was replaced by bone tissue. Interestingly, when $Ca(OH)_2$ was titrated, it lost its inductive activity, which suggests that $Ca(OH)_2$, and not calcium salts, was the inductive factor. *In vitro* studies have shown that $Ca(OH)_2$ stimulated proliferation of pulp cell fibroblasts [14]. HTR polymer is a very negatively charged material coated with polyhydroxyl methacrylate. When grafted with $Ca(OH)_2$ solution, it is possible that it produces a surface potential for cell-specific attachment and interaction with the cells present in the gap, so that the cells proliferate and differentiate into bone matrix.

Since HTR polymer by itself is a very inert plastic material, the authors suggest that this material be used in combination with $Ca(OH)_2$ where bone growth and regeneration are needed. Investigations of the biomechanical properties of the callus are presently being performed.

References

[1] Editorial, "Bone Harvesting and Transplantation," *Lancet,* Vol. 11, 1981, p. 730.
[2] Bassett, C. A. L., Mitchell, S. N., and Schink, M. M., "Treatment of Therapeutically Resistant Nonunions with Bone Grafts and Pulsing Electromagnetic Fields," *Journal of Bone and Joint Surgery,* Vol. 64A, 1982, pp. 1182–1220.

[3] Brighton, C. T., Black, J., Friedenberg, Z. B., Esterhai, J. L., Day, L. J., and Connolly, J. F., "A Multicenter Study of the Treatment of Nonunion with Constant Direct Current," *Journal of Bone and Joint Surgery,* Vol. 63A, 1981, pp. 2–13.

[4] Ashman, A. and Bruins, P., "Prevention of Alveolar Bone Loss Postextraction with HTR Grafting Material," *Oral Surgery,* Vol. 60, 1985, pp. 146–153.

[5] Moss, M. L. and Ashman, A., "Implantation of Porous Polymethylmethacrylate Resin for Tooth and Bone Replacement," *Journal of Prosthetic Dentistry,* Vol. 37, 1977, pp. 657–665.

[6] Salama, R., Burwell, R. G., and Dickson, I. R., "Recombined Grafts of Bone and Marrow," *Journal of Bone and Joint Surgery,* Vol. 55B, 1973, pp. 402–417.

[7] Salama, R. and Weissman, S. L., "The Clinical Use of Combined Xenografts of Bone and Autologous Red Marrow," *Journal of Bone and Joint Surgery,* Vol. 60B, 1978, pp. 111–115.

[8] Connolly, J. F., "Electrical Treatment of Nonunions: Its Use and Abuse in 100 Consecutive Fractures," *Orthopedic Clinics of North America,* Vol. 15, 1984, 89–106.

[9] Elves, M. W. and Salama, R., "A Study of the Development of Cytotoxic Antibodies Produced in Recipients of Xenografts (Heterografts) of Iliac Bone," *Journal of Bone and Joint Surgery,* Vol. 56B, 1974, pp. 331–339.

[10] Cohen, S. and Burns, R. C., *Pathways of the Pulp,* C.V. Mosby, St. Louis, 1976, pp. 578–582.

[11] Gutmann, J. L. and Heaton, J. F., "Management of the Open (Immature) Apex 1 Vital Teeth," *International Endodontic Journal,* Vol. 14, 1981, pp. 166–172.

[12] Schroder, U., "Effect of an Extra-Pulpal Blood Clot on Healing Following Experimental Pulpotomy and Capping with Calcium Hydroxide," *Odontology Review,* Vol. 24, 1973, pp. 257–268.

[13] Spangberg, L., "Biologic Effects of Root Canal Filling Materials: VII—Reaction of Bony Tissue to Implanted Root Canal Filling Material in Guinea Pigs," *Odontology Journal,* Vol. 77, 1969, pp. 501–527.

[14] Sumitra, D., "Effects of Certain Dental Materials on Human Pulp in Tissue Culture," *Oral Surgery,* Vol. 52, 1981, pp. 76–83.

Praphulla K. Bajpai,[1] Cathy M. Fuchs,[1] and Dale E. McCullum[2]

Development of Tricalcium Phosphate Ceramic Cements

REFERENCE: Bajpai, P. K., Fuchs, C. M., and McCullum, D. E., **"Development of Tricalcium Phosphate Ceramic Cements,"** *Quantitative Characterization and Performance of Porous Implants for Hard Tissue Applications, ASTM STP 953,* J. E. Lemons, Ed., American Society for Testing and Materials, Philadelphia, 1987, pp. 377–388.

ABSTRACT: Slowly resorbable and relatively nonresorbable bone cements have been developed to repair various types of defects in bone. A relatively fast-resorbing, porous, biocompatible beta tricalcium phosphate (TCP) cement would be ideal for replacing bone lost as a result of nonresorptive processes. In this investigation the following materials were tested as setting agents for developing a resorbable TCP cement: bovine serum albumin, Carbopol, Hespan, monomers of low-molecular-weight carboxylic acids, orthophosphoric acid, polyvinyl pyrrolidone, and water.

Initial experiments consisted of determining the amounts of setting agent or water or both required to gel the TCP powder. The gel time was monitored at 30-s intervals. The setting hardness was determined by mixing the optimal proportions of TCP and setting agent or water, or both, and monitoring the hardness at 15-min intervals at room temperature for 105 min. Mixtures of TCP and monomers of low-molecular-weight carboxylic acids yielded the most efficient cements. By varying the quantity of acid or combining different acids, or combining both methods, setting times from 30 s to 8 min were achieved for the TCP/polyfunctional acid cements. These cements have been used successfully to repair experimentally traumatized tibiae and femora in rabbits and rats, respectively.

KEY WORDS: porous implants, tricalcium phosphate, hydroxyapatite, tricalcium phosphate cements, aluminum-calcium-phosphorus oxide (ALCAP) cements, hydroxyapatite grouts, plaster of paris, malic acid, alpha-ketoglutaric acid, fumaric acid, polyfunctional carboxylic acids, orthophosphoric acid, bone repair, resorbable cements, nonresorbable cements, biodegradable cements, trauma

Various types of ceramic cements have been developed for replacing bone lost as a result of trauma or disease. Nonresorbable cements composed of calcium-depleted aluminum fluorosilicate glass powder [1] or calcium fluoroaluminosilicate glass powder and polycarboxylic acids [2] have been developed for use as bandages and splints, for joining two bones, and for attaching a device to bone. However, these cements are nonporous and cannot be replaced by endogenous bone. Aluminum-calcium-phosphorus oxide (ALCAP) cements resorb slowly [3], and the hydroxyapatite (HA) in hydroxyapatite grouts is nonresorbable [4]. Four weeks after surgery, ingrowth of both periosteal and endosteal trabecular bone was observed in ALCAP/alpha-ketoglutaric acid cement and HA/alpha-ketoglutaric acid grout filled holes in the tibiae of rabbits [5]. Bone ingrowth in the ALCAP/polyfunctional

[1] Professor of physiology and technical assistant, respectively, Biology Department, University of Dayton, Dayton, OH 45469.
[2] Ceramic specialist, University of Dayton Research Institute, University of Dayton, Dayton, OH 45469.

acid cement occurred around and into the periphery of the implant, especially where re-sorption of the cement had taken place. In holes filled with hydroxyapatite/polyfunctional acid grout, bone ingrowth was observed around and between the nonresorbable particles of HA.

Beta tricalcium phosphate (TCP) ceramics resorb relatively rapidly and are biocompatible. Beta tricalcium phosphate has been used in several forms in animals and humans to repair many types of defects in bone. Beta tricalcium phosphate blocks have been implanted in the tibia in rats [6] and in the femur, mandible, iliac crest, and inferior orbital rim in dogs [7–9]. Beta tricalcium phosphate has also been used to repair surgically created cleft palates [10] and periodontal osseous defects in dogs [11,12]. In monkeys, chipped or powdered TCP has been used effectively to repair defects on premolars and molars [13] and in apex-ification of teeth [14,15] and capping of pulp [16]. According to Metsger and co-workers, TCP powder or granules have been used to repair open apexes, periapical and marginal periodontal defects, and alveolar bony defects in humans [17]. Beta tricalcium phosphate by itself has not been reported to cause any adverse reactions in animal or human tis-sues [17].

Recently a mixture of tricalcium phosphate and plaster of paris (1:2) was used to recon-struct alveolar rims [18]. Since plaster of paris can cause renal osteodystrophy [19], syn-ergistic resorption of both tricalcium phosphate and plaster of paris (calcium sulfate) could enhance the chances of renal osteodystrophy. In contrast to TCP/plaster of paris mixtures, grouts of nonresorbable hydroxyapatite, with or without plaster of paris (20%), seem to be ideal for replacing bone lost from resorption. Therefore, a relatively fast-setting, resorbable, porous ceramic cement is needed for replacing bone that is lost because of nonresorptive processes. This investigation was designed to develop a TCP bone cement for replacing bone lost as a result of nonresorptive processes.

Procedure

The TCP used in this study was prepared by an aqueous precipitation and sulfate catalysis procedure [20]. The theoretical calcium/phosphorus ratio of 1.5 for tricalcium phosphate was used in preparing the reaction solutions. The precipitate was dried after centrifugation at 75°C and sintered in alumina crucibles by being heated at a rate of 9°C per minute to 1150°C in a Leco box furnace (Laboratory Equipment Corp., Saint Joseph, Michigan). Sintering was completed by soaking the precipitate at 1150°C for 1 h and then air quenching it to room temperature. Large pieces of the dried precipitate sludge were broken up before sintering to aid in the rapid air quenching of the material.

The sintered precipitate was prepared for X-ray diffraction analysis by grinding approx-imately 2 g of the material to pass a 325-mesh screen and then pressing the powder into a pellet 1.6 cm in diameter. The pellet was then placed in a diffractometer and exposed to X-rays from a high-intensity copper K-alpha source filtered by a monochrometer utilizing single-crystal graphite. The sample pellet was scanned over a range from 20 to 40° (2θ). X-ray diffraction performed on the first precipitation batch of material indicated that the sintered precipitate was pure beta tricalcium phosphate.

The sintered precipitate was crushed and ground in a laboratory-scale alumina ball mill apparatus and screened to the desired -400 mesh (-37 μm) on a Tyler Ro-Tap. All of the sintered material produced was crushed and screened in a single lot to homogenize the separate precipitate runs. Ball milling and screening operations were conducted repeatedly until all but a small fraction of the material (14.3 g) had passed a -400 mesh screen. A final X-ray diffraction sample was taken to confirm the purity of the -400 mesh beta tricalcium phosphate.

The beta tricalcium phosphate powder was mixed with 16 different setting agents. The setting agents tested with it were alpha-ketoglutaric acid, bovine serum albumin, Carbopol, citric acid, fumaric acid, Hespan, lactic acid, malic acid, orthophosphoric acid, oxaloacetic acid, polyvinyl pyrrolidone, pyruvic acid, succinic acid, and water.

The maximum viscosity of each mixture of TCP and setting agent in a water medium was estimated by observing the consistency of the mixture. The ratios of TCP and various solutions or powders were varied to obtain maximum viscosities. Gel time and set hardness were determined by allowing the mixture to dry at room temperature for 105 min. The gelling rate was evaluated at 30-s intervals and the setting hardness was evaluated on a scale from one to ten at 15 min intervals. During the mixing of the acids with TCP, care was taken to observe any exothermic reactions since the initial experiments conducted with ALCAP and orthophosphoric acid generated substantial amounts of heat.

To determine gel time and set hardness, triplicates of 1-g samples of TCP powder were mixed with either solutions or powders of various setting agents by means of a spatula in deep-well glass slides. Distilled and deionized water was added to each solid mixture to initiate the gelling reactions. All observations were made by the same person and all triplicates had a similar rating; therefore, the standard deviations for all values were zero.

Scanning electron micrographs of TCP/alpha-ketoglutaric acid and TCP/malic acid cements were taken after water had been added and the cements had set. One-quarter-gram mixtures of TCP and alpha-ketoglutaric acid or malic acid were set overnight in 14 by 5 by 3-mm rubber embedding molds. The cements were then removed from the molds and fractured by hand. The fractured specimens were mounted on an aluminum base and coated with a 1000-nm layer of gold in an Efa vacuum evaporator equipped with a sputtering head. (E. F. Fullam, Inc., Latham, New York). The cements were then examined in an Autoscan scanning electron microscope (Perkin-Elmer ETEC, Inc., Hayward, California) at 20 kV. Photomicrographs of the sections were taken with Polaroid 55N instant film.

Histological responses and bone ingrowth in the TCP cements were observed by implanting TCP/alpha-ketoglutaric acid and TCP/malic acid cements in holes drilled in the tibial cortex (not communicating with the marrow cavity) of rabbits and in surgically traumatized femurs of anesthetized albino male rats. TCP/alpha-ketoglutaric acid and TCP/malic acid cements were utilized because they yielded the best setting times and hardnesses without releasing heat. The rabbits were sacrificed four weeks after implantation and injection of tetracycline. The tissue sections were embedded in plastic, polished, and examined under ultraviolet light using a blue filter fluorescence for ingrowth of bone.

Each TCP cement was implanted in three rats. A groove approximately 1 mm wide and 5 mm long and just entering the medullary cavity was drilled in the proximal half of one femur. The groove was then filled with one of the cements and saline was added to facilitate gelling. Grooves of similar size were allowed to heal without the cement in sham-operated animals (controls). Urine was also collected for 24 h prior to surgery and immediately following surgery for 24 h. The urinalysis measurements included volume, specific gravity, pH, protein, glucose, ketones, bilirubin, blood, nitrates, and urobilinogen. The animals were sacrificed 70 days after surgery and the tissues were decalcified and embedded in paraffin. Five-micrometre-thick sections of decalcified tissue were stained with hematoxylin and eosin for histopathologic examination.

Results and Discussion

Maximum Viscosity

Maximum viscosity was attained with a 1:1 mixture of TCP with citric or succinic acid, after the addition of water (Table 1). A 1:1 ratio of TCP and Carbopol attained maximum

TABLE 1—*Ratios of calcined beta tricalcium phosphate ceramic powder and setting agent mixtures that attained maximum viscosity.*

Mixture of TCP and Setting Agent	Ratio
TCP/citric acid	1:1
TCP/succinic acid	1:1
TCP/Carbopol	1:1
TCP/orthophosphoric acid (30%)	5:4
TCP/water	5:3
TCP/polyvinyl pryrrolidone (20%)	5:3
TCP/bovine serum albumin (6%)	5:3
TCP/Hespan (6%)	5:3
TCP/lactic acid (85%)	5:3
TCP/oxaloacetic acid	4:1
TCP/pyruvic acid	2:1
TCP/α-ketoglutaric acid	2:1
TCP/fumaric acid	4:1
TCP/malic acid	2:1

viscosity. Maximum viscosity was also attained by a 5:4 mixture of TCP and orthophosphoric acid (30%). Beta tricalcium phosphate in combination with water, polyvinyl pyrrolidone (20%), bovine serum albumin (6%), Hespan (6%), or lactic acid (85%), in a ratio of 5:3, achieved maximum viscosity. Powders of TCP mixed with pyruvic, alpha-ketoglutaric, or malic acid reached maximum viscosity in a ratio of 2:1 after the addition of water. Maximum viscosity was attained by mixing TCP and oxaloacetic or fumaric acids in a ratio of 4:1 on the addition of water (Table 1). If maximum viscosity were the only criterion for developing a bone cement, a toothpaste-like mixture could be produced by varying the ratio of TCP and setting agents tested in this investigation.

Gel Time and Set Hardness

Mixtures of TCP and alpha-ketoglutaric acid gelled 30 s after the addition of water. However, by mixing different proportions of citric acid and alpha-ketoglutaric acid with TCP, gelling times of 2.5 to 3.5 min were achieved after the addition of water. The combination of TCP and 30% orthophosphoric acid gelled 45 s after water was added. Eight minutes after the addition of water the gelled TCP/malic acid paste could not be stirred with a spatula.

The TCP/malic acid cement reached a maximum hardness of ten within 15 min. Mixtures of TCP and 30% orthophosphoric acid or alpha-ketoglutaric acid reached a hardness of ten within 30 and 45 min, respectively. The TCP/orthophosphoric acid mixture generated substantial amounts of heat on mixing and, therefore, was not implanted in animals. All other TCP and setting agent mixtures failed to achieve maximum hardness within the 105-min test period (Table 2). The data suggest that mixtures of TCP and alpha-ketoglutaric acid or malic acid can be used as cements to replace bone lost as a result of nonresorptive processes.

The compressive strength at failure for TCP/alpha-ketoglutaric acid cylindrical plugs was 4.2 MPa and for TCP/malic acid plugs was 6.2 MPa [5]. However, the compressive strength at failure for ALCAP/alpha-ketoglutaric acid cylindrical plugs was 0.65 MPa and for ALCAP/malic acid cylindrical plugs was 1.3 MPa. Hydroxyapatite/alpha-ketoglutaric acid plugs failed at 0.7 MPa. These data show that TCP cements have a much higher strength than either ALCAP or HA cements.

Scanning electron microscopic examination of TCP/alpha-ketoglutaric acid and TCP/

TABLE 2—*Setting hardness of beta tricalcium phosphate ceramic powder and setting agent mixtures at 15-min intervals over a duration of 105 min.*

Setting Agent or Agents	Water Content, mL/g TCP	Degree of Hardness[a]						
		15 min	30 min	45 min	60 min	75 min	90 min	105 min
α-Ketoglutaric acid	0.5	8	9	10	10	10	10	10
Bovine serum albumin (6%)	…	2	2	2	4	5	6	7
Carbopol	…	1	1	2	2	2	3	3
Citric acid	0.3	7	8	8	8	8	9	9
Composition 1—α-ketoglutaric acid (0.35 g) and citric acid (0.15 g)	0.4	7	8	8	8	8	9	9
Composition 2—α-ketoglutaric acid (0.40 g) and citric acid (0.10 g)	0.4	8	8	8	9	9	9	9
Fumaric acid	0.6	2	2	3	4	5	6	6
Hespan (6%)	…	2	3	3	4	4	6	7
Lactic acid	…	1	1	1	1	2	2	2
Malic acid	0.4	10	10	10	10	10	10	10
Orthophosphoric acid (30%)	…	9	10	10	10	10	10	10
Oxaloacetic acid	0.5	1	1	1	1	2	2	2
Polyvinyl pyrrolidone	…	2	2	3	3	4	4	4
Pyruvic acid	0.6	2	3	4	6	7	8	9
Succinic acid	0.7	2	3	4	5	5	5	6
Water	0.6	1	2	3	5	6	7	8

[a] The degree of hardness was determined on a scale of 1 to 10, with 10 being the hardest.

FIG. 1a—*Scanning electron micrograph of the fractured surface of a TCP/alpha-ketoglutaric acid cement. The scale mark indicates 10 μm.*

FIG. 1b—*Scanning electron micrograph of the fractured surface of a TCP/malic acid cement. The scale mark indicates 10 μm.*

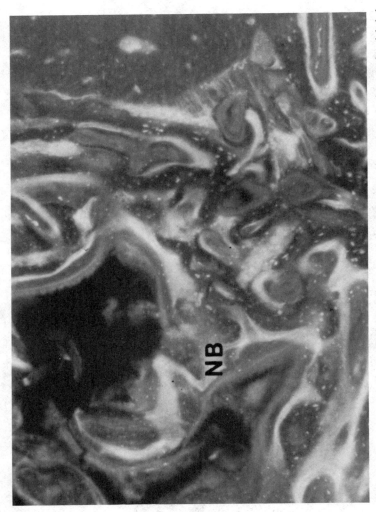

FIG. 2a—Photomicrograph of an undecalcified section of rabbit tibia showing ingrowth of tetracycline-labeled (white) new bone (NB) in a control drilled hole (magnification, ×125).

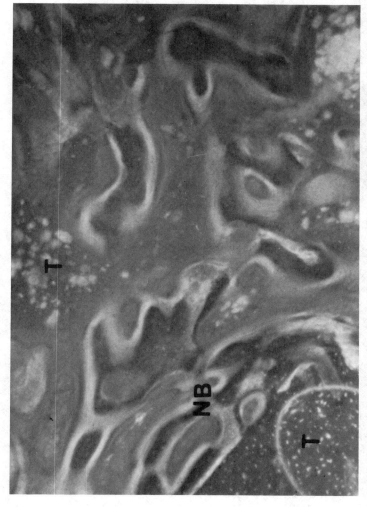

FIG. 2b—Photomicrograph of an undecalcified section of rabbit tibia showing ingrowth of tetracycline-labeled (white) new bone (NB) in a TCP/alpha-ketoglutaric acid (T) filled hole (magnification, ×125).

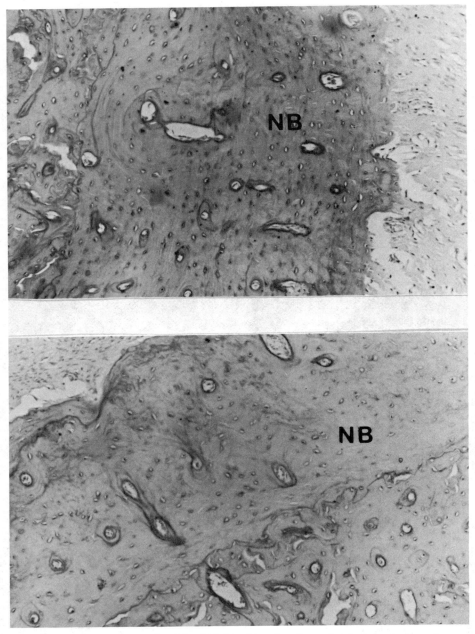

FIG. 3—*Photomicrographs of decalcified sections of rat femur showing growth of new bone (NB) at the* (top) *sham-operated site and* (bottom) *TCP/malic acid filled site (magnification, ×135).*

malic acid cements revealed ceramic particles held together by fused crystalline structures (Figs. 1a and 1b). The crystal morphology was dependent on the polyfunctional acid used as a setting agent.

Animal Studies

Initial application of the cement powder to a bone wound acts as a hemostatic agent and sets on reacting with blood. During surgery it was observed that mixtures of either TCP/alpha-ketoglutaric acid or TCP/malic acid can be applied directly to the bone and allowed to set. TCP/malic acid mixtures can also be premixed and then applied to the traumatized bone. Four weeks after surgery TCP/alpha-ketoglutaric acid cement filled holes in tibiae of rabbits showed ingrowth of large amounts of tetracycline-labeled periosteal and endosteal trabecular bone and extensive resorption of TCP cement (Figs. 2a and 2b). The bone ingrowth extended into the core of the specimen. With the ALCAP cements, however, the new bone did not penetrate the implant core and only peripheral resorption of the implant was observed [5]. After 70 days, rat femurs implanted with TCP/alpha-ketoglutaric acid or TCP/malic acid cement showed complete replacement of cement by bone (Fig. 3). The use of alpha-ketoglutaric acid or malic acid as a setting agent did not have any noticeable deleterious effects on surrounding tissues in either rabbits or rats. Urinalysis for the rats showed that the alpha-ketoglutaric acid resorbed from the TCP/acid cement during the first 24 h did not influence the metabolic status of the animals. Preliminary animal studies conducted to date suggest that surgically induced bone wounds in the rabbit tibia or rat femur can be filled with ease with TCP/alpha-ketoglutaric acid or TCP/malic acid cement.

General Conclusions

Data obtained on the gelling time and setting hardness suggest that mixtures of TCP with alpha-ketoglutaric acid or malic acid can be used successfully as bone cements. The healing pattern of bone wounds filled with either TCP/alpha-ketoglutaric acid or TCP/malic acid cement suggests that TCP/polyfunctional acid cements provide a scaffold for bone growth and can be used successfully to repair traumatized cortical bone.

Acknowledgments

The authors would like to thank Dr. J. Ricci for providing data on the mechanical strength of TCP cements in rabbit tibial implantation studies. We also acknowledge Dr. D. R. Mattie, AAMRL, WPAFB, Dayton, Ohio, for providing scanning electron micrographs of TCP cements. This study was supported in part by Orthomatrix, Inc., Dublin, California.

References

[1] Schmitt, W., Purrman, R., Jochum, P., and Gasser, O., U.S. Patent No. 4,376,835, 1983.
[2] Potter, W. D., Barclay, A. C., Dunning, R., and Parry, R. J., U.S. Patent No. 4,123,416, 1978.
[3] Bajpai, P. K., Fuchs, C. M., and Strnat, M. A. P. in Biomedical Engineering IV, Recent Developments, B. W. Sauer, Ed., Pergamon Press, New York, 1985, pp. 22–25.
[4] Bajpai, P. K. and Fuchs, C. M. in Proceedings, First Annual Scientific Session of the Academy of Surgical Research, Vol. 1, 1985, pp. 50–54.
[5] Ricci, J. L., Berkman, A., Bajpai, P. K., Alexander, H., and Parsons, J. R., Transactions, Twelfth Annual Meeting of the Society for Biomaterials, Vol. 9, 1986, p. 132.
[6] Bhaskar, S. N., Brady, J. M., Getter, L., Gomer, M. F., and Driskell, T., Oral Surgery, Vol. 32, No. 2, 1971, pp. 326–346.

[7] Driskell, T. D., Hassler, C. R., and McCoy, L. R., *Proceedings,* Annual Conference on Engineering Medicine and Biology, Vol. 15, 1973, p. 199.

[8] Cameron, H. U., Macnab, I., and Pilliar, R. M., *Journal of Biomedical Materials Research,* Vol. 11, No. 2, 1977, pp. 179–186.

[9] Ferraro, J. W., *Plastic and Reconstructive Surgery,* Vol. 63, No. 5, 1979, pp. 634–640.

[10] Mors, W. A. and Kaminski, E. J., *Archives of Oral Biology,* Vol. 20, No. 5, 1975, pp. 365–367.

[11] Levin, M. P., Getter, L., Cutright, D. E., and Bhaskar, S. N., *Oral Surgery,* Vol. 38, No. 3, 1974, pp. 334–351.

[12] Levin, M. P., Getter, L., and Cutright, D. E., *Journal of Biomedical Materials Research,* Vol. 9, No. 2, 1975, pp. 138–195.

[13] Levin, M. P., Getter, L., Adrian, J., and Cutright, D. E., *Journal of Clinical Periodontology,* Vol. 1, No. 4, 1974, pp. 197–205.

[14] Koenigs, J. F., Heller, A. L., Brilliant, J. D., Melfi, R. C., and Driskell, T. D., *Journal of Endodontics,* Vol. 1, No. 3, 1975, pp. 102–106.

[15] Koenings, J. F., "Induced Apical Closure in Pulpless Teeth on Monkeys Using Tricalcium Phosphate," M.Sc. thesis, Ohio State University, Columbus, OH, 1974.

[16] Heller, A. L., Koenigs, J. F., Brilliant, J. D., Melfi, R. C., and Driskell, T. D., *Journal of Endodontics,* Vol. 1, No. 3, 1975, pp. 102–106.

[17] Metsger, D. S., Driskell, T. D., and Paulsrud, J. R., *JADA,* Vol. 105, 1982, pp. 1035–1038.

[18] Vieco, E. B., *Implantologist,* Vol. 2, No. 2, 1981, pp. 56–62.

[19] Michalk, D., Klare, B., Manz, F., and Scharer, K., *Clinical Nephrology,* Vol. 16, No. 1, 1981, pp. 8–12.

[20] Jarcho, M., Salsbury, R. L., Thomas, M. B., and Doremus, R. H., *Journal of Materials Science,* Vol. 14, 1979, pp. 142–150.

Praphulla K. Bajpai,[1] George A. Graves, Jr.,[2] David R. Mattie,[3] and Frank B. McFall III[4]

Resorbable Porous Aluminum-Calcium-Phosphorus Oxide (ALCAP) Ceramics

REFERENCE: Bajpai, P. K., Graves, G. A., Jr., Mattie, D. R., and McFall, F. B. III, "Resorbable Porous Aluminum-Calcium-Phosphorus Oxide (ALCAP) Ceramics," *Quantitative Characterization and Performance of Porous Implants for Hard Tissue Applications, ASTM STP 953,* J. E. Lemons, Ed., American Society for Testing and Materials, Philadelphia, 1987, pp. 389–398.

ABSTRACT: A porous aluminum-calcium-phosphorus oxide (ALCAP) ceramic has been developed for potential use in corrrecting bone defects. The material is at least partly resorbed and appears to have acceptable biocompatibility and toxicity characteristics for its intended applications. ALCAP ceramics are fabricated by calcining mixtures of aluminum, calcium, and phosphorus oxide powders and sintering the compressed blocks from calcined particles of the desired size. By varying the dimensions of the die, particle size, sintering time, and temperature, ALCAP ceramics have been fabricated that may be useful for repairing a wide variety of bone defects. The fate of ALCAP ceramics in both *in vitro* and *in vivo* environments has been studied by means of chemical analysis of tissues, chemical and enzyme analysis of implants, radiography, radioactive isotope uptake, scanning electron microscopy, energy-dispersive X-ray analysis, and histology. ALCAP ceramics were implanted in experimentally induced defects in long bones of rats and mandibles of rabbits. Results of these studies suggest that the ALCAP ceramics resorb and allow ingrowth of new bone.

KEY WORDS: porous implants, bone repair, aluminum-calcium-phosphorus oxide (ALCAP) ceramics, mutagenicity, spinal fusion, resorption, oral surgery, and maxillofacial surgery

Bioceramics can be divided into three categories: according to their chemical reactivity with the environment, bioceramics have been classified as resorbable, surface reactive, or inert [1–5]. Inert bioceramics maintain their physical and mechanical properties for the life of the implant. Surface-reactive ceramics form chemical bonds with the surrounding tissue. Resorbable ceramics serve as templates to aid in the construction or replacement of the damaged tissue and are eventually absorbed or dissolved and replaced by living tissue. All three types of ceramics can release components into the host but in varying amounts and at vastly different rates. Since surface-reactive and resorbable ceramics release their chemical constituents and are modified primarily by the body fluids and tissues, they should be classified as sustained-release devices [2]. Since most of the resorbable ceramic implants used to date have been composed of combinations of aluminum, calcium, or phosphorus oxides, they are expected to release varying amounts of aluminum, calcium, or phosphorus in the host.

[1] Professor of physiology, Biology Department, University of Dayton, Dayton, OH 45469.
[2] Research ceramist, University of Dayton Research Institute, University of Dayton, Dayton, OH 45469.
[3] Research biologist, AAMRL, Wright-Patterson Air Force Base, OH 45433.
[4] Laboratory manager, Research and Development, C. R. Bard Inc., Billerica, MA 01821.

For a ceramic to be selected as a biomaterial it must have acceptable biological compatibility. A material has acceptable biocompatibility when it exists within the physiological environment of its intended application without adversely and significantly affecting the body or the environment of the body or adversely and significantly affecting the material [1,5–9]. The bioceramic should be biologically functional, which means that the material should function as intended in the body during the life of the item. A bioceramic must also have toxicity characteristics acceptable in the intended application. Toxicological manifestations include adverse local tissue, systemic, allergenic, and carcinogenic responses. Release of harmful chemical components from the biomaterials can induce these toxicological manifestations [10].

Historical Aspects

The development of aluminum-calcium-phosphorus oxide (ALCAP) ceramics has included both material characterization and performance testing in both *in vitro* and *in vivo* environments. Graves and co-workers developed porous polymorphic calcium aluminate and calcium aluminum phosphate ceramics [11]. They implanted the ceramics into rhesus monkeys for 32 weeks. The ceramics became completely impregnated with mineralized bone while undergoing resorption. The microstructural alterations observed in the polymorphic ceramics were due to loss of one or more soluble resorbable phases [10]. As the ceramic is resorbed, the pore size increases and encourages greater bone infiltration and formation. In follow-on studies in which 30 rhesus monkeys were implanted with calcium-aluminum-phosphate ceramics for up to three years no untoward effects were observed [11,12].

Studies in rats demonstrated that approximately 70% of the calcium from porous calcium aluminate ceramics implanted in femurs was resorbed by 75 days. By 150 days, the calcium content of the ceramic implants was significantly higher than the calcium content of unimplanted rat femurs. The phosphorus content of implanted calcium aluminate ceramics steadily increased with the duration of implantation [13].

Preliminary work conducted with bone enzymes suggested that the addition of phosphate to calcium aluminate ceramics would enhance their resorbability without decreasing their strength. Bajpai et al. [14] implanted calcium aluminate ceramics containing 20% phosphorous pentoxide into the femora of 28 male rats. Seven implanted rats plus seven sham-operated rats were sacrificed at 30, 60, 90, and 120 days [14]. The implant sites were radiographed prior to sacrifice and indicated complete compatibility between the ceramic and the surrounding bone. Alkaline phosphatase activity was higher in the ceramic implants than in the sham bone implants. This suggests that the osteoblastic activity in ceramic implants is higher than the osteoblastic activity in bone autografts.

Fabrication and Material Characterization

ALCAP ceramics are fabricated by mixing aluminum oxide, calcium oxide, and phosphorus pentoxide powders in a ratio of 50:34:16 by weight. The mixture of powders is calcined for 12 h at 1315°C. The calcined material is ground into particles in a ball mill and separated into three different ranges of particle sizes: 34 to 45, 45 to 60, and 60 to 75 μm. Using dry polyvinyl alcohol as a binder, cylindrical ceramics are made by pressing the calcined particles in a die at 27.6 MPa, using a hydraulic press. The green shape is sintered for 36 h at 1455°C [15]. ALCAP ceramics of various densities (Table 1), porosities, strengths, and degradabilities can be fabricated by varying the dimensions of the die, the particle size (Figs. 1a through 1c), the duration of sintering, and the maximum sintering temperature [15].

TABLE 1—*Effect of time, temperature, and particle size on the sintered density of aluminum-calcium-phosphorus oxide (ALCAP) ceramics.*

Particle Size Range, μm	Sintering Time, h	Density, gm/cm³, at Four Sintering Temperatures			
		1316°C	1371°C	1427°C	1454°C
60 to 75	18	1.15 ± 0.01	1.26 ± 0.02	1.26 ± 0.02	1.40 ± 0.01
	24	1.26 ± 0.02	1.31 ± 0.01	1.35 ± 0.01	1.42 ± 0.01
	36	1.29 ± 0.02	1.28 ± 0.01	1.34 ± 0.01	1.35 ± 0.02
45 to 60	18	1.26 ± 0.02	1.26 ± 0.01	1.38 ± 0.02	1.58 ± 0.02
	24	1.33 ± 0.03	1.35 ± 0.05	1.45 ± 0.03	1.64 ± 0.02
	36	1.34 ± 0.01	1.35 ± 0.01	1.53 ± 0.01	1.58 ± 0.01
35 to 45	18	1.31 ± 0.02	1.34 ± 0.02	1.41 ± 0.01	1.60 ± 0.01
	24	1.33 ± 0.03	1.41 ± 0.01	1.48 ± 0.01	1.69 ± 0.02
	36	1.35 ± 0.01	1.36 ± 0.01	1.51 ± 0.01	1.61 ± 0.01

In Vitro Dissolution Studies

Before ceramics were implanted in animals, *in vitro* degradation studies were conducted to determine how the material might behave in a physiological environment. The amount and rate at which the components are released from the implant helps to predict the toxic potential of the implant in the host and the host's potential to influence the mechanical properties of the implant.

Dissolution studies in human plasma were conducted to examine the solubility of ALCAP ceramics in a physiological medium. ALCAP ceramics were suspended in a fixed volume of plasma (static system) for seven days [16] and in a continuous flow-through system for twelve days at 37°C [17]. Significant amounts of aluminum, calcium, and inorganic phosphorus were released from ALCAP ceramics suspended in human plasma (Fig. 2). The hydrogen ion concentration was higher in plasma surrounding ALCAP ceramics than in control plasma in both static and flow-through systems. Since bone is formed by precipitation of calcium and phosphate salts in an alkaline medium, ALCAP ceramics implanted in bone should provide a favorable environment for formation of new bone [16,17]. The results of dissolution studies demonstrated that ALCAP ceramics contain materials extractable in human plasma and that ALCAP ceramics are composed of resorbable phases.

Biocompatibility Testing

A series of experiments was conducted to evaluate selected acute biocompatibility and toxicity characteristics of ceramics. The fate of aluminum, calcium, and phosphorus released during biodegradation was assessed in several studies. This was important for two reasons: (1) aluminum is extractable from ALCAP ceramics, and (2) it has been implied that aluminum acts as a toxic material in dialysis encephalopathy and osteodystrophy.

ALCAP disks containing an average of 274 ± 21 mg aluminum, 259 ± 32 mg calcium, and 77 ± 12 mg phosphorus were implanted subcutaneously in male rats for 1, 4, 8, and 12 weeks. The total amounts of aluminum, calcium, and inorganic phosphorus released from the ceramics were 93 ± 7, 57 ± 7, and 20 ± 3 mg, respectively. Tissue analysis showed that the materials released did not accumulate in the brain, bone, heart, liver, or spleen of

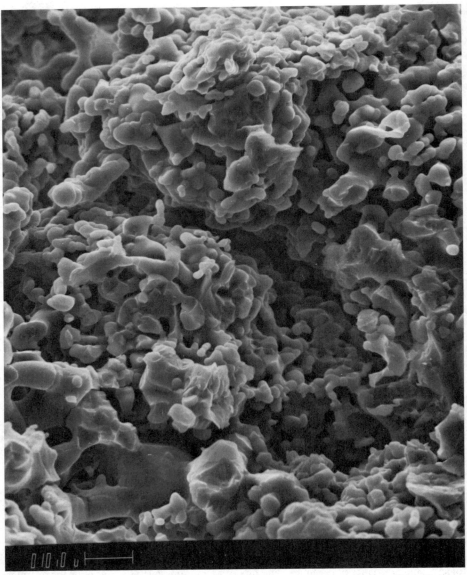

FIG. 1a—*Scanning electron photomicrograph of an ALCAP ceramic having a 35 to 45-μm particle size, sintered for 36 h at 1455°C.*

implanted rats but were excreted in the urine [*18*]. The resorbed components of ALCAP ceramics appear to pose no threat to the body's organs as they are removed from the body.

ALCAP ceramics were negative for mutagenicity, when tested by the Ames test (Table 2), and were not cytotoxic, when tested by the clonal cytotoxicity assay [*19*]. Based on the Ames test, ALCAP ceramics can be considered nonmutagenic.

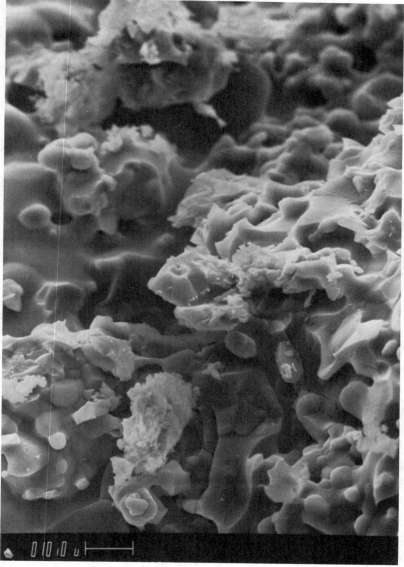

FIG. 1b—*Scanning electron photomicrograph of an ALCAP ceramic having a 45 to 60-μm particle size, sintered for 36 h at 1455°C.*

The hemolysis test is a relatively simple and rapid blood compatibility test that has been used to assess hemolytic properties of biomaterials [20]. ALCAP ceramics fabricated from 45 to 60 and 60 to 75-μm particle sizes were not hemolytic to human erythrocytes incubated in physiological saline at 37°C [21].

The rabbit muscle implant test is an *in vivo* biological test for evaluating the acute local

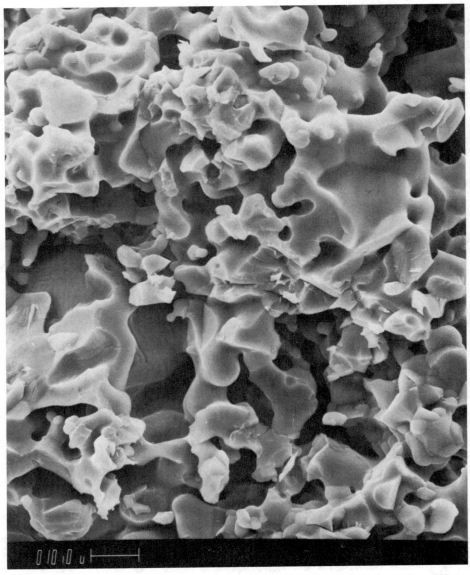

FIG. 1c—*Scanning electron photomicrograph of an ALCAP ceramic having a 60 to 700-μm particle size, sintered for 36 h at 1455°C.*

tissue response of a biomaterial [22]. ALCAP ceramics fabricated from all three particle size ranges were implanted into the paralumbar muscles in albino rabbits for one week. Sham procedures, negative controls (inert plastic, obtained from Persplex CQ, Higher Schulte, California), and ALCAP ceramics fabricated from all three particle size ranges induced the same acute inflammatory response in adjacent muscle tissue, caused by the surgery and

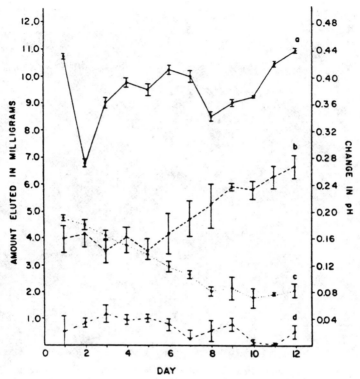

FIG. 2—*Changes in the pH* (a) *and in the total amount of aluminum* (b), *calcium* (c), *and inorganic phosphorus* (d) *eluted each day from ALCAP ceramics exposed to a continuous flow of human plasma.*

TABLE 2—*Plate incorporation test of the Ames/Salmonella assay performed on ALCAP ceramics.*[a]

Amount of ALCAP Ceramic per Plate, mg	Revertants per Plate, mean ± SEM[b]			
	Strain TA 98		Strain TA 100	
	−S9 Activation	+S9 Activation	−S9 Activation	+S9 Activation
DMSO solvent control	17 ± 3	29 ± 2	82 ± 1	103 ± 9
10	10 ± 2	24 ± 2	83 ± 5	108 ± 6
3	15 ± 2	23 ± 2	76 ± 6	92 ± 10
1	16 ± 1	28 ± 3	93 ± 7	108 ± 8

[a] No sample values were significantly greater ($P < 0.01$) than the solvent control values.
[b] SEM = standard error of the mean; the number of observations was three.

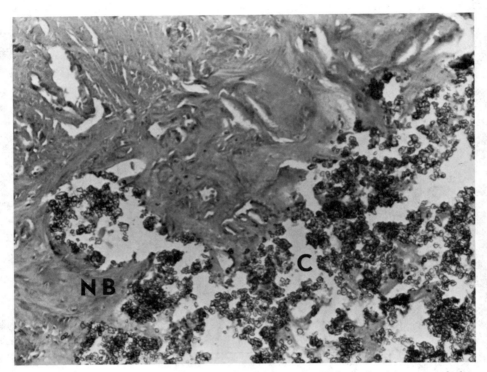

FIG. 3—*ALCAP ceramic implanted into a rat femur for 30 days, showing nonresorbed ceramic particles (C) and ingrowth of new bone (NB).*

healing process [21]. No significant adverse reactions were detected in these acute biocompatibility and toxicity tests.

Performance Testing

Semicylindrical ALCAP ceramics (0.82 by 0.82 cm) were implanted below the angle of the left mandible in experimentally induced defects in the mandibles of rabbits to evaluate the potential use of ALCAP ceramics in reconstruction of maxillofacial bones, a long-standing problem in both functional and cosmetic surgery. Radiographs of mandibles 30 and 60 days after implantation of ceramics showed continuous bone density at the interface of the implant and adjacent bone. Energy-dispersive X-ray analysis of the ceramic, bone, and ceramic-bone interface revealed that aluminum did not migrate beyond the interface of the ceramic while calcium and phosphorus were being resorbed [23,24].

ALCAP ceramics were implanted into the femurs of rats for 30 days. Radiographs, implant histology (Fig. 3), and scanning electron microscopy demonstrated osseous union of bone and implant as well as ingrowth of new bone into ALCAP ceramic implants [25].

A cement consisting of ALCAP plus carboxylic acid has been developed in an attempt to facilitate the implantation of ceramics into bone and to expand the potential uses of ALCAP ceramics. ALCAP/alpha-ketoglutaric acid (ALCAP-AKA) and ALCAP/malic acid (ALCAP-MA) cements have shown the best potential as cements for repairing bone defects. ALCAP cements can set at rates between 30 s and 8 min. The compressive strength at failure for ALCAP-AKA was 0.65 ± 0.07 MPa ($n = 8$) and for ALCAP-MA 1.3 ± 0.1 MPa ($n = 7$). Four weeks after surgery, ingrowth of trabecular bone was observed in ALCAP-AKA cement-filled holes in tibias of rabbits [26]. Dense cortical bone surrounded ALCAP particles after twelve weeks of implantation in femurs of rats [27].

Conclusions

A ceramic derived from aluminum oxide, calcium oxide, and phosphorus pentoxide has given good implant results in terms of compatibility and gradual replacement of the ceramic with endogenous bone. Experiments conducted so far suggest that ALCAP ceramic implants in bone biodegrade and have acceptable biocompatibility for potential use in correcting various defects in mandibular, maxillofacial, and long bones.

References

[1] Anderson, J. M. in *Fundamental Aspects of Biocompatibility*, Vol. II, D. F. Williams, Ed., CRC Press, Boca Raton, FL 1981, pp. 205–218.

[2] deGroot, K. in *Biocompatibility of Clinical Implant Materials*, Vol. I, D. F. Williams, Ed., CRC Press, Boca Raton, FL, 1981, pp. 199–222.

[3] Drummond, J. L. in *Biomaterials in Reconstructive Surgery*, L. R. Rubin, Ed., C. V. Mosby & Co., St. Louis, 1983, pp. 102–108.

[4] Hench, L. L. and Ethridge, E. C., *Biomaterials: An Interfacial Approach*, Academic Press, New York, 1982, pp. 1–25.

[5] Williams, D. F. *Fundamental Aspects of Biocompatibility*, Vol. I, CRC Press, Boca Raton, FL, pp. 1–7.

[6] Autian, J. in *Fundamental Aspects of Biocompatibility*, Vol. II, D. F. Williams, Ed., CRC Press, Boca Raton, FL, 1981, pp. 63–93.

[7] Bigi, A., Incerti, A., Roveri, N., Foresti-Serantoni, E., Mongiorgi, R., Rova di Sarseverino, L., Krajewski, A., and Ravgioli, A., *Biomaterials*, Vol. 1, 1980, p. 140.

[8] Galleti, P. M. and Boretos, J. W., *Journal of Biomedical Materials Research*, Vol. 17, 1983, pp. 539–555.

[9] Autian, J., *Transactions*, Fourth Annual Meeting of the Society for Biomaterials, Vol. 2, 1977, p. 42.

[10] Graves, G. A., Jr., Hentrich, R. L., Stein, H. G., and Bajpai, P. K., *Journal of Biomedical Materials Research Symposia*, Vol. 2, 1971, pp. 91–115.

[11] Graves, G. A., Jr., Noyes, F. R., and Villanueva, A. R., *Journal of Biomedical Materials Research Symposia*, Vol. 6, 1975, pp. 17–22.

[12] Graves, G. A., Jr., Bajpai, P. K., McCullum, D. E., Stein, H. G., and Noyes, F. R., "Bone Strength and In-flight Mechanical Stress," Technical Report AFOSR-F44620-71-C-0083, Aerospace Medical Research Laboratory, Wright-Patterson Air Force Base, Ohio, 1975, pp. 1–100.

[13] Carvalho, B. A., Graves, G. A., Jr., and Bajpai, P. K., *IRCS Medical Science*, Vol. 3, 1975, p. 185.

[14] Bajpai, P. K., Wyatt, D. F., Gilles, N. M., Stull, P. A., and Graves, G. A., *Clinical Research*, Vol. 24, 1976, p. 524A.

[15] Bajpai, P. K., Khot, S. N., Graves, G. A., Jr., and McCullum, D. E., *IRCS Medical Science*, Vol. 9, 1981, pp. 696–697.

[16] Mattie, D. R., Graves, G. A., Jr., Ritter, C. J. and Bajpai, P. K., *Proceedings*, Ninth North East Bioengineering Conference, Pergamon Press, New York, 1981, pp. 39–42.

[17] McFall, F. B., Ritter, C. J., Mattie, D. R., and Bajpai, P. K. in *Biomedical Engineering II, Recent Developments*, C. W. Hall, Ed., Pergamon Press, New York, 1983, pp. 357–360.

[18] Mcfall, F. B. and Bajpai, P. K., *Transactions*, Tenth Annual Meeting of the Society for Biomaterials, Vol. 7, 1984, p. 354.

[19] Mattie, D. R., McFall, F. B., Bajpai, P. K., and Gridley, J., *Federation Proceedings*, Vol. 43, 1984, p. 327.

[20] Wennberg, A. and Hensten-Petterson, A., *Journal of Biomedical Materials Research*, Vol. 15, 1981, pp. 433–435.

[21] Mattie, D. R., Lattendresse, J. R., and Bajpai, P. K., *IRCS Medical Science*, Vol. 13, 1985, pp. 420–421.

[22] Autian, J., *Artificial Organs,* Vol. 1., 1977, pp. 53–60.
[23] Freeman, M. J., Bajpai, P. K., Graves, G. A., Jr., and McCullum, D. E., *Ohio Journal of Science,* Vol. 80, 1980, p. 42.
[24] Bajpai, P. K., Graves, G. A., Jr., Wilcox, L. G., and Freeman, M. J., *Transactions,* Tenth Annual Meeting of the Society for Biomaterials, Vol. 7, 1984, p. 217.
[25] Mattie, D. R., Ritter, C. J., and Bajpai, P. K., *Transactions,* Tenth Annual Meeting of the Society for Biomaterials, Vol. 7, 1984, p. 353.
[26] Bajpai, P. K., Fuchs, C. M., and Strnat, M. A. P. in *Biomedical Engineering IV, Recent Developments,* B. W. Sauer, Ed., Pergamon Press, New York, 1985, pp. 22–25.
[27] Sutor, S. D., Strnat, M. A. P., and Bajpai, P. K., *Transactions,* Twelfth Annual Meeting of the Society for Biomaterials, Vol. 9, 1986, p. 113.

Akiyoshi Yamagami,[1] *Shuhei Kotera,*[1] *and Haruyuki Kawahara*[2]

Studies on a Porous Alumina Dental Implant Reinforced with Single-Crystal Alumina: Animal Experiments and Human Clinical Applications

REFERENCE: Yamagami, A., Kotera, S., and Kawahara, H., **"Studies on a Porous Alumina Dental Implant Reinforced with Single-Crystal Alumina: Animal Experiments and Human Clinical Applications,"** *Quantitative Characterization and Performance of Porous Implants for Hard Tissue Applications, ASTM STP 953,* J. E. Lemons, Ed., American Society for Testing and Materials, Philadelphia, 1987, pp. 399–408.

ABSTRACT: A new type of porous alumina dental implant has been designed. A biological seal is provided by single-crystal alumina on its cervical portion, and fixation capability in bone is provided by a porous alumina layer with an average interconnecting pore size of approximately 130 μm.

The mandibular premolar and molar extraction sites of 15 male rhesus monkeys (3 to 6 years old) were used in the animal experiments. All the implants and superstructures that were not retained by attachment to proximate teeth participated in functional occlusion at all times except during the first two to four weeks.

After four, six, and eight months, histological examinations, including light microscopy (LM), scanning electron microscopy (SEM), and electron probe microanalysis (EPM), were conducted. The surrounding gingival tissue strongly adhered to the single-crystal alumina on the cervical portion of the implant, and prolific bone ingrowth was observed within the porous network of the root component.

The human clinical evaluations showed a high success rate of 94% in 45 cases. Long-term clinical cases were included, the longest implant having survived for eight years in free-standing form. These cases included 23 free-standing cases and 22 bridge cases; 3 of these failed in the first stage of human clinical trials.

KEY WORDS: porous implants, dental implants, single-crystal alumina, artificial tooth root, porous alumina, free-standing dental implants

Artificial dental root implants have been made of Type 316 stainless steel, Co-Cr-Mo alloys, pure titanium, titanium alloys, tantalum, alumina (Al_2O_3) ceramics, hydroxyapatite, and other materials. The designs of endosseous dental implants have consisted of screws, blades, and plates.

Recently, porous materials, particularly Al_2O_3 porous ceramics, have been used in many animal studies and clinical applications. While many significant advancements have been made in recent years, some problems affecting the success or failure of the implant remain unresolved. One of these is a problem connected with tissue compatibility of these materials; it is difficult to prevent their contamination with oral microorganisms. The other significant

[1] Director and vice chairman, respectively, Kyoto Institute of Implantology, Kyoto 600, Japan.
[2] Professor, Department of Biomaterials, Osaka Dental University, Osaka 540, Japan.

problem is that insufficient ingrowth of new bony tissue into the pores makes the implants loosen under oral functioning. [1–18].

These Al_2O_3 porous implants were expected to become closely attached to the gingival tissue and prevent contamination because they have a smooth and biocompatible surface in the cervical portion. And, with suitable pore sizes for ingrowth of bone, this implant was expected to become fixed to the alveolar bone rapidly.

Purpose

The purpose of the animal studies was to confirm the anticontamination properties of this Al_2O_3 porous implant and to evaluate qualitatively the new bone ingrowth into the pores.

For the human clinical studies, which used the results of the animal studies, the purpose was to investigate free-standing implants with no retention by attachment to proximal teeth.

Materials and Methods

As shown in Fig. 1, a porous alumina implant was designed by H. Kawahara, Y. Yamagami, and M. Hirabayashi. The implant was composed of a core of cylindrical single-crystal alumina, an outer porous layer, and a smooth apex made of polycrystalline alumina. The porosity, interconnecting pore size, and range of pore sizes were evaluated by water-replacement and mercury intrusion porosity analyses, which produced the following results: volume porosity = 35%; average interconnecting pore size = 30 μm; and range of pore sizes = 10 to 300 μm.

For the animal studies, specially designed porous alumina implants were used. These implants had a 7-mm-long porous portion, a 4-mm diameter, and a total length of 20 mm. The porous layer was 1 mm wide.

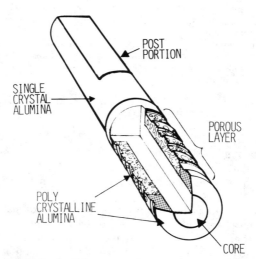

POST
PORTION

SINGLE
CRYSTAL
ALUMINA

POROUS
LAYER

POLY
CRYSTALLINE
ALUMINA

CORE

FIG. 1—*Diagrammatic sketch of the porous alumina dental implant used in this study. The root portion has a porous layer and apex made of polycrystalline alumina, and is designed as a self-tapping screw. The porous layer (1 mm in width and 4 mm in diameter) has interconnected pores approximately 130 μm in size and a volume porosity of 35%. The post portion and the core, which pierces the porous portion, were made of single-crystal alumina.*

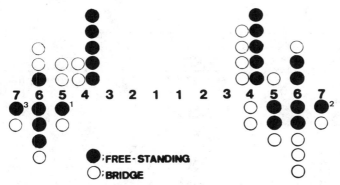

FIG. 2—*Distribution of the 45 implantion sites in human clinical cases operated on between 1977 and September 1985. The horizontal serial numbers indicate the tooth numbers. Each circle corresponds to one implant. The cases in the bridge form are marked with open circles and the cases in the free-standing form with closed circles. In 22 cases the prosthetic appliance was in a bridge form; in 23 cases a single crown was applied and functioned in a free-standing form. The smaller numbers 1, 2, and 3 indicate clinical cases No. 1, 2, and 3 in Fig. 3.*

For the human clinical studies, the basic implant, which had a total length of 24 mm and a 4.2 mm diameter at the root portion, was used.

Animal Studies

The implants were placed in extracted premolar and molar sites in 15 male rhesus monkeys (average age, 5.1 years old). The implants were placed three months after tooth extraction. All the opposing teeth were natural and the implants were free-standing at all times. A gold-silver-palladium alloy crown was placed on the implant.

The implantations were followed up by clinical examination, X-ray examination, assessment of the pocket depth, and a diagnosis of mobility and gingival color. After four, six, and eight months the monkeys were sacrificed randomly. Tissue reaction was evaluated by light microscopy (LM), scanning electron microscopy (SEM) and electron probe microanalysis (EPM).

Human Clinical Study

The human clinical evaluations involved implanting one or two implants into an intermediate long-span bridge or as a free-end and free-standing appliance. The distribution of the implantation sites can be seen in Fig. 2. The age, sex, year of surgery, and follow-up data are presented in Fig. 3.

Results

Animal Experiments

The animal experiments showed no gingival inflammation and the presence of epithelial tissue tightly adhering to the cervical portion of the implant, except in two cases of failure. The pocket depth was less than 3 mm in all 13 cases. Nearly normal gingival color was observed in all cases. The excellent histological reaction can be seen in Fig. 4.

Analysis of the porous root component revealed prolific bone tissue ingrowth, which

NO.	NAME	SEX	AGE	OPE. DATE	F/B	
1	T.O.	M	36	3.1977	F	Removed in 2.1980
2	T.O.	M	36	10.1977	F	Removed in 9.1978
3	M.H.	M	52	10.1977	B	Removed in 5.1980
4	T.K.	F	35	3.1978	B	
5	K.H.	F	51	11.1978	B	
6	T.N.	M	35	11.1978	B	
7.	M.T.	M	55	8.1979	B	
8	S.E.	F	28	3.1980	B	
9	R.A.	M	49	1.1981	B	
10	K.M.	F	42	10.1981	B	
11	M.M.	F	51	3.1982	B	
12	T.S.	M	34	8.1982	F	
13	A.N.	F	29	4.1983	F	
14	K.Y.	F	45	4.1983	B	
15	F.H.	F	60	5.1983	F	
16	T.M.	F	46	11.1983	F	
17	I.I.	F	24	3.1984	B	
18	F.I.	F	23	3.1984	B	
19	K.I.	F	32	3.1984	F	
20	K.M.	F	25	4.1984	F	M:+ → M:o
21	K.K.	M	52	4.1984	B	
22	T.Y.	M	31	6.1984	F	
23	I.T.	M	44	7.1984	F	
24	T.F.	M	47	8.1984	B	
25	I.I.	F	24	8.1984	F	
26	T.K.	M	68	9.1984	F	M:+ → M:o
27	H.M.	M	50	9.1984	F	
28	T.K.	F	57	9.1984	F	
29	M.S.	F	61	10.1984	B	
30	Y.N.	F	37	10.1984	B	
31	A.T.	M	54	11.1984	F	
32	H.H.	F	51	11.1984	F	
33	T.N.	M	42	11.1984	F	
34	T.N.	M	42	11.1984	F	
35	N.M.	M	59	3.1985	B	
36	N.M.	M	59	3.1985	B	
37	K.H.	F	58	4.1985	B	
38	T.I.	M	56	6.1985	B	
39	T.I.	M	56	6.1985	B	
40	T.K.	F	42	6.1985	B	
41	T.H.	M	50	7.1985	F	
42	A.Y.	M	27	8.1985	F	
43	T.F.	M	48	8.1985	B	
44	I.T.	F	22	8.1985	F	
45	T.T.	F	27	9.1985	F	

FIG. 3—*Personal data in the human clinical cases: the initials, sex, age, date of surgery, and prosthetic form (F/B stands for free-standing or bridge) are presented. Cases No. 1, 2, and 3, the initial trial cases in the human clinical study, were failures. The implants survived for 3, 1, and 2.5 years, respectively. Cases No. 20 and 26, which exhibited a slight mobility (indicated by M:+) have recovered and no inflammation is observed now (December 1985). The other cases show healthy gingival color and no mobility.*

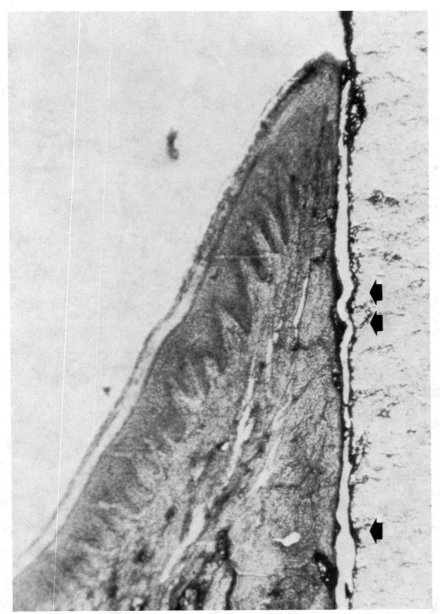

FIG. 4—*Light microscope photomicrograph of a four-month-old implant in the rhesus monkey, showing the cervical portion (hematoxylin and eosin stain, ×35). The arrows indicate single-crystal alumina fragments adhering to the epithelial tissue. The gingiva does not exhibit any inflammation.*

FIG. 5—Scanning electron photomicrograph of a longitudinal section of the porous root, illustrating the interconnected pore structure of the porous layer.

tightly adhered to the porous matrix (Figs. 5 and 6). This was confirmed by the EPM analysis, which showed an increase in and an abundance of calcium and phosphorus atoms within the pores of the root component. The EPM and histological analyses also revealed no gap between the surface and the new bone tissue.

Human Clinical Results

Patients examined between three months and eight years after surgery showed good clinical results. There were three implant failures, two due to infection and one as a result of the separation of the single-crystal core from the porous component. X-rays of the other implants showed no evidence of bone absorption due to infection and exhibited good adhesion of the bone to the porous root (Fig. 7).

Discussion

This porous alumina implant was fabricated to function in an alveolar bone in free-standing form. To form a potential biological seal, single-crystal alumina was used for the post portion, which is adjacent to the gingiva. The porous layer was fabricated with polycrystalline alumina that had interconnected pores approximately 130 μm in size; this pore size was chosen as a result of previous fundamental studies [1,5,19,20]. Consequently, our animal experiments showed firm gingival tissue adhering to the cervical portion of the implant, which was made of single-crystal alumina, and also showed calcified or formative bone tissue growing along the porous root and into the pores.

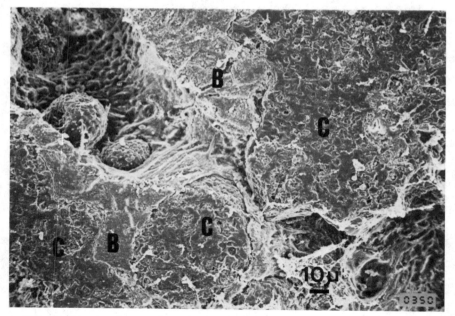

FIG. 6—*Scanning electron photomicrograph showing a cross-sectional view of the porous root six months after implantation. Osteogenesis in the pores can be seen. B = bone tissue; C = ceramic.*

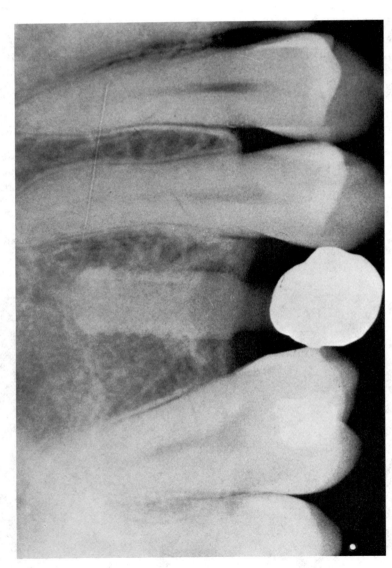

FIG. 7—*Radiograph of a free-standing implant in a human clinical application (Case No. 12 in Fig. 3) three years after surgery.*

Generally speaking [21,22], failures in previous applications of porous materials that pierce the gingival tissue, which interfaces with both the oral cavity and the underlying connective tissue (or bone tissue), were caused by bacterial contamination as a result of saliva adsorption in the dry porous component during the operation and a postoperative inflammatory reaction that prevented the formation of an effective biological seal between the two environments. These failures were primarily due to the surface microstructure of the post portion of the implant, to the biological properties of the material, or to both; for example, biodegradable material dissolves in the living body and diffuses its molecules, ions, and their derivatives, which may irritate the tissues. These causes of failure directly or indirectly produce alveolar bone absorption and soreness around the implant.

In the authors' animal experiments, two cases exhibited a gross inflammatory reaction and a high level of bone absorption. One of these failures was probably due to contamination ascribed to a projection of the porous portion out of the gingiva; the other was due to surplus bite pressure loading just after the insertion. It may be inferred that these two accidents occurred because of iatrogenic mistakes and not because of the properties of the materials or the design of the implant.

In the human clinical evaluations, two implants failed because of infection and one because of separation of the porous layer from the cylindrical core three years after the insertion; all three cases resulted in removal of the implant. The two cases of infection occurred very early in the authors' clinical evaluations and resulted in a change in the porous implant handling technique. This new procedure requires soaking the implant in sterile saline with an antibiotic. The incidence of infection after this technique was instituted was zero. And, since measures were taken to prevent failure due to mechanical properties, no breakage occurred.

The material properties and the fabrication technique will be reported at a later date.

Conclusions

Animal experiments revealed the potential of these porous alumina implants to function in a free-standing form. An average interconnecting pore size of approximately 130 μm appears to provide an adequate matrix for bony ingrowth and stability. It has also been observed that establishment of an effective biological seal at the cervical portion is necessary to prolong the duration of implantation.

Clinical experience with this porous dental implant of up to eight years has shown a success rate of 94%. The 21 free-standing cases are still functioning under proper bite pressure now.

To prevent bacterial contamination in the pores, the porous implant should be immersed in a sterile saline with antibiotics after sterilization prior to being implanted.

This implant appears to possess the important requirements for a long-term endosseous dental implant:

(a) long-term stability provided by prolific bone ingrowth,
(b) excellent biocompatibility, and
(c) a smooth cervical surface for a biological seal.

References

[1] Klawitter, J. J. and Hulbert, S. F., *Journal of Biomedical Materials Research,* Vol. 5, No. 2, 1971, pp. 161–229.
[2] Kawahara, H., Yamagami, A., and Nakamura, M., *International Dental Journal,* Vol. 18, No. 2, 1968, pp. 443–467.

[3] Griss, P., Von Andrian-Werbung, H., and Krempien, B., *Journal of Biomedical Materials Research,* Vol. 7, No. 4, 1973, pp. 453–462.

[4] Griss, P., Krempien, B., Von Andrian-Werbung, H., Heimke, G., Flemer, R., and Diehm, T., *Journal of Biomedical Materials Research,* Vol. 8, No. 5, 1974, pp. 34–49.

[5] Hulbert, S. F., Matthews, J. R., Klawwitter, J. J., Sauer, B. W., and Leonard, R. B., *Journal of Biomedical Materials Research,* Vol. 8, No. 5, 1974, p. 85.

[6] Selting, W. J. and Bhaskar, S. N., *Journal of Dental Research,* Vol. 52, 1973, p. 91.

[7] Kawahara, H., Yamagami, A., and Shibata, K., *Transactions,* Vol. 1, Third Annual Meeting of the Society for Biomaterials, New Orleans, LA, 1977, p. 133.

[8] Kawahara, H., *Journal of Oral Implantology,* Vol. 8, 1979, pp. 411–432.

[9] Yamagami, A. and Kawahara, H., *Transactions,* Fifth Annual Meeting of the Society for Biomaterials, Clemson, SC, 1979.

[10] Kawahara, H. in *First Proceedings of the Japan Society of Implant Dentistry,* 1975, pp. 187–196.

[11] Kawahara, H., *Transactions,* Tenth Annual Biomaterials Symposium, San Antonio, TX, 1978.

[12] Hammer, W. G. and Klawitter, J. J., *Journal of Dental Research,* Vol. 55B, 1976.

[13] Hammer, W. B., Topazian, R. G., Mackinney, R. V., and Hulbert, S. F., *Journal of Dental Research,* Vol. 52, 1973, pp. 356–361.

[14] Steflik, D. E., Mackinney, R. V., Jr., and Koth, D. L., *Transactions,* Society for Biomaterials, San Diego, CA, 1985.

[15] Yamagami, A., Kawahara, H., Yokota, J., Kobayashi, H., Sokawa, K., Maehara, S., Yokokawa, A., Nishimura, M., and Hirabayashi, M., *Transactions,* Third Annual Meeting of the Japanese Society for Biomaterials, Kyoto, Japan, 1981, pp. 141–142.

[16] Yamagami, A., Kawahara, H., and Hirabayashi, M., *Transactions,* 61st General Session of the International Association for Dental Research, Sydney, Australia, 1983, p. 679.

[17] Yamagami, A., Kawahara, H., and Hirabayashi, M., *Transactions,* Second Biosimposium, Lingnano Sabbiadore, Italy, 1982.

[18] Yamagami, A., Kotera, S., Hirabayashi, M., and Kawahara, H., *Transactions,* Second World Congress on Biomaterials, Washington, DC, 1984, p. 250.

[19] Topazian, R. G., Hammer, W. B., Boucher, L. J., and Hulbert, S. F., *Journal of Oral Surgery,* Vol. 29, 1971, p. 29.

[20] Imai, K., *Journal of Oral Implant Research,* Vol. 4, No. 4, 1982, pp. 26–48.

[21] Young, J. F., *Journal of the American Ceramic Society,* Vol. 53, 1970, pp. 65–69.

[22] Klawitter, J. J., Weinstein, A. M., Cooke, F. W., Peterson, L. J., Pennel, B. M., and Mackinney, R. V., *Journal of Dental Research,* Vol. 56, 1977, No. 7, pp. 768–776.

Indexes

Author Index

Subject Index